# Fermentation and Algal Biotechnologies for the Food, Beverage and Other Bioproduct Industries

# Multidisciplinary Applications and Advances in Biotechnology: Contributions from the Biotechnology Society of Nigeria Working Groups


*Series Editors:*

Benjamin Ewa Ubi
*Department of Biotechnology / Biotechnology Research & Development Centre, Ebonyi State University Abakaliki, Nigeria*

Abdulrazak B. Ibrahim
*Department of Biochemistry, Ahmadu Bello University, Zaria, Nigeria / Forum for Agricultural Research in Africa (FARA), Accra, Ghana*

James Chukwuma Ogbonna
*Department of Microbiology, University of Nigeria, Nsukka, Nigeria*

Emeka Godfrey Nwoba
*Algae Research & Development Centre, Murdoch University, Western Australia*

Sylvia Uzochukwu
*Biotechnology Centre, Federal University, Oye-Ekiti, Nigeria*

Charles Oluwaseun Adetunji
*Edo State University Uzairue, Nigeria*

Nwadiuto (Diuto) Esiobu
*Microbial Biotechnology Laboratory, Florida Atlantic University, Boco Raton, FL., USA / Applied Biotech International Nigeria Ltd., Wuye, Abuja, Nigeria*


Fermentation and Algal Biotechnologies for the Food, Beverage and Other Bioproduct Industries

*James Chukwuma Ogbonna, Sylvia Uzochukwu, Emeka Godfrey Nwoba, Charles Oluwaseun Adetunji, Nwadiuto (Diuto) Esiobu, Abdulrazak Ibrahim, and Benjamin Ewa Ubi*

Agricultural Biotechnology, Biodiversity and Bioresources Conservation and Utilization

*Olawole Obembe, Emmanuel Olufemi Ekundayo, Arinze Stanley Okoli, Abubakar Gidado, Charles Oluwaseun Adetunji, Abdulrazak Ibrahim, and Benjamin Ewa Ubi*

Medical Biotechnology, Biopharmaceutics, Forensic Science and Bioinformatics
*Hajia Mairo Inuwa, Ifeoma Maureen Ezeonu, Charles Oluwaseun Adetunji, Abubakar Gidado, Emmanuel Olufemi Ekundayo, Abdulrazak Ibrahim, and Benjamin Ewa Ubi*

Biosafety and Bioethics in Biotechnology: Policy, Advocacy, and Capacity Building
*Sylvia Uzochukwu, Arinze Stanley Okoli, Nwadiuto (Diuto) Esiobu, Emeka Godfrey Nwoba, Christpeace Nwagbo Ezebuiro, Charles Oluwaseun Adetunji, Abdulrazak Ibrahim, and Benjamin Ewa Ubi*

Microbiomes and Emerging Applications
*Nwadiuto (Diuto) Esiobu, James Chukwuma Ogbonna, Charles Oluwaseun Adetunji, Olawole Obembe, Ifeoma Maureen Ezeonu, Abdulrazak Ibrahim, and Benjamin Ewa Ubi*

Bioenergy and Environmental Biotechnology for Sustainable Development
*Akinola Rasheed Popoola, Emeka Godfrey Nwoba, James Chukwuma Ogbonna, Charles Oluwaseun Adetunji, Nwadiuto (Diuto) Esiobu, Abdulrazak Ibrahim, and Benjamin Ewa Ubi*

For more information about this series, please visit: https://www.routledge.com/Multidisciplinary-Applications-and-Advances-in-Biotechnology/book-series/MAABNIG

# Fermentation and Algal Biotechnologies for the Food, Beverage and Other Bioproduct Industries

Edited by
James Chukwuma Ogbonna, Sylvia Uzochukwu,
Emeka Godfrey Nwoba, Charles Oluwaseun Adetunji,
Nwadiuto (Diuto) Esiobu,
Abdulrazak B. Ibrahim, and Benjamin Ewa Ubi

CRC Press
Taylor & Francis Group
Boca Raton London New York

CRC Press is an imprint of the
Taylor & Francis Group, an **informa** business

First edition published 2022
by CRC Press
6000 Broken Sound Parkway NW, Suite 300, Boca Raton, FL 33487-2742

and by CRC Press
4 Park Square, Milton Park, Abingdon, Oxon, OX14 4RN

CRC Press is an imprint of Taylor & Francis Group, LLC



*Library of Congress Cataloging-in-Publication Data*
Names: Ubi, Benjamin Ewa, 1968- editor. | Biotechnology Society of Nigeria, issuing body.
Title: Multidisciplinary applications and advances in biotechnology : contributions from the Biotechnology Society of Nigeria working groups, six-volume set /
edited by Benjamin Ewa Ubi [and fourteen others].
Description: First edition. | Boca Raton : CRC Press, 2022. | Includes
bibliographical references and index. |
Contents: Volume 1. Fermentation and Algal Biotechnologies for the Food, Beverage and Other Bioproduct Industries—Volume 2. Agricultural biotechnology, biodiversity and bio-resources conservation and utilization—Volume 3. Medical Biotechnology, Biopharmaceutics, Forensic Science and Bioinformatics—Volume 4. Biosafety and Bioethics in Biotechnology—Volume 5. Microbiomes and Emerging Applications—Volume 6. Bioenergy and environmental biotechnology for sustainable development, | Summary: "This six volume book set examines a range of topics and applications related to biotechnology. Volumes include fermentation and algal biotechnologies; agricultural biotechnology; medical biotechnology, biopharmaceutics; biosafety, bioethics, biotechnology policy; microbiomes; bioenergy and environmental biotechnology for sustainable development"—Provided by publisher.
Identifiers: LCCN 2021047039 (print) | LCCN 2021047040 (ebook) |
ISBN 9780367766948 (v. 1 ; hardback) | ISBN 9781032013763 (v. 1 ; paperback) |
ISBN 9780367766955 (v. 2 ; hardback) | ISBN 9781032014982 (v. 2 ; paperback) | ISBN 9780367766962 (v. 3 ; hardback) | ISBN 9781032015026 (v. 3 ; paperback) | ISBN 9780367767020 (v. 4 ; hardback) | ISBN 9781032015781 (v. 4 ; paperback) | ISBN 9780367767044 (v. 5 ; hardback) | ISBN 9781032018331 (v. 5 ; paperback) | ISBN 9780367767051 (v. 6 ; hardback) | ISBN 9781032018416 (v. 6 ; paperback) | ISBN 9781003178378 (v. 1 ; ebook) | ISBN 9781003178880 (v. 2 ; ebook) | ISBN 9781003178903 (v. 3 ; ebook) | ISBN 9781003179177 (v. 4 ; ebook) | ISBN 9781003180241 (v. 5 ; ebook) | ISBN 9781003180289 (v. 6 ; ebook)
Subjects: LCSH: Biotechnology—Nigeria.
Classification: LCC TP248.195.N6 M95 2022 (print) |
LCC TP248.195.N6 (ebook) | DDC 660.609669—dc23
LC record available at https://lccn.loc.gov/2021047039
LC ebook record available at https://lccn.loc.gov/2021047040

ISBN: 978-0-367-76694-8 (hbk)
ISBN: 978-1-032-01376-3 (pbk)
ISBN: 978-1-003-17837-8 (ebk)

DOI: 10.1201/9781003178378

Typeset in Times
by codeMantra

# Contents

# Foreword

Biotechnology is an essential broad field of study that encompasses the use of living organisms, biomolecular processes and biological derivatives for the advancement of multiple technologies and the development of end-products in an effort to significantly improve our well-being and the environment. The rapid growth in human population coupled with the current global food security has tremendously necessitated the development of innovative technologies and breakthrough products within the field of biotechnology. Furthermore, biotechnology-derived solutions are also essential to overcome various challenges associated with food production, tackling diseases (e.g. COVID-19 pandemic), cleaning up pollution and to produce a wide range of bioproducts.

Aligning well with the recent interest and need for biotechnology advancement, this book series focuses on a wide range of current multidisciplinary applications and advances within the field of biotechnology. As such, the aim of this book edition is to review the most important aspects of fermentation and algal-based bioprocesses for successful commercial application within food, beverage and bioproduct industries, especially in the African continent. This book is developed in a way to guide and provide necessary resources for general readers and newcomers, as well as experienced researchers within this field. For the ease of the reader, this book is conveniently divided into two sections: Section A – Dairy and Fermented Foods and Section B – Microalgae and Novel Products.

In the first two chapters of Section A, the readers are introduced to a wide range of native fermented foods in the locality of Africa with special emphasis on their importance and potential as well as the integration of innovative technologies to improve their quality and production. Subsequent chapters in this section focus on other recent biotechnological advancements within the food and beverage industry such as enzyme production (Chapter 3), dairy production (Chapter 4), soy-based products (Chapter 5) and caffeine-based products such as cocoa (Chapter 6). This section concludes with the in-depth discussion of upcoming biotechnological tools such as metabolic engineering for the development of novel biological pathways, designs and systems (Chapter 7) as well as the potential role of intelligent feedback-based control systems for improving the productivity and quality of bio-fermented products (Chapter 8).

Section B begins with an introduction to the history of algal biotechnology R&D and its current application for the production of a wide range of novel food products in Africa (Chapters 1 and 2). Among the key attractions of microalgae is their ability to produce various commodity-based products. Chapter 3 details some of the latest research and development in microalgae cultivation and production technologies to produce a range of biofuels and high-value products.

Overall, I am thrilled and amazed at the completeness and amount of valuable discussion that is being included in this book on the ongoing advances in both fermentation- and algal-biotechnology. I trust the readers will share the level of enthusiasm as much as I did and most importantly gain a better understanding of the role of biotechnology in shaping our future.

**Dr. Ashiwin Vadiveloo**
*Algae Research & Development Centre*
*Murdoch University, South Street*
*Murdoch WA 6150, Australia*

# Preface

In recognition of the pivotal role of biotechnology as a burgeoning technology in fostering a knowledge-driven bio-based economy, the Biotechnology Society of Nigeria (BSN) has – in the about 40 years of its founding – made enormous contributions towards expanding the frontiers of biotechnology in Nigeria for sustainable livelihoods. This book is one of the multi-volume series on the Multidisciplinary Applications and Advances in Biotechnology: Contributions from the BSN, with each volume covering specific contemporary topics targeted at providing scientists in the academia, industry, research institutes/agencies, etc. with recent advances in the field so as to bridge the knowledge gap in the crucial areas of biotechnology.

This book is designed to advances in biotechnology in Nigeria over the past decades and the emerging biotech sector as a key driver of the bioeconomy. It is a timely knowledge product documenting key issues on recent advances in the field for use by a variety of readers including postgraduate students, professionals in the field, industry players, policy makers, science advocacy groups, etc. and engenders a deeper understanding of underlying mechanisms, technologies, processes and science-policy nexus that have placed Nigeria as a leader in biotechnology in Africa.

This volume covers 2 of the 16 Working Groups of the Society, namely the Dairy and Fermented Foods (WG01) and Microalgae and Novel Products (WG03), and contains a total of 12 chapters which bring to light the applications of available knowledge and deliverables in the thematic areas covered in the topics by the array of seasoned contributing authors. The chapters cover a wide range of topics including adopting biotechnologies/fermentation technologies for improving the quality of dairy and African fermented and underutilized/novel foods towards combating malnutrition in sub-Saharan Africa, industrial enzyme production for the food and beverage industries, algae cultivation systems and use as a novel food source and other high-value products/applications, mushroom from Africa as a functional and nutraceutical food source, nutraceutical beverages derived from cocoa (an important tropical crop) and other caffeine products, metabolic engineering of yeasts from palm wine (a yeast-enriched fermented traditional alcoholic drink obtained from the sap of palm trees) for the production of some food ingredients, etc. Each chapter addresses the relevant research area and discusses each of the thematic areas concisely and with emphasis on the important points, largely illustrating research processes beautifully with clear schematic diagrams to ease comprehension of the presented subject areas. Many of the chapters are co-authored by a team of different scientific experts in the field, thereby producing in-depth scholarly output in a high level of interdisciplinary collaborative manner, using the pool of expertise available in the BSN Working Groups, which would go a long way in ensuring a higher level of sustainability on the deliverables expected from these thematic research areas towards fostering a sustainable, knowledge-driven, bio-based economy.

This book is a useful reference material, a comprehensive description of the recent advances in the field, and a useful guide to the original literature in the subject matter, being specifically targeted at scientists and researchers working in the field of food biotechnology, enzyme biotechnology, biopharmaceuticals, nutraceuticals, fermentation technology, algae biotechnology, bioenergy, bioremediation, etc. in the universities, research institutions, industry or government agencies.

Overall, this volume is an excellent piece of scientific contributions that targets a wide range of audience including global leaders working on strategies at combating malnutrition particularly in sub-Saharan Africa, industrialists with interest in valorization of the abundant bioresources in Africa and biotech start-ups, the food industry, the agricultural and fisheries sectors, professionals involved in biopharmaceutics and therapeutics, investors, farmers, policy makers, educators and researchers, and educational and research institutions (most especially for postgraduate students). Readers will find this book a valuable resource containing recent advances contributed by seasoned and diverse array of scholars within this professional society.

The carefully selected editorial team and the authors of individual chapters drawn from several key institutions were chosen for their recognized international expertise and contributions to the various subject areas. Their willingness to make these invaluable contributions in their respective fields of expertise reflecting the different working groups and in a collaborative way to foster research networking is gratefully acknowledged. The editors wish to thank the reviewers of the different manuscripts for their unceasing dedication and great input. We thank Prof. Malachy O. Akoroda of the Department of Agronomy, University of Ibadan, who shared the initial idea that inspired this book initiative with me just after the conclusion of the 32nd Annual Conference of the Society held at the International Institute of Tropical Agriculture (IITA) Ibadan, Nigeria. Furthermore, we highly appreciate the foresight, diligent support and understanding of the publisher, Taylor and Francis Group (USA), for publishing *Multidisciplinary Applications and Advances in Biotechnology- I: Fermentation and Algal Biotechnologies for the Food, Beverage and Other Bio-product Industries with their international hallmark of excellence. We return special thanks to Dr. Marc Gutierrez and Mr. Nick Mould, whose great and painstaking efforts, constant suggestions and guidance at all stages saw to the successful publication of this book.*

**Prof. Benjamin Ewa Ubi**
*President, Biotechnology Society of Nigeria (BSN)*
*& Chief Editor*
*15 February, 2021*

# Editors

**James Chukwuma Ogbonna,** Ph.D., is a Professor of Industrial Microbiology and Biotechnology. He obtained a B.Sc. (First Class Hons in Botany) from the University of Jos, Nigeria (1982); M.Eng. (Fermentation Technology) from Yamanashi University, Japan (1988); and Ph.D. (Applied Biochemistry) from the University of Tsukuba, Japan (1991). He is currently the Director, National Biotechnology Development Agency, South East Zonal Biotechnology Centre, University of Nigeria, Nsukka, and the immediate past Deputy Vice Chancellor (Academic), University of Nigeria, Nsukka. He was the National President, Biotechnology Society of Nigeria (2010–2014) and the current President, Foundation for African Development through International Biotechnology. His current research interests include PBR design and optimization, development of microbial cell culture systems and applications, and bioenergy production. He has published three books, co-edited three books, contributed 13 book chapters, published 111 peer-reviewed journals articles, and presented nine conference proceedings and over 120 papers in national and international conferences in over 11 countries. His current H-index is 34 with over 3,760 citations in Google Scholars. Prof. Ogbonna is a Fellow of the Biotechnology Society of Nigeria, the Foundation for African Development through International Biotechnology, and the Institute of Corporative Administration.

**Sylvia Uzochukwu,** Ph.D. is a Professor of Food Science and Biotechnology at the Department of Plant Science and Biotechnology, Federal University, Oye-Ekiti, Nigeria. She holds a B.Sc. in Botany from the University of Nigeria Nsukka, and M.Sc. and Ph.D. in Food Science and Technology, from the same University. She was the Director of the Biotechnology Centre of the Federal University of Agriculture, Abeokuta from 2002 to 2009, Head of Department of Biotechnology of the Federal University Oye-Ekiti from August to November 2013, and Dean of the Faculty of Science at the same University, 2013–2017. She has been involved in capacity building in biotechnology in Nigeria, and has organized up to 20 short training courses in Molecular Biology and Biosafety between 2000 and 2018, and has attended about eight short training courses in Molecular Biology in Europe, Asia and America. She has won many International and National research grants for biotechnology research, and currently works on unintentional effects of genetic modification on GM foods approved for use in Nigeria. She is currently the Director of the Biotechnology Centre, Federal University Oye-Ekiti, Nigeria.

**Dr. Emeka Godfrey Nwoba,** Ph.D., is currently a research fellow at the Algae Research & Development Centre, Murdoch University, Western Australia, Australia. His research areas include industrial & environmental biotechnology, microalgae biotechnology, applied phycology/algology, energy-efficient photobioreactors (PBRs) and biostatistics. He has over 6 years of research and development experience in algae (microalgae and macroalgae) biotechnology including algae-based wastewater treatment, bio-based products from algae, and design, optimization and scale up of PBRs. His academic training was at Ebonyi State University, Abakaliki, Nigeria, where he received his B.Sc. and M.Sc. degrees, in Biochemistry/Industrial Biotechnology. Dr. Nwoba obtained his Ph.D. in Biotechnology/Energy from Murdoch University, Australia. He has published several peer-reviewed research and review articles in Q1 Journals and book chapters. His current research focuses on partnering engineering, material science and biology to develop high efficiency/next-generation algal PBRs (i.e. autonomous/standalone PBRs) with the capability to co-produce chemical (algal biomass) and electrical energy towards freshwater-based cooling and grid-electricity independent operation. He is a member of Australasian Society of Phycology and Aquatic Botany, International Society for Applied Phycology, Phycological Society of Nigeria and Biotechnology Society of Nigeria.

**Charles Oluwaseun Adetunji,** Ph.D., is an Associate Professor of Microbiology and Biotechnology at the Microbiology Department, Faculty of Science, Edo State University Uzairue (EDSU), Edo State, Nigeria, where he utilized the application of biological techniques and microbial bioprocesses for the actualization of sustainable development goals and Agrarian revolution, through quality teaching, research and community service. He is currently the Acting Director of Intellectual Property and Technology Transfer, Sub-Dean for Faculty of Science, and the Head of Department of Microbiology at EDSU. He is a Visiting Professor and the Executive Director for the Centre of Biotechnology, Precious Cornerstone University, Ibadan. He has won several scientific awards and grants from renowned academic bodies like Council of Scientific and Industrial Research (CSIR), India; Department of Biotechnology (DBT), India; the World Academy of Science (TWAS), Italy; Netherlands Fellowship Programme (NPF), the Netherlands; the Agency for International Development Cooperation; Israel; and Royal Academy of Engineering, UK, among many others. He has filed several scientific patents on Bioherbicides, Biopesticides, Nanobiosurfactants, Nanobiopesticdes and many more. He has published over 300 scientific articles in refereed journals and conference proceedings both international and local journals. He was ranked among the top 500 prolific authors in Nigeria between 2019 till date by SciVal/SCOPUS. He is currently editing several biotechnology textbooks with Elsevier. He was recently appointed as the President and Chairman Governing Council of the Nigerian Bioinformatics and Genomics Network Society. He is presently a series editor with Taylor and Francis, USA, editing several textbooks on Agricultural Biotechnology, Nanotechnology, Pharmafoods and Environmental Sciences. The breadth of his scholarly contributions to research is evident from his contributions which cover several topics relating to Food Security, Agriculture and Environmental Sustainability. His research interest includes microbiology, biotechnology, bioresources and nanotechnology. He is an editorial board member of many international journals and serves as a reviewer to many double-blind peer-review journals like Elsevier, Springer, Taylor and Francis, Wiley, PLOS One, Frontiers, Nature, American Chemistry Society, Bentham Science Publishers, etc. He is a member of many scientific and professional bodies including the American Society for Microbiology, Nigerian Young Academy, Biotechnology Society of Nigeria and Nigerian Society for Microbiology. He has won a lot of international recognition and also acted as a keynote speaker delivering invited talks/position papers at various universities, research institutes and several centres of excellence which span across several continents of the globe. Over the last 15 years, he has built strong working collaborations with reputable research groups in numerous and leading universities across the globe. He is the convener for Recent Advances in Biotechnology, which is an annual international conference where renowned Microbiologists and Biotechnologists come together to share their latest discoveries (https://raibconference.wixsite.com/website). He is the Founder & CEO of BECTIK Biotechnology and Nanotechnology Company. His Biotechnology Company consists of a team of leading academics in their fields and thrives to deliver solutions all around Bio- and Nanotechnology. The company also delivers solutions that are tailored to your business needs and for any scale.

**Prof. Nwadiuto (Diuto) Esiobu** (Ph.D., University of Louvain, Belgium; Post Doc. Massachusetts Institute of Technology, Cambridge, USA) is currently a Professor of Microbiology and Biotechnology at Florida Atlantic University, Boca Raton, FL, USA, where her microbial biotech laboratory integrates environmental genomics and health risk assessment with rapid pathogen detection using synthetic biology. Her research interests in agriculture and health include molecular ecology of plant invasion, soil fertility and the applications of microbiomes in humans and plants. She has served in senior professional leadership roles in academia, industry and policy, including selection panel for the US President's award for excellence in Math and Science (Washington, DC) and more than 11 years on Committees of the Board of the American Society for Microbiology. A Fellow of the Biotechnology Society of Nigeria and a Jefferson Science Fellow, she was selected by the National Academy of Sciences as a senior science advisor to Secretary Clinton's Office of Global Food Security (GFS) and the Office of International Health and Biodefense (OES/IHB), US Department of State, Washington, DC, and named world-class Faculty by the City of Fort Lauderdale for her contributions to the scientific enterprise.

**Abdulrazak B. Ibrahim,** Ph.D., is an agricultural biotechnologist with experience and skills on plant biotechnology stewardship, stakeholder engagement, foresight analysis, development of transgenic crops, laboratory and field management of GM crops, biosafety analysis, adoption and discontinuation of events. He was trained in the Brazilian Agricultural Research Cooperation (EMBRAPA Center for Genetic Resources), from where he obtained a doctorate degree in Molecular Biology from Universidade de Brasilia (UnB) and M.Sc. from Universidade Fedeal do Ceara (UFC), Fortaleza, and acquired best practices in agricultural biotechnology leading to the development and patenting of a transgenic technology for the control of whitefly using RNA interference technology within EMBRAPA, and the development of virus-resistant cowpea, under the Africa-Brazil Innovation Platform, implemented by EMBRAPA and Forum for Agricultural Research in Africa (FARA), with support from Department for International Development (DFID), Bill & Melinda Gates Foundation (BMGF), International Fund For Agricultural Development (IFAD) and Foundation Arthur Bernardes (FUNARBE). Dr. Ibrahim has worked as a Research Officer in Jigawa Research Institute (JRI), Nigeria, where he set up the agricultural biotechnology unit of the institute, before joining the Department of Biochemistry of Ahmadu Bello University (ABU), Nigeria, and rose to the rank of Senior Lecturer (currently under external assessment for Associate Professorship). Currently, he is a capacity development expert at the FARA, where he coordinates the Agricultural Research and Innovation Fellowship for Africa (ARIFA). He is a member of Nigeria's Tertiary Education Trust Fund's (TETFund) Research and Development Standing Committee (RDSC). He has taught several courses in biology, biochemistry, plant science, agricultural biotechnology, nutrition, genetic engineering and biosafety at undergraduate and postgraduate levels, is a published author and is the Editor-in-Chief of the *Nigerian Journal of Biotechnology* (NJB).

**Benjamin Ewa Ubi,** Ph.D., is a Professor of Plant Breeding and Biotechnology. He obtained his B.Sc. Honours (Agric.) Agronomy from the University of Calabar (1990) and M.Sc. Agronomy – Crop Science (Plant Breeding and Tissue Culture) (1994) and Ph.D. Agronomy (Plant Breeding and Biotechnology) (1998) from the University of Ibadan, Nigeria. His Ph.D. research was completed at the then Biotechnology Laboratory of the International Institute of Tropical Agriculture (IITA), Ibadan, where he was a graduate research fellow and became the President of the International Association of Research Scholars and Fellows (IARSAF) in 1996. Among several other research engagements, Prof. Ubi served as Science and Technology Agency (STA) Postdoctoral Fellow under the Japan International Science and Technology Exchange Centre (JISTEC) 1999–2001 at the Plant Breeding Department (Biotechnology Laboratory) of the National Institute for Livestock and Grassland Science (NARO, National Agricultural Research Organization), Nishinasuno, Japan; and as a three-term Visiting Scientist at the NARO National Institute of Fruit Tree Science (NIFTS), Tsukuba, Japan, under the auspices of the Inoue Research Foundation of Japan Fellowship (April 2003–April 2004) and the Japanese Society for the Promotion of Science (JSPS) visiting fellowships (October 2008–October 2009; May 2012–March 2013). Prof. Ubi served as a Visiting Professor at the Arid Land Research Centre (The Global Centre of Excellence for Dryland Science) (Molecular Breeding Laboratory), Tottori University, Japan; and Visiting Professor, University of Tokyo, Japan. Among several other academic engagements, Prof. Ubi has served as the Director of the Biotechnology Research and Development Centre, Ebonyi State University, Abakaliki (August 2006–March 2008) and the Head of the Department of Biotechnology, Ebonyi State University for several years; Dean, Faculty of Science, Ebonyi State University Abakaliki and Pioneer Editor-in-Chief of *Science Journal* <https://sciencejournal.ng>, a publication of the Faculty of Science, Ebonyi State University, Abakaliki; President, Biotechnology Society of Nigeria (BSN) and Leader of the Crops Genetics, Breeding and Genetics Working Group of the Biotechnology Society of Nigeria (BSN).

# Contributors

**Tolulope Adeleye**
Department of Microbiology
Federal University of Agriculture
Abeokuta, Nigeria

**Taofikat A. Adesalu**
Department of Botany
University of Lagos
Lagos, Nigeria

**Abosede T. Adesalud**
Department of Botany
University of Lagos
Lagos, Nigeria

**Charles Oluwaseun Adetunji**
Applied Microbiology, Biotechnology and Nanotechnology Laboratory, Department of Microbiology
Edo State University Uzairue
Okpella, Nigeria

**Juliana Bunmi Adetunji**
Nutritional and toxicological Research Laboratory, Department of Biochemistry Sciences
Osun State University
Osogbo, Nigeria

**Mathias A. Chia**
Department of Botany
Ahmadu Bello University
Zaria, Nigeria

**David Chuka-Ogwudeb**
Algae R&D Centre
Murdoch University
Perth, Australia

**Wadazani Palnam Dauda**
Crop Science Unit, Department of Agronomy
Federal University Gashua
Gashua, Nigeria

**Daniel Ingo Hefft**
University Centre Reaseheath
Nantwich, UK

**John N. Idenyi**
Department of Biotechnology
Ebonyi State University
Abakaliki, Nigeria

**Saher Islam**
Institute of Biochemistry and Biotechnology, Faculty of Biosciences
University of Veterinary and Animal Sciences
Lahore, Pakistan

**Sarafadeen Kareem**
Department of Microbiology
Federal University of Agriculture
Abeokuta, Nigeria

**Samuel Makinde**
Informatics and CyberPhysical Systems Unit, Department of Computer Science
Edo State University Uzairue
Okpella, Nigeria

**Florence U. Masajuwa**
Faculty of Law
Edo State University Uzairue
Okpella, Nigeria

**Augusta Anuli Nwachukwu**
Department of Biotechnology, School of Biological Sciences
Federal University of Technology
Owerri, Nigeria

**Wilson Nwankwo**
Informatics and CyberPhysical Systems Unit, Department of Computer Science
Edo State University Uzairue
Okpella, Nigeria

**Emeka Godfrey Nwoba**
Algae R&D Centre
Murdoch University
Perth, Australia

**Priscilla Amaka Ogbodo**
Department of Biotechnology, School of Biological Sciences
Federal University of Technology
Owerri, Nigeria

**Innocent Ogbonna**
Department of Microbiology
Federal University of Agriculture
Makurdi, Nigeria

**James Chukwuma Ogbonna**
Department of Microbiology
University of Nigeria
Nsukka, Nigeria

**Toochukwu Ekwutosi Ogbulie**
Department of Biotechnology, School of Biological Sciences
Federal University of Technology
Owerri, Nigeria

**Philippa C. Ojimelukwe**
Department of Food Science and Technology
Michael Okpara University of Agriculture
Umudike, Nigeria

**Olugbemi Tope Olaniyan**
Laboratory for reproductive Biology and Developmental Programming, Department of Physiology
Edo State University Uzairue
Okpella, Nigeria

**Julius Kola Oloke**
Department of Pure and Applied Biology
Ladoke Akintola University of Technology
Ogbomoso, Nigeria

**Olusola Olawale Olaleye**
Nigerian Institute for Trypanosomiasis Research
Ibadan, Nigeria

**Christiana N. Opara**
Department of Microbiology
Federal University of Technology
Otueke Bayelsa State, Nigeria

**Osarenkhoe O. Osemwegie**
Department of Biological Sciences, Microbiology Unit
Landmark University
Omu-Aran, Nigeria

**Olusola Oyewole**
Department of Food Science and Technology
Federal University of Agriculture
Abeokuta, Nigeria

**Devarajan Thangadurai**
Department of Botany
Karnatak University
Dharwad, India

**Benjamin Ewa Ubi**
Department of Biotechnology
Ebonyi State University
Abakaliki, Nigeria
and
Arid Land Research Center
Tottori University
Tottori, Japan

**Kingsley Eghonghon Ukhurebor**
Climatic/Environmental/
Telecommunication Physics Unit,
Department of Physics
Edo State University Uzairue
Okpella, Nigeria

# 1 Improving Traditionally Fermented African Foods through Biotechnology
## *Need to Translate Art to Science*

*Philippa C. Ojimelukwe*
Michael Okpara University of Agriculture

**CONTENTS**



## 1.1 INTRODUCTION

Food fermentation is an age-long culture in Africa. Howbeit, the pace of development of fermented food products is slow due to several factors: Lack of infrastructure and funds for research, inability of African researchers to take their own development as a priority and changes in food culture that fail to promote the consumption of indigenous fermented foods have constituted impediments

DOI: 10.1201/9781003178378-1

to the development of traditionally fermented African foods. Poor implementation of existing policies and creation of innovative policies to drive research in African food fermentations have also slowed down the pace of development of fermented African foods. Biotechnology may be defined as "the use of living organisms (and their derivatives), to formulate or modify products or processes for specific use that will benefit man" (FAO, 2010). Biotechnology manipulates living organisms and biological systems to make products that advance agriculture, health and environmental control by modifying things to benefit humans. During fermentation, enzymes found in microorganisms metabolize food for enhanced nutritive value, sensory properties and shelf stability (FAO, 2010). Many biotechnological techniques are available for the effective improvement of traditionally fermented African foods. Classical mutagenesis and conjugation may be used for improving the suitability of microorganisms and the yield and concentration of metabolites (Turpin, 2012). They are traditional methods of genetic improvement (Turpin et al. 2012). Recombinant gene technology may be used for strain improvement in microorganisms used for food fermentations (Adrio and Damain, 2010). The capacity for improved enzyme production and improved production of food ingredients (such as monosodium glutamate) by microbial cultures may be modified through recombinant gene technology. Using metagenomics, information about the microbiomes that will enhance fermentation may be deduced (Mallikarjuna and Yellama, 2019). Enzyme-linked immunoabsorbent assays are valuable diagnostic tools for the design of starter cultures (Macwana and Muriana, 2012). Polymerase chain reaction (PCR) is used for detecting microorganisms in assays (Ouoba et al., 2004). Tools for genetic manipulations and methodologies for improving starter cultures are regulated and are sensitive to consumer perceptions (https://www.reatch.ch/de/content/regulation-genetically-modified-organisms-present-and-future).

Modern biotechnology diagnostic tools are required for developing and designing starter cultures for African food fermentations. They direct appropriate metabolic transformation of traditional foods and innovations in the design of bioreactors (and other equipment) for controlled fermentation (Brandt, 2014). In Asia, fermentation inoculants are available in granular form or as pressed cake. Fermentation biotechnology is more advanced in Asia than in Africa (Swain et al., 2014). Bioreactors are used more frequently for starter culture production. Biotechnological methods are more accurate and they reduce fermentation time. Developments in biotechnology have provided low-cost rapid identification with lower costs and less time requirements than conventional techniques. Biotechnological standards as defined by the FAO could be used as evaluation criteria for further development of traditional African food fermentations. Have the key fermenting microorganisms been identified? Are the functions of the key microorganisms known? Has genetic improvement been carried out for the key microorganisms involved? Are starter cultures suitable for the fermentation process available? Have the raw materials most suited for the product been identified? Are improved technologies developed for the fermentation process? Has any of such technologies been adopted? Has the fermented product been produced on the pilot scale? Has it been produced on an industrial scale? (FAO, 2010).

## 1.2 METHODOLOGY

Internet searches were conducted using several search engines for primary, secondary and tertiary sources of literature on African food fermentations. The key words used were food fermentations; African food fermentations; microorganisms and food fermentations; biotechnology; and microbial improvement of food fermentations. Information on food fermentations were obtained using different search engines (Google Scholar, WebCrawler, MetaCrawler, America Online Library, etc.). Abstracts obtained were scrutinized for their relevance, and subsequently, original journal articles, chapters in books, books, conference proceedings and documentaries containing relevant information were used. Relevant information was collated, analysed and organized into the review paper.

## 1.3 RESULTS AND DISCUSSION

### 1.3.1 Microorganisms That Ferment Traditional African Foods

Many microorganisms are involved in natural fermentations. The microbial flora of each environment (soil, water and air) naturally initiates the fermentation of food products in that environment. Some microbes transform the substrate more than others. Such microorganisms are potential starter cultures for the controlled fermentation and process standardization of the food products. Yeasts and lactic acid bacteria (LAB) are predominant in cereal fermentations (Oguntoyinbo et al., 2011). Most natural legume fermentations are predominated by *Bacillus* species and are alkaline fermentations (Battcock and Azam Ali, 2001). *Leuconostoc* species and LAB dominate vegetable fermentations (Swain et al., 2014). Table 1.1 shows the microorganisms that have been identified in traditionally fermented African foods.

### 1.3.2 Improvements in Microorganisms Used for African Food Fermentations

Microorganisms used in food fermentations can be improved for better efficiency using several genetic improvement strategies (Adewumi et al., 2013). Mutations may be induced using mutagenic agents like ultraviolet light and various chemicals may be employed. This has to be followed up by selections to screen off undesirable mutants and further development of strains with improved activity for the food fermentation processes (Nam et al., 2012). Transduction is a method of natural gene transfer. A bacteriophage facilitates the genetic modification process. Bacteriophages facilitated the genetic modification of *Escherichia coli* but have not been employed for the modification of many other food fermentation microorganisms because gene transfer between unrelated strains is difficult and the process is not very efficient (Nam et al., 2012). This could constitute an innovative research focus. In addition, bacteriophages for most bacterial strains are not well characterized. Conjugation is a natural gene transfer method involving close physical contact between donor and recipient. Improvement of microbial strains by the use of conjugation requires good understanding of plasmid biology, but only very few conjugative plasmid encoding genes have been characterized (Gonzalez et al., 2016). During transformation (which

**TABLE 1.1**
**Microorganisms Associated with Selected Traditional African Fermented Foods**

| Fermented Product | Microbial Flora | References |
|---|---|---|
| **Legume products** | | |
| *Ugba Pentaclethra macrophylla* | *B. subtilis*; *B. pumilus*; *B. licheniformis*; *Staph. saprophyticus* | (Ahaotu et al., 2013) |
| *Ogiri/Ogili* from melon/ castor oil/ pumpkin/bean/ sesame seeds | *B. subtilis*; *B. licheniformis*; *B. megaterium*; *B. pumilus*; *B. rigui*; *Pediococcus* spp.; *Staph. saprophyticus*; *Lb. plantarum* | (Odunfa and Oyewole, 1998) |
| Okpehe seeds from *Prosopis africana* | *B. subtilis*; *B. cereus*; *B. amyloliquefaciens*; *B. licheniformis* | (Oguntoyinbo et al., 2001) |
| Soumbala from African locust bean seeds | *B. subtilis*; *B. pumilus*; *B. atrophaeus*; *B. amyloliquefaciens*; *B. mojavensis*; *Lysininbacillus sphaericus*; *B. thuringiensis*; *B. licheniformis*; *B. cereus*; *B. badius*; *B. firmus*; *B. megaterium*; *B. mycoides*; *B. sphaericus*; *Peanibacillus alvei*; *Brevibacillus laterosporus* | (Ouoba et al., 2004) |
| Dawadawa (African Locust bean) | *B. pumilus*; *B. licheniformis*; *B. subtilis*; *B. firmus*; *B. atrophaeus*; *B. amyloliquefaciens*; *B. mojavensis*; *Lysininbacillus sphaericus* | (Amoa-Awua et al., 2006; Meerak et al., 2008) |
| Kinda (African Locust bean) | *B. subtilis*; *B. amyloliquefaciens*; *B. pumilus*; *B. licheniformis*; *B. atrophaeus*; *B. mojavensis*; *Lysininbacillus sphaericus* | (Meerak et al., 2008) |
| African yam bean (*Sphenostylis stenorcarpa*) | *Lb. jensenii*; *B. coagulans*; *Aerococcus viridans*; *Pediococcus cerevisiae*; *Aspergillus niger* | (Jeff-Agboola et al., 2007) |
| **Cereal products** | | |
| Ogi/Ogi baba/ Akamu (maize, millet, sorghum) | *Lb. plantarum*; *Lb. fermentum*; *Leuc. mesenteroides;* *Corynebacterium* spp.; *Acetobacter*; *L. brevis*; *L. pentosus*; *L. cellibiosus*; *Lactococcus rafinolactis*; *Pediococcus*; *P. pentosaceus*; *L. vacinostercus*; *C. krusei*; *S. lactis* | (Olasupo et al., 2000) |
| Chibuku | *Lactobacillus*; *S. cerevisiae* | Mohammed et al., 1991; Nwachukwu et al., 2010) |
| Umqombothi | | |
| Mahewu | *Lb. bulgaricus*; *Lb. brevis* | (Fadahunsi and Soremekun, 2017) |
| Munkoyo: Chibwantu | *Lactobacillus* spp.; *Weissella* spp. | (Schoustra et al., 2013) |
| Borde (from maize) | *Lactobacillus* spp.; aerobic mesophilic bacteria, LAB; yeasts | (Ketema et al., 1998) |
| Kenkey or Koko (maize) | *Lb. fermentum*; *Lb. brevis*; *Lb. plantarum*; *L. reuteri*; *Saccharomyces*; *Candida* spp.; *L. confusa*; *L. salivarius*; *Pediococcus* spp.; *E. cloacae*; *Acinetobacter* spp.; *S. cerevisiae*; *C. mycoderma* | (Oguntoyinbo et al., 2011) (Halm et al., 1996) |

(*Continued*)

**TABLE 1.1 (*Continued*)**
**Microorganisms Associated with Selected Traditional African Fermented Foods**

| Fermented Product | Microbial Flora | References |
|---|---|---|
| Mawe (maize) | *Lb. fermentum*; *L. reuteri*; *Lb. paraplantarum*; *L. brevis* | (Hounhouigan et al., 1993) |
| Kunun zaki (maize, sorghum) | *Lb. plantarum*; *L. pantheris*; *S. cerevisiae*; *L. fermentum*; *B. subtilis*; *L. vaccinostercus*; *Corynebacterium* spp. *Aerobacter* spp.; *P. pentosaceus*; *C. mycoderma*; *S. gallolyticus* subsp. *macedonicus*; *W. confusa*; *Rhodotorula* spp.; *Cephalosporium* spp.; *Fusarium* spp.; *Aspergillus* spp.; *Penicillium* spp. | (Olasupo et al., 2000; Oguntoyinbo et al., 2011) (Oguntoyinbo and Nabad., 2012; Inyang and Dabot, 1997) |
| Mbege (finger millet/banana) | *Lb. plantarum*; *S. cerevisiae*; *Schizosaccharomyces pombe*; *Leuc. mesenteroides* | (Odunfa and Oyewole, 1997) |
| Fura | *Lb. fermentum*; *L. reuteri*; *Lb. paraplantarum*; *P. acidilactici*; *W. confusa*; *L. salivarius* | Olasupo et al., 2000) |
| *Enjera/Injera* | *Lb. plantarum*; *Lb. pontis*; *Leuc. mesenteroides*; *Ped. cerevisiae*; *S. cerevisiae*; *Cand. glabrata* | (Olasupo et al., 2010) |
| Ben-saalga | *Lb. plantarum*; *Lb. fermentum*; *Lactobacillus* spp.; *Pediococcus* spp.; *Leuconostoc* spp.; *Weissella* spp.; yeasts | (Songré-Quattara et al., 2008; Turpin et al., 2012; Tou et al., 2006) |
| Gowe (sorghum) | *Lb. fermentum*; *P. acidilactici*; *W. confusa*; *Pichia anomala*; *L. mucosae*; *Lb. fermentum*; *Lb. reuteri*; *Lb. brevis*; *Lb. confuse*; *Lb. curvatus*; *Lb. buchneri*; *Lb. salivarius*; *Lact. lactis*; *Ped. pentosaceus*; *Ped. acidilactici*; *Leuc. mesenteroides*; *Candida tropicalis*; *C. krusei*; *Kluyveromyces marxianus* | (Abegaz et al., 2002) (Greppi et al., 2013a,b) |
| Malted sorghum grain/flour | *Lb. fermentum*; *P. acidilactici*; *W. confuse* | (Sawadogo-Lingani et al., 2010) |
| Mawe (maize) | *Lb. fermentum*; *Lb. reuteri*; *Lb. brevis*; *Lb. Lb. curvatus*; *Lb. buchneri*; *Leuc. mesenteroides*; *Candida glabrata*; *S. cerevisiae*; *Lb. salivarius*; *Lact. lactis*; *Ped. pentosaceus*; *Ped. acidilactici*; *Kluyveromyces marxianus*; *Clavispora lusitaniae* | (Greppi et al., 2013a,b) |
| Pito (maize, sorghum) | *G. candidum*; *Lactobacillus* spp.; *Candida* spp. | (Odunfa and Oyewole, 1997) |
| Potopoto (maize) | *Lb. plantarum*; *Lb. reuteri*; *Lb. gasseri*; *Lb. paraplantarum*; *Lb. acidophilus*; *Lb. delbrueckii*; *Lb. casei*; *Bacillus* spp.; *Enterococcus* spp.; yeasts | (Abriouel et al., 2006) |
| Uji (maize; sorghum; millet) | *Lb. mesenteriodes*; *Lb. plantarum* | (Odunfa and Oyewole, 1997) |
| Busa (maize, millet, sorghum) | *S. cerevisiae*; *Schizosacchromyces pombe*; *Lb. brevis*; *Lb. plantarum*; *Lb. helveticus*; *Lb. salivarius*; *Lb. casei*; *Lb. buchneri*; *Leuc. mesenteroides*; *Ped. damnosus* | (Odunfa and Oyewole, 1997) |
| Kisra (sorghum) Koko (maize porridge) | *Ped. pentosaceus*; *Lb. confusus*; *Lb. brevis*; *Erwinia ananas*; *Klebsiella pneumoniae*; *Ent. cloacae*; *Cand. intermedia*; *Deb. hansenii*; *Aspergillus* spp.; *Penicillium* spp.; *Fusarium* spp.; *Rhizopus* spp.; *Ent. cloacae*; *Acinetobacter* spp.; *Lb. plantarum*; *Lb. brevis*; *S. cerevisiae*; *Cand. mycoderma* | (Mohammed et al., 1991; Blandino et al., 2003) Hamad et al. (1997) |

(*Continued*)

**TABLE 1.1 (*Continued*)**
**Microorganisms Associated with Selected Traditional African Fermented Foods**

| Fermented Product | Microbial Flora | References |
|---|---|---|
| Togwa (maize, sorghum, millet) | *Lb. brevis*; *Lb. cellobiosus*; *Lb. fermentum*; *Lb. plantarum*; *Ped. pentosaceus*; *Candida pelliculosa*; *C. tropicalis*; *Issatchenkia orientalis*; *S. cerevisiae* | (Odunfa and Adeleye, 1985) |
| Arrow; Cere or Cakry | *Lb. plantarum*; *Leuc. mesenteroides*; *Lb. plantarum*; *B. subtilis*; *Candida* spp. | (Totté et al., 2003) |
| **Roots and tubers** | | |
| Cassava fufu | *L. plantarum*; *L. brevis*; *L. casei*; *Bacillus* spp.; *Leuc. mesenteroides*; *L. cellobiosus*; *L. coprophilus*; *Leuc. lactis*; *L. bulgaricus*; *Klebsiella* spp.; *Leuconostoc* spp.; *Corynebacterium* spp.; *Candida* spp.; *L. delbrueckii* | (Oyewole and Odunfa, 1990) |
| Garri (from Cassava) | *Corynebacterium manihoti*; *Geotrichum* spp.; *Lb. planetarium*; *Lb. buchneri*; *Leuconostoc* spp.; *Streptococcus* spp. | (Oyewole et al., 2004) |
| Chikwangue | *Corynebacterium*; *Acinetobacter*; *Bacillus*; *Lactobacillus*; *Micrococcus*; *Pseudomonas*; *Moraxella* | (Odunfa and Oyewole, 1998) |
| Cingwada | *Corynebacterium*; *Bacillus*; *Lactobacillus*; *Micrococcus* | (Odunfa and Oyewole, 1997) |
| Lafun/Konkonte | *Bacillus* spp.; *Klebsiella* spp.; *Candida* spp.; *Aspergillus* spp.; *Lb. mesenteroides*; *Corynebacterium manihoti*; *Lb. plantarum*; *Micrococcus luteus*; *Geotrichum candidum* | (Odunfa and Oyewole, 1997) |
| Yam (Elubo) | *L. plantarum; L. brevis; L. delbrueckii; B. subtilis*; *Lactococcus lactis; Klebsiella pneumoniae; Citrobacter freundii;* yeasts – *Pichia burtonii*; *Candida krusei* | (Achi and Akubor, 2000) |
| **Milk-based products** | | |
| Kindirmo | *Bacillus* spp.; *Streptococcus* spp.; *Micrococcus* spp.; *Lb. casei*; *L. lactis*; *L. brevis*; *L. bulgaricus* | (Esiobu et al., 2014) |
| Nono | *Lb. fermentum*, *L. plantarum*; *L. helveticus*; *Leuc. mesenteroides*; *Ent. faecium*; *Ent. italicus*; *Weissella confusa*; *Candida parapsilosis*; *C. rugose*; *C. tropicalis*; *Galactomyces geotrichum*; *Pichia kudriavzevii*; *S. cerevisiae* | (Akabanda et al., 2013) |
| Labanrayeb | *Candida* spp.; *Saccharomyces* spp.; *Lactobacillus* spp.; *Leuconostoc*; *L. casei*; *L. plantarum*; *L .brevis*; *Lact. lactis*; *Leuconostoc* spp. | (Odunfa and Oyewole, 1997; Ezeronye et al., 2005) |
| Mabisi (Amasi) | *L. plantarum*; *Lactococcus*; *Streptococcus*; *Leuconostoc* | (Heita et al., 2013) |
| Ambere (Amarunanu) | *Strept. thermophiles*; *L. plantarum*; *Leuc. mesenteroides*; *yeasts*; *Lactococcus lactis*; *Leuc. mesenteroides* subsp. *dextranicum*, *Lb. curvatus*, *Leuc. paramesenteroides*; *Lb. plantarum* | (Muigei et al., 2013) |

is a natural gene transfer method), naked DNA is taken up from the surrounding medium. The technique is not commonly used for the modification of food fermentation microorganisms. In electroporation, high-voltage short-time electric pulses are used to form transient pores in cell walls and cell membranes of microorganisms. Under properly controlled conditions, DNA in the surrounding medium will enter through the pores (Assad-Garcia et al., 2008). Strain improvement in microorganisms is being revolutionized by genetic engineering and is positively advancing food fermentation. It is a tool for identifying appropriate hosts and cloning vectors and for determining gene transfer procedures for *Escherichia coli*, *Bacillus subtilis*, some LAB and yeasts (Gonzalez et al., 2016).

## 1.4 METABOLIC PROCESSES ASSOCIATED WITH FOOD FERMENTATIONS

In microbial fermentations, biochemical reactions are used to transform substrates to products (such as organic acids, alcohols, aldehydes and ketones) (Weckx et al., 2010). Microbes such as *Lactobacillus*, *Lactococcus*, *Leuconostoc*, *Enterococcus*, *Streptococcus*, *Penicillium* and *Saccharomyces* could be involved in traditional food fermentations (Thakur et al., 2015). If fermentation is not controlled, it will contribute to food spoilage. Temperature, water activity and hydrogen ion concentration (pH) are factors which determine fermented product quality (Qiao et al., 2010). The substrate and the amount of oxygen available for fermentation are other important factors. In alcohol fermentation, sugar is converted to mainly ethanol, while carbon dioxide is often given off as a by-product (Kedia et al., 2007). Acid fermentation may yield lactic acid, acetic acid, formic acid, succinic acid, etc. (Ray and Panda, 2007). Types of acid fermentation include homolactic (yielding mainly lactic acid), heterolactic (which yields lactic acid, acetic acid, ethanol and $CO_2$) and mixed acid fermentation (yielding mixtures of acetic acid, lactic acid, formic acid, succinic acid and malic acid) (Di Cagno et al., 2013). LAB are extensively found in the traditionally fermenting cereals (maize, sorghum and millet) in the African setting (Sengun and Karabiyiklii, 2011).

### 1.4.1 Innovations in Metabolite Manipulations for Improving Traditional Fermentations

A good knowledge of the biochemistry and metabolic activities of a microorganism is required for the development of appropriate genetic manipulation strategies. The metabolic pathways and regulatory mechanisms for numerous microorganisms are not yet known. This information is, however, very valuable for the genetic improvement of microbial strains and for understanding of the dynamics of microbial communities. For instance, bacteriocins are metabolites which can improve food safety (Settanni and Corsetti, 2008). They can be transferred to microbes being used as starter cultures. Increased acid production and improved heat tolerance are examples of other factors that can be conferred on food fermentation microorganisms after a good understanding of their metabolic processes.

## 1.5 USE OF MIXED CULTURES FOR FOOD FERMENTATIONS

Mixed microbial strains yield better quality products than single strain fermentations (Greppi, 2013a). The most predominant microorganisms in the succession of natural fermentations are usually responsible for the major metabolic changes that convert the substrate to the fermented product. Many factors influence the growth and development of microorganisms. The most important factors relevant to food fermentations include the levels of acid, alcohol, oxygen, use of starters, temperature and amount of salt. These factors also influence the shelf stability of the fermented products (Smid and Lacroix, 2012).

To standardize the fermentation process, inoculants containing high concentrations of live microorganisms which can direct the course of fermentation (starter cultures) are developed. Many research efforts have been used to characterize the microorganisms commonly used in the processing of food products (Owusu-Kwarteng et al., 2012; Adimpong et al., 2012; Oguntoyinbo et al., 2011; Ekwem, 2014; Vieira-Dalode et al., 2007; Nwachukwu et al., 2010; Hounhouigan et al., 1993; Oguntoyinbo and Narbad, 2012; Sawadogo-Lingani et al., 2010; Turpin et al., 2012; Halm et al., 1993; Songré-Quattara et al., 2010; Obinna-Echem, 2015). Such studies demonstrate that natural fermentation involves mixed cultures of microorganisms (bacteria, yeasts and moulds). Microorganisms in the environment are often associated with uncontrolled traditional fermentation. Isolated microbial strains may be used to achieve the following: reduction of dry matter loss; better control over the fermentation steps; enhanced acid production; flavour enhancement; improved nutritional quality; improved product acceptability; and reduction of anti-nutrients and contaminants (Enwa et al., 2011; Ekwem, 2014; Halm et al., 1996; Hounhouigan et al., 1993; Lei and Jacobsen, 2004; Annan et al., 2003; Lardinois et al., 2003; Fandohan et al., 2005; Teniola et al., 2005; Agarry et al., 2010).

Phenotypic tests used for species identification may include the observation of phenotypic characteristics and gas production from glucose. Molecular typing techniques include pulsed-field gel electrophoresis, PCR-based methods and DNA sequencing. While a considerable level of species identification has been achieved for some bacterial species such as *Lactobacillus plantarum* (Zhang et al., 2009: Adeyemo and Onulide, 2014) and *Bacillus subtilis* (Figure 1.1), many food microorganisms have not been fully characterized using molecular techniques.

## 1.6 SCALING UP AND INDUSTRIALIZATION OF THE FERMENTATION PROCESS

Traditionally fermented foods are consumed in many African countries. The majority of these products are consumed as beverages, breakfast or as snack foods, while a few are consumed as staples and used as complementary foods (Odunfa and Oyewole, 1998). Generally, processing treatments such as drying, de-hulling, washing, soaking, grinding and sieving in addition to fermentation are possible steps applied during the processing of fermented foods. Many traditional African food fermentations are still cumbersome household activities that require further research and scaling up (Holzapfel, 2002). Successful industrialization has been achieved in Africa in beer production using different fermented substrates (Holzapfel, 2002). The products

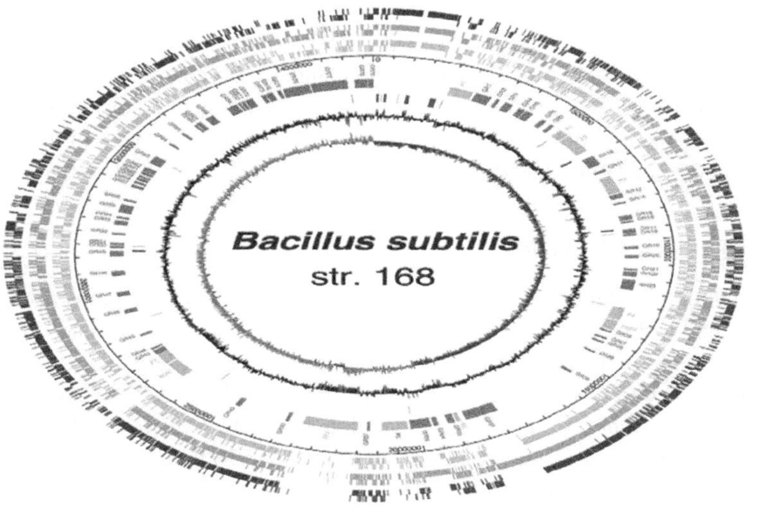


**FIGURE 1.1** Genome sequence of *Bacillus subtilis*. (Kunst et al. 1997. https://www.genome.jp/kegg-bin/show_organism?org=bsu.)

that are in high demand commercially should be scaled up and industrialized. The shelf stability of many traditional African foods can be enhanced by fermentation and combination fermentation with other unit processing operations. For example, pasteurization of Kunun followed by preservation using sodium metabisulphite gives products having longer shelf life. At 28°C–30°C, the samples could be stored for 4–5 weeks (Ojimelukwe et al. 2011a).

## 1.7 BIOTECHNOLOGICAL IMPROVEMENT OF THE BENEFICIAL EFFECTS OF AFRICAN FERMENTED FOODS ON HUMAN HEALTH

The beneficial effects of African fermented foods on human health still require a lot of research investigations (Tu et al., 2010). Health-promoting effects of Asian fermented foods are better documented in literature than that of African fermented foods (Lamsal and Faubion, 2009). Probiotic foods are gaining importance because they promote good health (Savadogo et al., 2016). Probiotic products contain mono or mixed cultures of live organisms which enhance microbial balance when ingested balance (FAO/WHO, 2001). Human probiotic foods usually contain *Lactobacilli* or *Bifidobacterium* (Herbel et al. 2013). *L. lactis*, *S. thermophilus* and *Enterococcus faecium* are common probiotic bacteria beneficial to humans and animals. Recent researches have shown that some fermented foods have probiotic potentials. Probiotic microorganisms associated with fermented foods are very beneficial. They degrade anti-nutrients like tannins and phytic acid. They improve digestion and promote the production of nutrients, like vitamins and minerals (Niu et al., 2013). Probiotic microbes inhibit food spoilage and pathogenic microorganisms. They are effective for the prevention and treatment of diarrhoea. They prevent tooth decay and help to manage type 2 diabetes. Some probiotic microorganisms reduce "bad" cholesterol. Probiotic fermentation improves shelf life, sensory properties and nutritional value, and detoxifies harmful substances (Jacques and Caseregola, 2008). African fermented foods have great potentials as probiotic foods. Biotechnological improvement of microorganisms and the unique properties of the probiotic foods will improve the health of consumers of African fermented foods.

## 1.8 STARTER CULTURES

Starter cultures are tailor-made organisms that accelerate the fermentation process and lead to improved product quality under controlled conditions. Developed starter cultures should be viable and active for them to be suitable for large-scale production. This implies that they should be stable during drying and long-term storage (Yao et al., 2009). It is important to ensure that starter cultures do not lose their viability and metabolic activity during packaging and storage (Yao et al., 2008; Coulibaly et al., 2009). Dried cultures should contain up to 95% dry matter. Ideally, they should be stored at low temperatures possibly sealed under vacuum. Probiotic products should contain up to $10^6$–$10^7$ cfu/g viable microorganisms during consumption (Agarry et al., 2010). Important factors that affect probiotic starter cultures are pH, post-acidification period, production of metabolites with preservative effects, oxygen level, temperature, and food matrix (Dave and Shah, 1998; Bachman et al., 2015). Microbial and biochemical dynamics studies of microbial communities are required to reveal causal links between key microorganisms and their role in the biochemistry and health benefits associated with the fermented product (Chakravorty et al., 2016). Genome sequence analysis (involving the screening of microorganisms and selection of strains with desired properties using automated laboratory processes) is currently used for the selection of strains with desired properties (Leroy and De Vuyst, 2004; Johansen et al., 2015)

## 1.9 CURRENT RESEARCH NEEDS

Fermented foods play significant roles in the African diet. Several microorganisms involved with fermentation are partially or fully identified phenotypically but not at the molecular levels. Based on research works published on fermented African foods, DNA-dependent characterization of microorganisms is limited. Genetic variations responsible for metabolic activities could exist and should be detected. Proper genetic identification of microorganisms is very important. Metabolic transformations that lead to desired end products of fermentation are not adequately monitored and documented. Uncontrolled fermentation processes lead to variations in quality and shelf stability of the products. It is very needful to apply modern biotechnology techniques for quality upgrade and to standardize the fermentation processes. The desirable quality characteristics of each product should be maintained while upgrading the fermentation processes. Nutrient composition and nutritional quality of standard fermented foods should be properly documented. Health benefits of commercial probiotic products and microorganisms generated from traditional African fermentations should be properly documented in literature. Such probiotic products should be developed for household consumption to improve the health and well-being of the populace.

Cell immobilization and the use of biofilms could be explored to upgrade the fermentation process (Shade, 2011). Biofilms are naturally formed microbial cells capable of adsorbing to a support without assistance by chemicals and polymers that entrap the cells. The bio-reactor environment, microorganisms and other environmental factors influence biofilm formation. Biofilms are resistant to toxicity. They have high biomass potentials. They have improved stability when compared with cells in suspension (Shade, 2011). These unique characteristics position biofilms for exceptional potentials for industrial fermentation. Another important and particularly relevant line of research is the investigation of the effect of continued fermentation on the nutrient quality and composition of fermented products especially after the optimum quality has been achieved and the mechanisms for terminating fermentation once the desired optimum nutrient quality has been achieved. Molecular tools based on DNA analysis will provide greater insights into many known alterations and also help to uncover new ones.

## 1.10 CONCLUSION

Fermented foods are very important components of the African diet. Most microorganisms involved with natural food fermentations have been identified phenotypically to a large extent. More comprehensive identification should be conducted using molecular tools. Starter cultures should be developed to standardize and optimize the production processes for optimal product quality. Unhygienic processing environment and poor packaging characterize traditional African food fermentation processes. Modern food processing equipment should be employed. Production should be industrialized for products consumed in large quantities, where the raw material supply is adequate. Fermented foods with additional health benefits should be developed. The FAO standard of 2010 is a useful guide for the improvement of traditional African fermented foods. The key microorganisms involved in fermentation of many traditional African foods are known. Genetic improvement has been limited to only few of the key microorganisms involved. Suitable starter organisms are not available for many of the traditional African fermentations. Only very few researches have been dedicated to identifying the raw materials most suited for the fermented product. Improved technologies need to be developed for fermentation processes with potentials for scaling up. Adoption of newly developed technologies and pilot- and industrial-scale productions of traditionally fermented African foods also require urgent research attention.

Author's contribution: Philippa C. Ojimelukwe conducted the relevant searches, organized the data, collated and interpreted them, and wrote up the review paper.

## REFERENCES

Adewumi G.A., Oguntoyinbo F.A., Keisa S., Romi W., Jeyaram K. (2013). Combination of culture-independent and culture dependent molecular methods for the determination of bacterial community of iru, a fermented *Parkia biglobosa* seeds. *Frontiers in Microbiology*, 3:436. doi:10.3389/fmicb.2012.00436.

Adeyemo S.M., and Onilude A.A. (2014).Molecular identification of *Lactobacillus plantarum* isolated from fermenting cereals. *International Journal of Biotechnology and Molecular Biology Research*, 5:59–67. doi:10.5897/IJBMBR2014.0184.

Adimpong B., Nielsen D.S., Sørensen K.I., Derkx, J.L. (2012). Genotypic characterization and safety assessment of lactic acid bacteria from indigenous African fermented products. *BMC Microbiology*, 12:75–89.

Agarry O.O., Nkama I., Akoma O. (2010). Production of Kunun-zaki (A Nigerian fermented cereal beverage) using starter culture. *International Research Journal of Microbiology*, 1(2):018–025.

Ahaotu I., Anyogu A., Njoku O.H., Odu N.N., Sutherland J.P., Ouoba, L.I.I. (2013). Molecular identification and safety of Bacillus species involved in the fermentation of African oil beans (*Pentaclethra macrophylla* Benth) for production of Ugba. *International Journal of Food Microbiology*, 162:95–104.

Amoa-Awua W.K., Terlabie N.N., Sakyi-Dawson E. (2006). Screening of 42 *Bacillus* isolates for ability to ferment soybeans into dawadawa. *International Journal of Food Microbiology*, 106:343–347.

Abriouel H., Omar N.B., López R.L., Martínez-Cañamero M., Keleke S. Gálvez A. (2006). Culture-independent analysis of the microbial composition of the African traditional fermented foods potopoto and dégué by using three different DNA extraction methods, *International Journal of Food Microbiology*, 111:228–233.

Adrio J., Demain A.L. (2010). Recombinant organisms for production of industrial products. *Bioengineered Bugs*, 1 (2):116–131.

Achi O.K., Akubor P.I. (2000). Microbiological characterization of yam fermentation for 'Elubo' (yam flour) production. *World Journal of Microbiology and Biotechnology*, 16(1):3–7.

Akabanda F., Owusu-Kwarteng J., Tano-Debrah K., Glover R. Nielsen, L.K., Jespersen L. (2013). Taxonomic and molecular characterization of lactic acid bacteria and yeasts in nunu, a Ghanaian fermented milk product. *Food Microbiology*, 34:277–283.

Annan N.T., Poll L., Sefa-Dedeh S., Plahar W.A., Jakobsen M. (2003). Volatile compounds produced by *Lactobacillus fermentum, Saccharomyces cerevisiae* and *Candida krusei* in single starter culture fermentations of Ghanaian maize dough. *Journal of Applied Microbiology*, 94(3):462–474.

Assad-Garcia J.S., Bonnin-Jusserand M., Garmyn D., Guzzo J., Alexandre W., Grandvalet C. (2008). An improved protocol for electroporation of *Oenococcus oeni* ATCC BAA-1163 using ethanol as immediate membrane fluidizing agent. *Letters in Applied Microbiology*, 47:333–338.

Bachman H., Prank J.T., Kleerbezem M., Teusink B. (2015). Evolutionary engineering to enhance starter culture performance in food fermentations. *Current Opinion in Biotechnology*, 32:1–7.

Battcock M., Azam Ali S. (2001). Fermented Foods and Vegetables. *FAO Agricultural Services Bulletin*, 134:96.

Blandinob A., Al-Aseeria M.E., Pandiellaa S.S., Canterob D., Webb C. (2003). Cereal-based fermented foods and beverages. *Food Research International*, 36:527–543.

Brandt M.J. (2014). Starter cultures for cereal based foods. *Food Microbiology*, 36(6):527–543.

Chakravorty S., Bhattacharya S., Chatzinotas A., Chakrabborty W., Bhattacharya D., Gachhui R. (2016). *International Journal of Food Microbiology*, 22(2):63–72.

Coulibaly I., Yao A.A., Lognay G., Destain J., Fauconnier M.L., Thonart P. (2009) Survival of freeze-dried *Leuconostoc mesenteroides* and *Lactobacillus plantarum* related to their cellular fatty acids composition during storage. *Applied Biochemistry and Biotechnology*, 157(1):70–84.

Dave R.I., Shah N.P. (1998). Ingredient supplementation effects on viability of probiotic bacteria in yogurt. *Journal of Dairy Science*, 81: 2804–2816.

Di Cagno R., Coda R., De Angelis M., Gobbetti M. (2013). Exploitation of vegetables and fruits through lactic acid fermentation, *Food Microbiology*, 33:1–10.

Enwa F.O., Beal J., Arhewoh M.I. (2011). Effect of maize and bacteria starter culture on maize fermentation process. *International Journal of Biomedical Research*, 2(11):561–567.

Ekwem O.H. (2014). Isolation of antimicrobial producing lactobacilli from akamu (a Nigerian fermented cereal gruel). *African Journal of Microbiology Research* 8 (7):718–720. doi:10.5897/AJMR2013.6251.

Esiobu E., Igwe E.C., Ojimelukwe P.C. (2014). Variations in the traditional starter culture for production of a Nigerian fermented milk product (Kindirmo). *Focusing on the Modern Food Industry*, 3(1):35–42.

Ezeronye O.U., Elijah A.I., Ojimelukwe P.C. (2005). Effect of *Saccoglottis gabonensis* and *Alstonia boonei* on the fermentation of fresh palm sap by *Saccharomyces cerevisae. Journal of Food Technology*, 3(4):586–591.

Fadahunsi I.F. Soremekun O.O. (2017). Production, nutritional and microbiological evaluation of mahewu - a South African traditional fermented porridge. *Journal of Advances in Biology and Biotechnology*, 14(4):1–10.

Fandohan P., Zoumenou D., Hounhouigan D.J., Marasas W.F., Wingfield M.J., Hell K. (2005). Fate of aflatoxins and fumonisins during the processing of maize into food products in Benin. *International Journal of Food Microbiology*, 98(3):249–259.

FAO (2010). FAOSTAT Food and Agriculture Organization of the United Nations.

Food and Agriculture Organisation/World Health Organisation (FAO/WHO) (2001) Health and nutritional properties of probiotics in food including powder milk with live lactic acid bacteria. Geneva: WHO.

Gonzalez R., Tronchoni J., Quiros M., Morales P. (2016). Genetic improvement and genetically modified microorganisms. Wine Safety, Consumer Preference and Human Health, pp. 71–96.

Greppi A., Rantsiou K., Padonou W., Hounhouigan J., Jespersen L., Jakobsen M., Cocolin L. (2013a). Determination of yeast diversity in ogi, mawè, gowé and tchoukoutou by using culture-dependent and-independent methods. *International Journal of Food Microbiology*, 165:84–88.

Greppi A., Rantsiou K., Padonou W., Hounhouigan J., Jespersen L., Jakobsen M. (2013b) Cocolin L. Yeast dynamics during spontaneous fermentation of mawè and tchoukoutou, two traditional products from Benin. *International Journal of Food Microbiology*, 165:200–207.

Halm M., Lillie A., Sorensen A.K., Jackobsen M. (1993) Microbiological and aromatic characteristics of fermented maize dough for 'Kenkey' production in Ghana. *International Journal of Food Microbiology*, 19(2):135–43.

Halm M., Osei-Yaw A., Hayford A., Kpodo K.K.A., Amoa-Awua W.K.A. (1996). Experiences with the use of a starter culture in the fermentation of maize for 'kenkey' production in Ghana. *World Journal of Microbiology and Biotechnology*, 12(5):531–536.

Hamad S.H., Dieng M.C., Ehrmann M.A., Vogel R.F. (1997). Characterization of the bacterial flora of Sudanese sorghum flour and sorghum sourdough. *Journal of Applied Microbiology*, 83:764–770.

Heita L., Cheikhyoussef I.A., Shikongo M., Nambabi M.S. (2013). Microbiological and physicochemical analysis of traditional fermented milk from North central and North-eastern Namibia. *Conference: V International Conference on Environmental, Industrial and Applied Microbiology – BioMicroWorld2013* At: Madrid, Spain.

Herbel S.R., Vahjen W., Wieler L.H., Guenther S. (2013). Timely approaches to identifying probiotic species of the genus Lactobacillus. *Gut Pathogens*, 5:27–40

Holzapfel W.H. (2002). Appropriate starter culture technologies for small-scale fermentation in developing countries. *International Journal of Food Microbiology*, 75:197–212.

Hounhouigan D.J., Nout M.J.R., Nago C.M., Houben J.H., Rombouts F.M. (1993). Composition and microbiological and physical attributes of mawè, a fermented maize dough from Benin. *International Journal of Food Science and Technology*, 28(5):513–517.

Ketema B., Tetemke M., Mogessie A. (1998).The microbial dynamics of 'Borde' fermentation, a traditional Ethiopian fermented beverage. *SINET: Ethiopian Journal of Science*, 21:195–205.

Inyang C.U., Dabot Y.A. (1997). Storability and portability of pasteurized and sterilised kunun-zaki: A fermented sorghum beverage. *Journal of Food Processing and Preservation*, 21(1):1–7.

Jacques N., Casaregola S. (2008). Safety assessment of dairy microorganisms: The hemiascomycetous yeasts. *International Journal of Food Microbiology*, 126:321–326.

Johansen E., Oregaard G., Sorensen K.I., Derkx P.M.F. (2015). Modern approaches for isolation, selection and improvement of bacterial strains for fermentation applications. Advances in Fermented Foods and beverages. Improving quality, technologies and health benefits. Woodhead Publishing Series in Food Science, Technology and Nutrition, pp. 227–248.

Kedia G.R., Wang H., Patel S., Pandiella S. (2007). Use of mixed cultures for the fermentation of cereal-based substrates with potential probiotic properties. *Process Biochemistry*, 42:65–70.

Leroy F., De Vuyst L. (2004). Lactic acid bacteria as functional starter cultures for the food fermentation industry. *Trends in Food Science and Technology*, 15:67–78.

Mohammed S.I., Steenson L.R., Kirlies A.W. (1991). Isolation and characterization of microorganisms associated with traditional sorghum fermentation for production of Sudanese Kisra. *Applied Environmental Microbiology*, 57 (9):2529–2533.

Nwachukwu E., Achi O.K., Ijeoma I.O. (2010). Lactic acid bacteria in fermentation of cereals for the production of indigenous Nigerian foods. *African Journal of Food Science and Technology*, 1(2):021–026.

Jeff-Agboola A.A. (2007). Microorganisms associated with natural fermentation of African Yam Bean (*Sphenostylis sternocarpa* Harms) seeds for the production of otiru. *Journal of Microbiology*, 2 (11):816–823.

Kunst F., Ogasawara N., Dachin A. (1997). The complete genome sequence of the Gram-positive bacterium *Bacillus subtilis*. *Nature*, 390:249–256.

Lamsal B.P., Faubion J.M. (2009). The beneficial use of cereal and cereal components in probiotic foods, *Food Reviews International*, 25(2):103–114. doi:10.1080/87559120802682573.

Lardinois M., Totté A., Tounkara L., Mbaye C.T., Beye C., Thonart P., Ngom E.H.A. (2003). Conservation de produits locaux à travers un transfert de technologie de séchage: Cas de l'atomisation au Sénégal. In: *International Relief and Development (IRD), editor. Proceedings of the 2nd International workshop in Food-based Approaches for a Healthy Nutrition*; Ouagadougou, Burkina Faso. 23–28 November 2003; Arlington, US: IRD.

Lei V., Jackobsen M. (2004). Microbiological characterization and probiotic potential of koko and koko sour water, African spontaneously fermented millet porridge and drink. *Journal of Applied Microbiology*, 96 (2):384–397.

Mallikarjuna N., Yellamma K. (2019). Genetic and metabolic engineering of microorganisms for the production of various food products. *Recent Developments in Applied Microbiology and Biochemistry*, 67–182. doi:10.1016/B978-0-12-816328-3.00013-1.

Macwana S., Muriana P.M. (2012). Spontaneous bacteriocin resistance in Listeria monocytogens as a susceptibility screen for identifying different mechanisms of resistance and modes of action by bacteriocins of lactic acid bacteria. *Journal of Microbiology Methods*, 88 (1):7–13.

Meerak J., Yukphan P., Miyashita M., Sato H., Nakagawa Y., Tahara Y. (2008). Phylogeny of polyglutamic acid producing Bacillus strain isolated from a fermented locust bean product manufactured in West Africa. *Journal of General and Applied Microbiology*, 54:159–166.

Muigei S.C., Shitandi A., Muliro P., Bitonga O.R. (2013). Production of exopolysaccharides in the kenyan fermented milk. *Mursik International Journal of Scientific Research*, 2 (2):79–89.

Nam Y.D., Lee S.Y., Lim S.I. (2012). Microbial community analysis of Korean soybean pastes by next-generation sequencing. *International Journal of Food Microbiology*, 155: 36–42.

Niu L.Y., Jiang S.T., Pan L.J. (2013). Preparation and evaluation of antioxidant activities of peptides obtained from defatted wheat germ by fermentation. *Journal of Food Science and Technology*, 50 (1):53–61.

Nwachukwu E., Achi O.K., Ijeoma I.O. (2010). Lactic acid bacteria in fermentation of cereals for the production of indigenous Nigerian foods. *African Journal of Food Science and Technology*, 1(2):021–026.

Obinna-Echem P.C. (2015). Inhibitory activity of lactobacillus plantarum strains from Akamu - A Nigerian fermented maize food against Escherichia coli. *American Journal Food Science and Technology*, 3 (5):118–125.

Odunfa S.A., Oyewole O.B. (1997). *African Fermented Foods*. London: Blackie Academic and Professional.

Odunfa S.A., Oyewole O.B. (1998). African fermented foods. In: Wood B.J.B. (eds). *Microbiology of Fermented Foods*. Springer, Boston, MA. https://doi.org/10.1007/978-1-4613-0309-1_23.

Odunfa S.A., Adeyele S. (1985). Microbiological changes during the traditional production of ogi-baba, a West African fermented Sorghum Gruel. *Journal of Cereal Science*, 3 (2):173–180.

Oguntoyinbo F.A., Sanni A.I. Onilude, A.A. (2001). Microflora and proximate composition of okpehe, a fermented condiment from Prosopis africana. *Advances in Food Sciences* 23:165–171.

Oguntoyinbo F.A., Tourlomousis P., Gasson M.J., Narbad A. (2011). Analysis of bacterial communities of traditional fermented West African cereal foods using culture independent methods. *International Journal of Food Microbiology*, 145:205–210.

Oguntoyinbo F.A., Narbad A. (2012). Molecular characterization of lactic acid bacteria and in situ amylase expression during traditional fermentation of cereal foods. *Food Microbiology*, 31 (2):254–262.

Ojimelukwe P.C., Okudu H.O., Uche N.J. (2011a). Effect of pasteurization and preservation treatment on the quality of "kunun zaki" made from Sorghum (Sorghum biocolor). *Nigerian Journal of Nutritional Sciences*, 32 (2):42–45.

Ojimelukwe P.C., Ojinnaka M.C., Okechi P. (2011b). Physicochemical characteristics of fermenting castor seeds containing lime and NaCl as additive. *African Journal of Food Science*, 5 (14):754–760.

Olasupo N.A., Osikoya A.F., Odunfa S.A., Kuboye A.O., Olatunji O. (2000). An investigation on the preservation of kunun-zaki, an African Fermented Cereal-Based Food drink. *Acta Alimentaria*, 29(4):385–392.

Ouoba II L., Diawara B., Amoa-Awua W.K., Traoré A.S., Lange M.P. (2004). Genotyping of starter cultures of Bacillus subtilis and *Bacillus pumilus* for fermentation of African locust bean (*Parkia biglobosa*) to produce Soumbala. *International Journal of Food Microbiology*, 90:197–205.

Ouoba L.I., Nyanga-Koumou C.A.G., Parkouda C., Sawadogo H., Kobawila S.C., Keleke S., Diwara B., Louembe D., Sutherland J.P. (2010). Genotypic diversity of lactic acid bacteria isolated from African traditional alkaline-fermented foods. *Journal of Applied Microbiology*, 8 (6):2019–2029.

Owusu-Kwarteng J., Akabanda F., Nielsen D.S., Tano-Debrah K., Glover R.L.K. (2012) Identification of lactic acid bacteria isolated during traditional fura processing in Ghana. *Food Microbiology*, 32 (1):72–78.

Oyewole S., Odunfa S.A. (1990). Characterization and distribution of lactic acid bacteria in cassava fermented during fufu production. *Journal of Science of Food and Agriculture*, 20:265–271.
Oyewole A.O., Audu O.T., Amupitan J.O. (2004). A survey of chemical constituents and biological activities of some medicinal plants. *Journal of Chemical Society of Nigeria*, 4:162–165.
Qiao Z., Dong Chen X., Cheng Y.Q., Liu H., Liu Y., Li L. (2010). Microbiological and chemical changes during the production of acidic whey, a traditional Chinese tofu-coagulant. *International Journal of Food Properties*, 13(1):90–104.
Ray R.C, Panda S.H. (2007). Lactic acid fermented fruits and vegetables: An overview. In: Palino, M.V. (ed.) *Food Microbiology Research Trends*. Nova Science Publishers Inc., Hauppauge, New York, pp. 155–188.
Shade A. (2011). Kombucha biofilm: A model system for microbial ecology. Final report on research conducted during the microbial diversity carse. Marina Biological laboratories. Wood Hole. MA.
Savadogo A., Filbert G., Tapsoba F. (2016). Probiotic microorganisms involved in cassava fermentation for Gari and AttiÃokÃ production. *Journal of Advances in Biotechnology*, 6 (2):858–864.
Sawadogo-Lingani H., Diawara B., Glover R.K., Tano-Debrah K., Traoré A.S., Jakobsen M. (2010). Predominant lactic acid bacteria associated with the traditional malting of sorghum grains. *African Journal of Microbiology Research*, 4 (3):169–179.
Schoustra S.E., Kasase C., Toarta C., Kassen R., Poulain A.J. (2013). Microbial community structure of three traditional Zambian fermented products: Mabisi, Chibwantu and Munkoyo. *PLoS One*, 8(5). doi:10.1371/journal.pone.0063948.
Sengun I.Y., Karabiyikli S. (2011). Importance of acetic acid bacteria in food industry. *Food Control*, 22: 647–656.
Settanni L., Corsetti A. (2008). Application of bacteriocins in vegetable food biopreservation. *International Journal of Food Microbiology*, 121:123–138.
Smid E.J., Lacroix C. (2012). Microbe–microbe interactions in mixed culture food fermentations. *Current Opinion in Biotechnology*, 24: 1–7.
Songré-Ouattara L.T., Mouquet-Rivier C., Humblot C., Rochette I., Diawara B., Guyot J.P. (2010). Ability of selected lactic acid bacteria to ferment a pearl millet-soybean slurry to produce gruels for complementary foods for young children. *Journal of Food Science*, 75(5):261–269.
Songré-Ouattara L.T., Mouquet-Rivie C., Verniere C., Humblot C., Diawara B., Guyot J.P. (2008). Enzyme activities of lactic acid bacteria from a pearl millet fermented gruel (Ben-saalga) of functional interest in nutrition. *International Journal of Food microbiology*, 128(2):395–400.
Swain M.R., Anandharji M., Ray R.C., Rani R.P. (2014). Fermented fruits and vegatables of Asia: a potential source of probiotics. *Biotechnology Research International*, 1–19. doi:10.1155/2014/250424.
Teniola O.D., Addo P.A., Brost I.M., Farber P., Jany P.K., Alberts J.F., van Zyl W.H., Steyn P.S., Holzapfel W.H. (2005). Degradation of aflatoxin $B_1$ by cell-free extracts of Rhodococcus erythropolis and Mycobacterium fluoranthenivorans sp. nov. DSM44556. *International Journal of Food Microbiology*, 105:111–117.
Thakur K., Nanda D.K., Kumar N., Tomar S.K. (2015). Phenotypic and genotypic characterization of Indigenous Lactobacillus Species from Diverse Niches of India. *Current Trends in Biotechnology and Pharmacy*, 9:222–227.
Totté A., Tine E., Seye N., Mathiam J.M., Roblain D., Thonard P. (2003). Innovation et transfert de technologie: Cas du contrôle de la fermentation du mil par l'utilisation d'un starter lactique. In International Relief and Development (IRD) (ed.). *Proceedings of the 2nd International workshop in Food-based Approaches for a Healthy Nutrition*. Ouagadougou, Burkina Faso, 23–28 November 2003. Arlington, US: IRD.

Tou E.H., Guyot J.P., Mouquet-Rivier C., Rochette I., Counil E., Traoré A.S., Treche S. (2006). Study through surveys and fermentation kinetics of the traditional processing of pearl millet (Pennisetum glaucum) into Ben-saalga, a fermented gruel from Burkina Faso. *International Journal of Food Microbiology*, 106: 52–60.

Tu R.J., Wu H.Y., Lock Y.S., Chen M.J. (2010). Evaluation of microbial dynamics during the ripening of a traditional Taiwanese naturally fermented ham. *Food Microbiology*, 27(4): 460–467.

Turpin E. H., Soule A. M. (2012). Agriculture, its fundamental principles. Ulan Press (August 31, 2012). 338pages. https://www.amazon.com/Agriculture-Fundamental-Principles-Henry-Turpin/dp/1296811425.

Turpin W., Humblot C., Noordine M.L., Thomas M., Guyot J.P. (2012). Lactobacillaceae and cell adhesion: Genomic and functional screening. *PLoS ONE*, 7, e38034.

Vieira-Dalodé G., Jespersen L., Hounhouigan D.J., Moller P.L., Nago C.M., Jakobsen M. (2007). Lactic acid bacteria and yeasts associated with Gowé production from sorghum in Bénin. *Journal of Applied Microbiology*, 103(2):342–349. doi:10.1111/j.1365-2672.2006.03252.x.

Weckx S., Van der Muelen R., Maes D., Scheirlinck I., Huys G., Vandamme P., De Vuyst L. (2010). Lactic acid bacteria community dynamics and metabolite production of rye sourdough fermentations share characteristics of wheat and spelt sourdough fermentations. *Food Microbiology*, 27 (8):1000–1008.

Yao A.A., Coulibaly I., Lognay G., Fauconnier M.L., Thonart P. (2008). Impact of polyunsaturated fatty acid degradation on survival and acidification activity of freeze-dried Weissella paramesenteroides LC11 during storage. *Applied Microbiology and Biotechnology*, 79(6):1045–1052.

Yao A.A., Wathelet B., Thonart P. (2009). Effect of protective compounds on the survival, electrolyte leakage and lipid degradation of freeze-dried Weissella paramesenteroides LC11 during storage. *Journal of Microbiology and Biotechnology*, 19 (8):810–817.

Zhang Z.Y., Liu C., Zhu Y.Z., Zhong Y., Zhu Y.Q., Zheng H.J., Zhao G.P., Wang S.Y., Guo X.K. (2009). Complete genome sequence of Lactobacillus plantarum JDM1delta. *Journal of Bacteriology*, 191:5020–5021.

https://www.reatch.ch/de/content/regulation-genetically-modified-organisms-present-and-future 2007. Accessed 25/08/2019.

# 2 Indigenous Fermented and Underutilized/Novel Foods with Potentials for Combating Malnutrition in Sub-Saharan Africa

*Philippa C. Ojimelukwe*
Michael Okpara University of Agriculture

## CONTENTS



## 2.1 INTRODUCTION

Malnutrition is a major problem facing Africa despite its arable land mass and lush vegetation (Oladele, 2011). Poor nutrition causes 45% of deaths in children (0–5 years) (about 3.1 million children annually) (Baldermann et al., 2016). There is need to produce more food and to produce food in better ways. Africa needs to add value to her food crops, prevent seasonal scarcity of foods and diversify her sources of nutrients. Fermentation is a feasible food processing and preservation method for Africa. Malnutrition in Africa reflects a complex situation. Global challenges are emerging in the areas of climate change, increased demand for biofuels, unprecedented

DOI: 10.1201/9781003178378-2

financial crisis increased growth rate in developing countries and increased poverty. These issues should be addressed from a food and nutrition perspective, since food is a basic necessity of life. Food security should be placed at the centre of developmental policies. Insufficient calorie intake, the double burden of malnutrition and hidden hunger are on the increase worldwide. Anaemia, stunting and wasting depict the high level of food insecurity (FAO, 2018). Fermentation remains a plausible avenue for improved nutrition in Africa given the tropical climate in many parts and the limited use of refrigeration for food preservation. Exploring underutilized fermented foods with potentials for further development will help alleviate malnutrition in Africa. Sustainable diets may be developed from healthy cheap local foods. Fermentation provides safe nutritious foods with proven health benefits. Investments in research, education, training and policy are required for the promotion of underutilized local fermented foods.

## 2.2 UNDERUTILIZED CROPS

A lot of names are used to describe underutilized food crops. They may be referred to as abandoned, orphaned, neglected, underused, forgotten, underdeveloped, lost or minor crops. Underutilized crops share some common features:

- They have food and may have medicinal value.
- Their cultivation may be localized or they might have been cultivated in the past and later abandoned.
- They are currently receiving less attention than other crops.
- They are used traditionally in some areas.
- They have not received any significant research attention or technological attention.
- They usually have no formal system for propagation of their planting materials.

## 2.3 POTENTIAL ROLES OF UNDERUTILIZED FOOD CROPS IN FOOD AND NUTRITION SECURITY

Many underutilized crops are rich in nutrients. The neglect of the use of these crops undermines the food security of the populace. Their enhanced use will improve nutrition and health. For example, many underutilized cereals contain more vitamin $B_1$ and dietary fibre than widely available varieties. Exploring the potentials of underutilized crops is an effective way of diversifying diets to combat micronutrient deficiencies (hidden hunger). Some of these underutilized crops were used in the past but are now receiving very little attention or have very good potentials to promote food and nutrition security as fermented foods (Simatende et al., 2015). This chapter is dedicated to underutilized fermented African foods.

## 2.4 FERMENTATION

Fermentation is a method of food processing and food preservation which adds value to perishable raw materials through microbial transformations (Subramaniyam and

Vimala, 2012). Enzymes found in microorganisms (amylases, proteases and lipases) hydrolyse polysaccharides, proteins and lipids to easily digestible forms. New compounds are formed. Flavours, aromas and textures are also altered by the fermentation process. In many cases, detoxification reactions also take place. According to Steinkraus (2002), fermentation plays at least five roles in food processing:

- It diversifies the human diet through the development of appealing sensory properties.
- Fermentation leads to the development of metabolites that preserve foods (lactic acid; acetic acid; ethanol, bacteriocins; acetic, formic and propionic acids; etc.).
- It inhibits spoilage and pathogenic microorganisms.
- It makes food more digestible.
- It leads to the development of more nutrients, when properly controlled.
- It leads to the detoxification of foods.
- It reduces cooking time and energy.

## 2.5 TYPES OF FOOD FERMENTATION

Fermentation basically involves the transformation of substrates by enzymes found in microorganisms. Lactic acid bacteria (LAB) are the most predominant microorganisms in food fermentations and are generally regarded as safe (GRAS). The most common features of LAB fermentation are acid production and the lowering of pH (Ray and Panda, 2007). It not only produces lactic acid, which imparts flavour, but also preserves health promoting bioactive compounds (Shivashankara et al., 2004).

## 2.6 ALCOHOL FERMENTATION

In alcohol fermentation, pyruvate (from glucose) is converted to acetaldehyde which is eventually reduced to ethanol by alcohol dehydrogenase. Sugars such as glucose are converted to ATP, while ethanol and carbon dioxide are produced during the process (Zamora, 2009) (Figure 2.1).

## 2.7 ACID FERMENTATION

In homolactic fermentation, the keto group from pyruvate is converted to a hydroxyl group by lactate dehydrogenase yielding lactate. NADH is oxidized to NAD+.

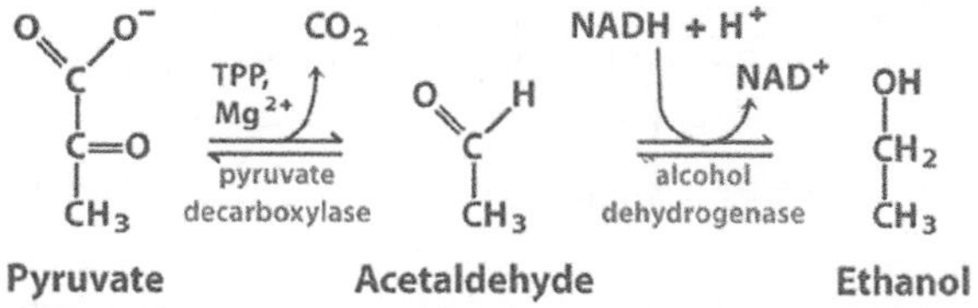


**FIGURE 2.1** Conversion of sugars to ethanol. (Zamora 2009.)

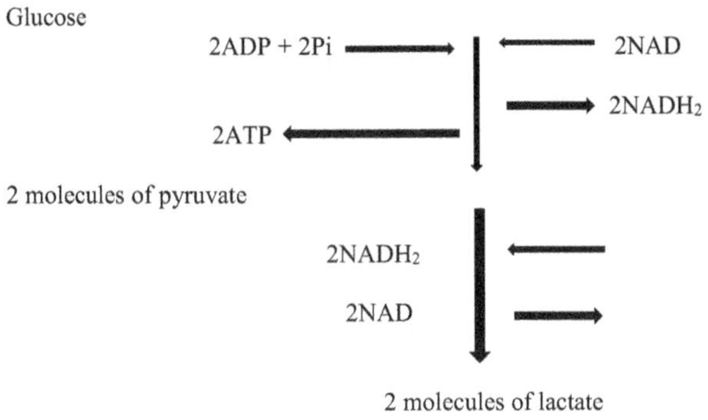


**FIGURE 2.2** Lactic acid fermentation pathway.

In homolactic acid fermentation, the two molecules of pyruvate are converted to lactate. One molecule of glucose will yield two molecules of lactate (Figure 2.2).

## 2.8 MIXED ACID FERMENTATION

In mixed acid fermentation, a hexose sugar such as glucose or mannose is converted to a mixture of different acids. This fermentation pathway differs from other fermentation pathways that lead to the production of fewer acids in fixed amounts. Mixed acid fermentation involves bacteria that utilize two or more pathways in their terminal stage of fermentation. These bacteria will ferment glucose to acetate, formate, lactate, succinate, etc. Many members of the *Enterobacteriaceae*, e.g. *Escherichia*, *Salmonella* and *Shigella*, undertake mixed acid fermentation. The various acids produced decrease the pH of the fermenting medium significantly (Figure 2.3).

## 2.9 ALKALINE FERMENTATION

Alkaline fermentation is common with protein-rich foods. Many legume fermentations are primarily alkaline fermentation. The dominant microorganism is *Bacillus* sp. which hydrolyses proteins to produce amino acids and ammonia. These products

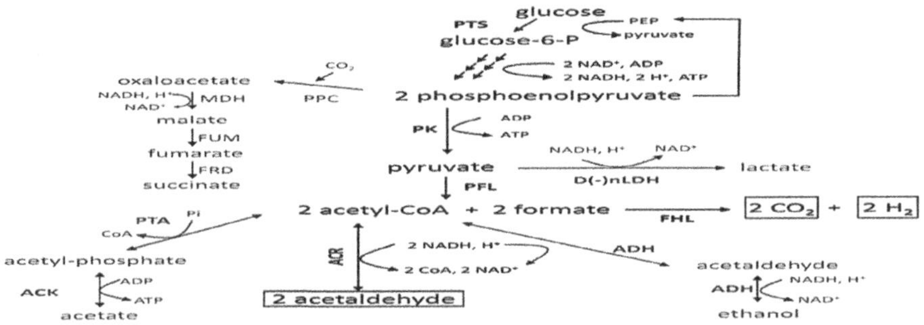


**FIGURE 2.3** Mixed fermentation pathway for microorganisms. (Doi 2019.)

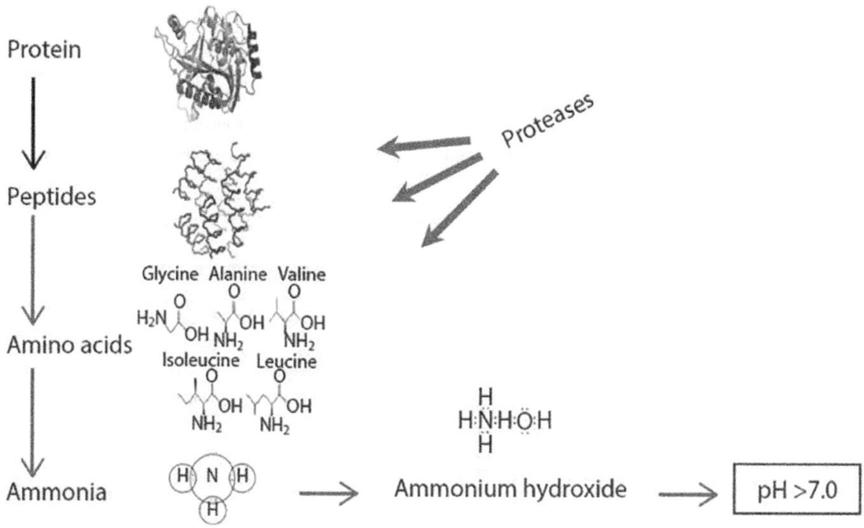


**FIGURE 2.4** The Alkaline fermentation process. (nattomatrika.wordpress.com.)

are responsible for a rise in pH. Fermentation of ugba and dawadawa and many other legumes follows this pattern (Jensen et al., 2009). Some dairy and vegetable products also undergo alkaline fermentation. Endogenous enzymes in microorganisms decompose proteins and free amino acids into peptides, amino acids, aldehydes and alcohols. The alkaline pH inhibits the growth and proliferation of pathogenic and spoilage microorganisms. It is also responsible for the development of the ammonia-like flavour of legume condiments (Figure 2.4).

## 2.10 SOME LOCALLY FERMENTED UNDERUTILIZED AFRICAN FOODS

Table 2.1 shows some fermented African leafy vegetables. Vegetables are plant products which may be used as accompaniment to main dishes. Both the leafy vegetables and other vegetables may be fermented to improve digestibility, health benefits and shelf life. Africa needs to acculturate the use of fermented vegetables for improved nutrition. "Ntobambodi" is a condiment prepared from cassava leaves (Sanni and Oguntoyinbo, 2014). Cassava leaves are sun-dried, de-stalked and washed (Kobawila et al., 2005). The washed leaves are wrapped in papaya leaves and fermented in a basket for 2–4 days. The pH of the solid substrate alkaline fermentation increases to up to 8.5. The microorganisms involved in the chance fermentation have been shown in Table 2.1. Kawal is a fermented product from *Cassia obtusifolia* L. and is consumed as a meat substitute in Sudan (Suliman et al., 1987). The leaves are cheap sources of proteins and amino acids (Dirar et al., 1985). Kawal is produced by solid-state fermentation of *C. obtusifolia* leaves by bacterial species such as *Bacillus subtilis*, *Propionibacterium* and *Staphylococcus sciuri*. LAB such as *L. plantarum* also take part in this fermentation (Dirar et al., 1985). In North Africa, several fruits

**TABLE 2.1**
**Fermented Underutilized Green Leafy Vegetables**

| Raw Materials | Fermented Products | Microorganisms Involved | Region/ Country | References |
|---|---|---|---|---|
| **Leaves** | | | | |
| *Cassia obtusifolia* | Kawal | *B. subtilis*, *Propionibacterium*, and *S. sciuri*, LAB such as *L. plantarum* | North Africa | Dirar et al. (1985) |
| Manihot esculenta (Cassava leaf) | Ntobambodo | *B. subtilis*, *Weissella confusa*; *B. macerans*, and *B. pumilus*; *S. xylosus* and *Erwinia* sp.; LAB such as *E. faecium*, *E. hirae*, *E. casseliflavus*, and *Pediococcus* sp. | Africa | Sanni and Oguntoyinbo (2014) |

and vegetables are preserved by brining which restricts the growth of spoilage microorganisms (Aayah et al., 2010).

Many green leafy and other vegetables have good potentials as fermented products. *Psophocarpus tetragonolobus* (winged bean) is a legume. All parts of winged bean are edible. The seeds and roots contain a lot of proteins (40% and 20%, respectively). It is also a good source of pro-vitamin A (300–900 IU). It is an underutilized legume with good potentials for easing the problem of protein-energy malnutrition, and fermentation will add value to its utilization. *Vigna angularis* (Adzuki bean) is another example. The germinated beans are very good sources of folic acid, the B group of vitamins and vitamin A. The seeds are also rich in protein. Both the seeds and leaves have medicinal properties (Duke, 1981). *Hibiscus sabdariffa* contains many phytochemicals which are beneficial to humans (Hibiscus acid and its glucosides and methyl esters; chlorogenic acid and its derivatives; quercetin and its derivatives, flavonoids and others). Baobab leaves contain 10–18 times more calcium than the common calcium-rich foodstuffs (e.g. milk, green leafy vegetables and pulses). Controlled fermentation of leafy vegetables and fruits will improve safety and prolong shelf life, and also enhance the availability of some trace minerals, vitamins and anti-oxidants (Di Cagno et al., 2013). Other green leafy vegetables that have potentials for improving their nutritional and medicinal value through fermentation include *Gnetum africanum*, *Telferia occidentalis*, *Talinum triangulare*, *Moringa oleifera*, *Amaranthus hybridus* and *Vigna unguiculata* (McKay et al., 2010). Various vegetables and fruits are preserved by pickling. Pickled cucumber is made in Africa, Latin America, Asia and Europe, and pickled olives are common in Asia and Europe. The cleaned vegetable is put in 5% brine until the pH reduces to 4.7–5.7. It is further inoculated with *L. plantarum* or *Pediococcus pentosaceus* or a combination of these organisms and left for 2 weeks when the pH reduces further to pH 3.4–3.6. Onions (*Allium cepa*) are pickled with LAB to fermented products. *Telferia occidentalis* and *Amaranthus hybridus* have been shown to possess good potentials as fermented green leafy vegetables (Koko and Yako, 2018). Microorganisms associated with the fermentation process were *Bacillus, Lactobacillus, Escherichia coli, Staphylococcus, Proteus, Serratia, Citrobacter* and *Pseudomonas*. Fermentation is known to increase the mineral and protein contents of vegetables (Ifesan et al., 2014; Koko and Yako, 2018).

Many traditional African condiments are produced by the alkaline fermentation of protein-rich legumes (see Table 2.2). The names and details of methods of production vary with location. Ogiri (owoh) and very similar condiments are prepared from the fermentation of many legumes and oil seeds as shown in Figure 2.5. It is used to add flavour to dishes traditionally. The fermentation of *Parkia biglobosa* (African locust bean), *Prosopis africana* (mesquite bean) *Citrullus vulgaris* (melon) seeds, *Ricinus communis* (castor bean) seeds, *Telferia occidentalis* (fluted pumpkin) seeds, *Glycine max* (soya bean) seeds, *Vigna subterranea* (Bambara groundnut) seeds is essentially similar (Ouoba et al., 2003; Omafuvbe et al., 2004; Ojokoh et al., 2013; Oguntoyinbo et al., 2010). They are solid-state alkaline fermentations characterized by proteolytic and deamination reactions. The initial processing of the seeds differs based on the locality, nature and composition of the raw material. The decorticated seeds are wrapped in banana, plantain (or other leaves) leaves and cooked for 2–4 hours. The boiled seeds are cooled, ground in a mortar and wrapped again, packed in a clean container and fermented at room temperature (30°C±2°C) for at least 4 days.

The cooking period may vary with the level of toxicants found in the legume. Some legumes such as *Pentaclethra macrophylla* are cooked more than once. "Ugba" from African oil bean is obtained through several processing steps followed by solid-state fermentation (Onwuliri et al., 2004). The hard shell is cut open to release the undecorticated seeds. In the wild, the hard outer shell of mature dry seeds explodes by themselves to release the seeds which are picked from the ground. After boiling for 3–4 hours, the seeds are dehulled, sliced to 0.5–1 cm thickness and boiled again for at least 1 hour. The sliced seeds are drained, washed and wrapped in banana leaves, packed in a clean container and allowed to ferment at room temperature for 2–3 days. If it is to be used as a condiment, fermentation of the sliced seeds is prolonged to achieve the desired softening.

Many legume fermentations lead to an increase in the vitamins (especially the B vitamins) due to microbial synthesis. Some of these substrates contain toxicants which are removed or reduced by fermentation. Flatulent oligosaccharides may be hydrolysed to simple sugars (e.g. in the fermentation of African locust bean to produce dawadawa). The levels of anti-nutritional factors such as phytates and tannins are also reduced. Many leaves are used to wrap the fermenting condiments and these include *Musa sapientum*; cocoa leaves. In some cases, traditional starter cultures are used. A starter culture "kuru" is made from *H. sabdariffa* and used in the processing of *Parkia biglobosa* (African locust bean seeds) to dawadawa. A lot of proteolysis occurs during alkaline fermentation of legumes (Enujiugha et al., 2002; Enujiugha, 2003). Generally, the amino acid concentration increases with fermentation period and culminates in a peak before it starts to decrease. In oil seeds, it may take a longer time to attain this peak which suggests further metabolism of the fatty acids by the fermenting microorganisms (Enujiugha et al., 2004; Enujiugha and Akanbi 2008). The flavour of the product is developed within this period.

**TABLE 2.2**
**Fermented Underutilized Legumes**

| Legumes | Product | Microorganisms | Region/Country | References |
|---|---|---|---|---|
| *Parkia biglobosa* (African locust bean) | dawadawa; iru-pete; iru-woro | *B. subtilis*; *Coryneform bacteria*, *B. amyloliquefaciens*; *Staphylococcus*; *Lactobacillus* sp.; *R. stolonifer*; *Streptococcus* sp.; *A. Fumigates*; *Pediococcus* sp. | West Africa | Oboh et al. (2008), Jensen et al. (2009), Omodara and Aderibigbe (2014, 2018), Adewumi et al. (2018) |
| *Hibiscus sabdariffa* (Roselle) seeds | Bikalgia; dawadawa botso (Niger), Datou (Mali), Furundu (Sudan), Mbuja (Cameroon) | *B. subtilis*, *B. licheniformis*, *B. cereus*, *B. pumilus*, *B. badius*, *Weissella confusa*, *Pediococcus pentasaseus*, *Weissella cibaria*, *L. plantarum*, *Enterococcus casseliflavus*, *E. faecium*, *E. faecalis*, *Brevibacillus bortelensis*, and *B. fusiformis* | Africa; Burkina Faso; Cameroon, Mali, Niger; Sudan | Bengely (2001), Ouoba et al. (2007, 2008, 2010) |
| *Prosopis africana* (African mesquite bean) | Okpiye; Okpeye; Okpehe | *B. subtilis*, *B. pumilus*, *B. licheniformis*; *B. circulans*; *LAB*; *B. amyloliquefaciens*; *B. cereus*; *Enterococcus* sp. | Nigeria | Odibo et al. (2008), Oguntoyinbo et al. (2007, 2010), Oguntoyinbo (2012, 2014) |
| *Telferia occidentalis* (Fluted pumpkin) | Ogiri-ugu | *Bacillus* sp.; *E. coli*, *S. aureus*, *Citrobacter* sp.; *Lactobacillus* sp.; *Serratia* sp.; *Penicillium* sp.; *Fusarium* sp. | West Africa | Barber et al. (1989), Koko and Yako (2018) |
| *Ricinus communis* (Castor bean) | Ogiri-isi | *Pseudomonas* sp.; *B. licheniformis*; *B. megaterium*; *Pseudomonas aeruginosa*; *Staphylococcus* sp.; *Micrococcus* sp.; *Corynebacterium* sp. | West Africa | Ojinnaka and Ojimelukwe (2012), Ojimelukwe et al. (2011a) |
| *Citrullus vulgaris* (Melon seeds) | Ogiri-egusi | *Lactococcus*, *Streptococcus*, *Pediococcus*, *Leuconostoc*, *Propionibacter* sp. | West Africa | Nwosu and Ojimelukwe (1993) |

(*Continued*)

**TABLE 2.2 (*Continued*)**
**Fermented Underutilized Legumes**

| Legumes | Product | Microorganisms | Region/Country | References |
|---|---|---|---|---|
| *Pentaclethra macrophyla* (African oil bean) | Ugba | *B. subtilis*, *B. licheniformis*, *B. megaterium*, *E. coli*, *B. pumilus*, *Staphylococcus* sp.; *Micrococcus* sp.; *Leuc mesenteroides*; *L. plantarum*; *S. lactis*, *Proteus* sp., *Enterobacter* sp. and *E. coli* | West Africa | Ogueke and Aririatu (2004), Ogueke et al. (2010) |
| *Treculia africana* (African bread fruit) | Ukwa | *B. subtilis*, *A. niger*, *A. flavus*, *B. pumulis*, *S. aureus*; *L plantarum*; *L. bulgaricus*; *Leuc. mesenteroides*, and *S. cerevisiae* | West Africa | Nwaneri et al. (2017), Osabo et al. (2009), Olapade and Umeonuorah (2014), Okafor (1981,1985), Ojokoh et al. (2013) |
| *Vigna subterranea* (Bambara groundnut) | Dawadawa type product | *B. subtilis*; *B. licheniformis* | West Africa | Amadi et al. (1999), Barimalaa et al. (1994) |
| *Gossypium hirsutum* L. (Cotton seed) | Owoh | *B. brevis*; *B. megaterium*; *B. polymyxa*; *Staphylococcus* sp.; *P. aeruginosa*; *B. subtilis*; *B. licheniformis* | West Africa | Omafuvbe et al. (2004), Sanni et al. (2000), Sanni and Ogbonna (1991) |
| *Sphenostylis stenocarpa* (African Yam Bean) | Owoh type product | *B. pumilus*; *B. subtilis*; *Staphylococcus* sp. | West Africa | Ogbonna et al. (2001) |
| *Glycine max* (Soybean) | Daddawa | *B. subtilis*, *E. faecium*; *B. pumilus*; *Candida parapsilosis*, *Geotrichum candidum*; *B. licheniformis*, *B. cereus*, *B. circulans*, *B. thuringiensis*, *B. sphaericus* | Africa | Omafuvbe et al. (2003) |
| *Adansonia digitata* (baobab) – Maari | Maari | *B. subtilis*; *E. faecium*; *E. casseliflavus*; *Pediococcus acidilactici* | Burkina Faso | Parkouda et al. (2009), Sanni and Oguntoyinbo (2014) |

Legume/Oil seed
↓
Dehull
↓
Wrap in clean banana or plantain leaves (or other leaves)
↓
Boil (2–4 hours: boiling time depends on the nature of the legume)
↓
Allow to cool
↓
Grind in a mortar
↓
Wrap in plantain/ banana leaves
↓
Pack in a clean container and ferment for about four days
↓
Ogiri (Owoh; Okpei)

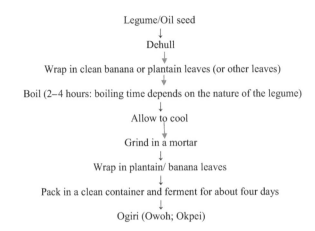

**FIGURE 2.5** Process flow for the production of "Ogiri" from legume/oil seed.

Nono, Kindirmo, Warankasi and Maishanu are some underutilized milk products that require research attention (Table 2.3). Nono is traditionally prepared from cow or goat milk by mixing a little leftover as starter into fresh milk and allowing fermentation to take place for 24 hours at room temperature (30°C±2°C). The product becomes sour due to the formation of lactic acid. After fermentation, milk butter is removed by churning and the residual sour milk is used as beverage. Bacteria, yeasts and moulds take part in the fermentation process. It may be taken with fura (fermented compressed

**TABLE 2.3**
**Fermented Underutilized Animal Products – Meat, Milk Fish and Crustaceans)**

| Animal Products | Fermented Product | Microorganisms Involved in the Fermentation Process | Region/ Country | References |
|---|---|---|---|---|
| Milk | Nono | *Streptococcus thermophilus*, *Leuconostoc* sp.; *Lactococcus* sp. (*L. cremoni* and *L. lactis*), *Saccharomyces* sp., *Lactobacilli* (*L. acidophilus* and *L. bulgaris*) | Nigeria, Mali, Chad | Okonkwo (2011) |
| Milk | Kindirmo; Warankasi; | LAB; *Bacillus* sp.; *Streptococcus* sp.; *Micrococcus* sp.; *L. casei*; *L. lactis*; *L. brevis*; *L. bulgaricus*; *L. brevis* | Nigeria, West Africa | Igwe et al. (2011), Okonkwo (2011) |
| Milk | Amasi | *L. plantarum*; *Lactococcus*; *Streptococcus*; *Leuconostoc* | South Africa | Hama et al. (2009) |
| Meat Fish | Sungu; Bala; Bonga fish | *Streptococcus* sp., *Escherichia coli*, *Yersinia* sp., *Enterobacter* sp., *S. aureus*, *Enterococcus* sp., *Vibrio cholerae*, *Proteus* sp., *Shigella* sp., *Salmonella* sp. and *Campylobacter* sp. | North Africa West Africa | Norredine et al. (2013) |

millet ball powder); as fura de nunu. Maishanu (local butter) may be obtained by boiling milk for about 3 hours after which it is allowed to stand for a day in a calabash. The maishanu floats, while the nono sediments. Wara (Yoruba soft cheese) can also be produced from goat or cow milk. The milk is coagulated with the juice extract of Sodom apple (*Calotropis pocera*). The mixture of the juice and the milk is heated for about 40 min and cooled to remove the skin and floating dirt. The white curd is further heated to obtain the desired firmness and flavour. The curd is poured into a basket mould which gives it a characteristic shape before it is dropped back into the whey while it is still hot (about 65°C) and sold. Kindirmo (sour yoghurt) and cuku (Fulani cheese) are other small-scale milk products that are produced traditionally.

Enriched fermented milk products may be developed from milk co-fermented with dietary fibres, starch, minerals, vitamins and vegetables. Sweet potato, French bean, lemon and soybean may be co-fermented with milk to produce enriched curds and yoghurts. Curd is popular in Asian countries, while yoghurt is popular in America and Europe (Younus et al., 2002). Banda is a low-quality meat product made from different livestock and wildlife (donkey, horse, camel, buffalo, etc.). Chance fermentation occurs before it is smoke-dried. Suya (esire; balangu) is produced from smoke-dried spiced salted slices of meat (usually beef) after natural fermentation. Spontaneous fermentations are associated with the processing of meat products prior to smoke drying. Ndariko and Jirge are other Nigerian traditionally processed meat products.

Several small fishes (such as *Ethmalosa fimbriata* – Bonga fish) and crustaceans found in fast flowing streams are harvested and left to spoil through uncontrolled fermentation during their peak season (Anihouvi et al., 2006, 2007). Better processing and utilization of these products through controlled fermentation and drying will promote food and nutrition security.

## 2.11 FERMENTED COCOYAM: SAPAL; KOKOBELE

Table 2.4 shows some underutilized fermented tubers. Taro (*Colocasia esculenta*) is fermented to produce sapal which is popular in the northern coast of Papua New Guinea and other regions of Africa. To make sapal, cooked grated taro corm is mixed with coconut cream in a ratio of 5:1 and allowed to ferment at ambient temperature. The product has a firm texture and a jelly-like consistency. LAB such as *Leuconostoc mesenteroides* or *Leuc. paramesenteroides* are responsible for the fermentation and may reach populations of 1.6 × 10' cfu/ml in the final product (Gubag et al., 1996). Traditionally, sapal is seasonal and is often used on social occasions.

Another variety of cocoyam (*Xanthosoma sagittifolium*) is fermented into *kokobele* (a common food in the Ondo State of Western Nigeria). Cocoyam tubers are processed by peeling, washing and slicing, and are steeped under water at room temperature to ferment for 2–3 days. At the end of fermentation period, the steep water is drained off and discarded, while the cocoyam slices are sun-dried for 3–5 days and milled into flour to produce kokobele. Preparation of kokobele involves reconstitution in water and boiling with pepper, tomatoes, palm oil, fish and spices to enhance flavour (Iwuoha and Eke, 1996). Fermented foods from roots and tubers are diverse in nature and offer nutritional and health benefits in a wide variety of flavours and textures. Fermented foods cassava, yam and sweet potato are part of the regular

**TABLE 2.4**
**Fermented Underutilized Roots and Tubers**

| Root and Tuber | Fermented Product | Microorganisms Involved | Region/ Country | References |
|---|---|---|---|---|
| *Colocasia* sp. (Taro) | Sapal | *Leuc. mesenteroides*; *Leuc paramesenteroides*; *Geotrichum candidum*; *Mycoderma vini* | Africa | Gubag et al. (1996), Cai and Nip (1990) |
| *Xanthosoma sagittifolium* | Kokobele | | Nigeria | Duru and Uma (2003) |
| Yam | Amala | | Nigeria | Iwuoha and Eke (1996) |

African, Latin American and Asian diets. An alcoholic beverage prepared by semi-solid fermentation of *Colocasia* using steamed or autoclaved fresh *Colocasia* grits has been also reported in literature (Cai and Nip, 1990).

Sweet potatoes are used industrially as a source of starch which could be fermented with *Lactobacillus cellobiosus*, *Streptococcus lactis* and *Corynebacterium* sp. As the inoculum improves starch extraction through the concentration of amylases is reduced (Ray and Ravi, 2005; Jyothi et al., 2005). In China, Japan, Korea and India, sweet potato is used for alcohol production. Shochu is an Asian alcoholic beverage made from sweet potato and other ingredients (Jyothi et al., 2005). *A. niger* and *S. cerevisiae* contribute significantly to the fermentation process.

## 2.12 NOVEL FERMENTED FOODS FROM SWEET POTATO

Fermentation of sweet potatoes using *L. plantarum* as starter culture yields lacto-pickles (Ray and Panda, 2007). It produces mainly acids increasing the sourness and decreasing sweetness as the sugars are converted to acids. Lacto-pickles have been prepared both from orange-fleshed (β-carotene-rich) and anthocyanin-rich sweet potato roots. Diced sweet potato roots are subjected to fermentation (after brining) using probiotic *L. plantarum* MTCC 1407 (for 28 days). The lacto-pickle has acceptable sensory properties (Panda et al., 2007). Sweet potato petioles can also be pickled (Panda et al., 2009).

**Lacto-juice** Lacto-juices may be produced by the lactic fermentation of a number of beverages. They have high nutritive value (Ray and Panda, 2007). β-Carotene and anthocyanin-rich sweet potato cultivars may be fermented by inoculating *Lb. plantarum* MTCC 1407 or other LAB to produce lacto-juice (Panda and Ray, 2007; Panda et al., 2008). Sweet potato roots (fresh or boiled) are fermented with *Lb. plantarum* at 28°C ± 2°C for 48 hours to make lacto-juice (Panda and Ray, 2007). Sweet potato curd and yoghurt are produced by lactic acid fermentation of milk. They are enriched milk products (Younus et al., 2002).

Some underutilized fermented cereal products are shown in Table 2.5. Kwunu, Fura and Tuwo are some of the underutilized cereal products with good potentials for improving food and nutrition security in Africa. Tuwo can be made from any cereal (rice, maize, millet, sorghum, hungry rice, etc.) (see Table 2.5). The cereal is steeped

**TABLE 2.5**
**Underutilized Fermented Cereals**

| Cereals | Fermented Product | Microorganisms Involved | Region/Country | References |
|---|---|---|---|---|
| Maize products | Amahewu; Fura; chibwantu; munkoyo; togwa | *L. fermentum*; *L. reuteri*; *Saccharomyces*; *Candida* sp; *L. fermentum*; *L. reuteri*; *L. paraplantarum*; *P. acidilactici*; *W. confusa* | Africa | Adimpong et al. (2012), Amenan et al. (2014), Enwa et al. (2011), Hama et al. (2009) |
| Sorghum products | Chibuku; Kunun; Kome; Fura; Dagnan; Ben-saalga; Coco-baca; Baca; Pito Burukutu Umqombothi | *L. fermentum*; *P. acidilactici*; *W. confusum*; *Pichia anomala*; *L. plantarum*; *B. subtilis*; *L. plantarum* and *B. subtilis*; *L. fermentum*, *L. plantarum*, *S. gallolyticus* subsp. *macedonicus*, *P. pentosaceus*, *W. confusa* | All parts of Africa | Odunfa and Adeleye (1985), Ekwem (2014), Guyot (2010), Ojimelukwe et al. (2011b), Vieira-Dalode et al. (2008) |
| Millet, sorghum and maize products | Aklui; Arraw; cakry; cere; Banku; Dagnan | *L. fermentum*; *P. acidilactici*; *W. confusum*; *LAB*; *fungi* | Different parts of Africa | Adimpong et al. (2012), Amenan et al. (2014), Oguntoyinbo and Narbad (2012), Vieira-Dalode et al. (2007) |

overnight in water, spread to dry in the sun and ground into powder. The flour is stirred into hot boiling water until a smooth paste is formed. It is left for some days (sometimes up to 4 days (for fermented flavour development)). Potash may be added as a preservative. Kwunu is made by initially grinding millet in a mortar to release the kernels. The decorticated cereal is further ground into flour, mixed with a little water and poured into boiling water, where it is stirred to form a thick paste. It is fermented for about 3 days prior to consumption. Fura is millet balls, prepared like Tuwo, made into balls and served as a combination dish with milk.

Both pito and burukutu are beverages made by germinating and fermenting cereals (Table 2.6). They can be made from maize sorghum or millet. The grain is soaked for 24 hours and allowed to germinate for 2 days. It is sun-dried and ground into flour. The flour is mixed with water and boiled for 3–4 hours. It is allowed to settle at room temperature (30°C). The floating water is decanted, while more water is added and the slurry of the sediment is re-cooked for 3 hours. The mixture is allowed to cool. The upper clearer layer is separated as pito while the sediment is stored in another container as burukutu. Both beverages are allowed to undergo natural fermentation for 24 hours before use. LAB and yeasts are involved in the fermentation process (Okoro et al., 2011).

The pulp of harvested cocoa is used as a nourishing fermented drink. Ripe cocoa pods are cut and placed in raffia baskets placed in a basement (like in a pot). They are covered with leaves and allowed to undergo natural fermentation. The dripping liquid from the fermenting cocoa pulp is collected and used as a fermented cocoa drink (De Vust and Weck, 2016). Kuru is a local starter culture fermented drink from the pulp of *Parkia biglobosa* (Omodara and Aderibingbe, 2014).

## 2.13 CONCLUSION

Crops used for most of the underutilized fermented foods are obtained from the wild or produced indigenously on a very small scale. They have traditional medicinal uses that have not received sufficient research attention. Fermentation as a processing and preservation method for leafy vegetables and fruits is grossly underutilized in Africa, although such fermented products could significantly contribute towards improving nutrition and food security. Plants like the baobab, winged bean and several local green leafy vegetables have potentials as probiotic foods but have not been scientifically explored. These underutilized food sources need scientific attention to boost their potentials for food and nutrition security in Africa. Suitable technology and packaging for these fermented foods have not received good research attention. Some of the underutilized crops like African locust bean trees in many countries such as Nigeria have been cut down through deforestation. There are no plantations for sustainable production. Concerted research efforts and policy changes are needed to harness the benefits of underutilized fermented foods.

The potentials of underutilized fermented foods need to be harnessed in the quest for food and nutrition security in Africa. They are long-term options for reducing hidden hunger, the double burden of malnutrition and food/nutrient diversification. Biotechnological upgrade of the fermentation of underutilized food crops needs priority research attention.

**TABLE 2.6**
**Underutilized Fermented Fruits and Beverages**

| Beverage Source | Product | Microorganisms | Region/Country | References |
|---|---|---|---|---|
| Oil palm/Rafia palm | Palm wine | LAB; yeasts | Africa | Ezeronye et al. (2005), Elijah et al. (2007) |
| Cereal-based beverages | Mabisi; Kome Mukoya; Chibwantu; Kunun-zaki (non-alcoholic beverages) | Acetic acid bacteria | Africa: Sudan, South Africa; Zambia | Hammes et al. (2005) |
| Barley | Borde; Shamita | Yeasts and LAB | North Africa; Ethiopia | Amenan et al. (2014), Abegaz et al. (2002) |
| | Ogogoro; | Yeasts and LAB | | |
| Maize, sorghum and millet | Burukutu; Dalaki; Kunun-zaki; Ben-saalga pito (alcoholic beverages) | *Acetobacter* sp., *Saccharomyces cerevisiae*, *Lactobacillus plantarum* and *L. fermentum*; *L. fermentum*, *L. delbrueckii*, *P. acidilactici*, *S. cerevisiae*, *C. tropicalis*, *K. apiculata*, *H. anomala*, *S. pombe*, *K. africanus* | Africa | Okoro et al. (2011), Sawadogo-Lingani et al. (2007) |
| Palm fruit | Eketeke | *S. cerevisiae*; *Leuconostoc*; *Lactobacillus*; *Acetobacter* sp.; *Zymomonas mobilis* | Nigeria | Iwuoha and Eke (1996), Njoku and Onwu (2010) |
| Cocoa pulp | Fermented cocoa drink | LAB, yeasts, and acetic acid bacteria | West Africa | De Vust and Weck (2016) |

## REFERENCES

Aayah, H., Ahansal, L., Guyot, J.P., Ibnsouda, S., Chevallier, I. and Boussaid, A. (2010). Characterization of the dry salted process for the production of the msayer, a traditional lemon aromatizing condiment. *LWT—Food Science and Technology* 43: 568–72.

Abegaz, K., Beyene, F., Langsrud, T. and Narvhus, J.A. (2002). Parameters of processing and microbial changes during fermentation of borde, a traditional Ethiopian beverage. *Journal of Food Technology* 7(3): 85–92. http://www.bioline.org.br/request?ft02021.

Adewumi, B., Germaine, G.A., Ogunwole, E., Akinola Ogunwole, E., Akingunsola, O.C., Falope, O.C. and Eniade, A. (2018). Effects of sub-lethal toxicity of Chlorpyrifos and DDforce pesticides on haematological parameters of *Clarias gariepinus. International Journal of Environmental Research and Public Health* 5(5): 62–71.

Adimpong, B., Nielsen, D.S., Sørensen, K.I. and Jespersen, D.L. (2012). Genotypic characterization and safety assessment of lactic acid bacteria from indigenous African fermented products. *BMC Microbiology* 12: 75–89.

Amadi, E.N., Barimalaa, I.S. and Omosigho, J. (1999). Influence of temperature on the fermentation of bambara groundnut (*Vigna subterranea*) to produce a dawadawa-type product. *Plant Foods for Human Nutrition* 54(1): 13–20. doi:10.1023/A:1008003118374.

Amenan, A., Kouakou, S.Y., BrouAmani, B.G., Thonart, A.P. and Djè, K.M. (2014). The use of lactic acid bacteria starter cultures during the processing of fermented cereal-based foods in West Africa: A review. *Tropical Life Science Research* 25(2): 81–100.

Anihouvi, V.B., Ayernor, G.S., Hoounhouigan, J.D. and Sakyi-Dawson, E. (2006). Quality characteristics of Lanhouin: A traditionally processed fermented fish product in the republic of Benin. *African Journal of Food, Agriculture, Nutrition and Development* 6(1): 1–15.

Anihouvi, V.B., Sakyi-Dawson, E., Ayernor, G.S. and Hounhouigan, J.D. (2007). Microbiological changes in naturally fermented cassava fish (*Pseudotolithus* sp.) for *lanhouin* production. *International Journal of Food Microbiology* 116(2): 287–291.

Baldermann, S., Blagojević, L., Frede, K., Klopsch, R., Neugart, S., Neumann, A., Ngwene, B., Norkeweit, J., Schröter, D., Schröter, A., Schweigert, F., Wiesner, J. and Schreiner, M. (2016). Are neglected plants the food for the future? *Critical Reviews in Plant Science* 35(2): 106–119. doi:10.1080/07352689.2016.1201399.

Barber, L., Ibiama, E.A. and Achinewhu, S.C. (1989). Microorganisms associated with fermentation of fluted pumpkin seeds (*Telfairia occidentalis*). *International Journal of Food Scienceand Technology* 24 (2): 189–193.

Barimalaa, I.S., Achinewhu, S.C., Yibatama, I. and Amadi, E.N. (1994). Studies on the Solid Substrate Fermentation of Bambara Groundnut (*Vigna subterranea* (L) Verde). *Journal of the Science of Food and Agriculture*, 66(4): 443–446. doi:10.1002/(ISSN)1097-0010.

Cai, T. and Nip, W.K. (1990). Biochemical changes in the development of alcoholic fermented products from taro (*Colocasia esculenta* L) Schott. *Tropical Science* 30 (4): 379–390.

De Vust and Weck, S. (2016). The cocoa bean fermentation process: From ecosystem analysis to starter culture development. *Journal of Applied Microbiology* 121 (1): 5–17.

Di Cagno, R., Coda, R., De Angelis, M. and Gobbetti, M. (2013). Exploitation of vegetables and fruits through lactic acid fermentation. *Food Microbiology* 33(1): 1–10.

Dirar, H.A., Harper, D.B. and Collins, M.A. (1985). Biochemical and microbiological studies on Kawal, a meat substitute derived by fermentation of *Cassia obtusifolia* leaves. *Journal of the Science of Food and Agriculture* 36: 881–892. doi:10.1002/jsfa. 2740360919.

Doi, Y. (2019). Glycerol metabolism and its regulation in lactic acid bacteria. *Applied Microbiology and Biotechnology* 103. doi:10.1007/s00253-019-09830-y.

Duke, J.A. (1981). Handbook of Legumes of World Economic Importance. Plenum Press, New York.

Duru, C. C. and Uma, N. U. (2003). Protein enrichment of solid waste from cocoyam (Xanthosoma sagittifolium (L.) Schott) cormel processing using Aspergillus oryzae obtained from cormel flour. *African Journal of Biotechnology* 2(8): 228–232.

Ekwem, O.H. (2014). Isolation of antimicrobial producing lactobacilli from akamu (a Nigerian fermented cereal gruel). *African Journal of Microbiology Research* 8(7): 718–720. doi:10.5897/AJMR2013.6251.

Elijah, A., Ojimelukwe, P.C. and Ezeronye, O.U. (2007). Preliminary investigations on the effect of incorporation of *Alstonia boonei* bark powder on the fermentation of palm wine. *Journal of Food Science and Technology* 44 (2) 190–194.

Enwa, F.O., Beal, J. and Arhewoh, M.I. (2011). Effect of maize and bacteria starter culture on maize fermentation process. *International Journal of Biomedical Research* 2(11): 561–567. http://scholar.google.com/citations?user=Fw0INyAAAAAJ&hl=en.

Enujiugha, V.N., Amadi, C. and Sanni, T. (2002). Amylase in raw and fermented African oilbean seed (*Pentaclethra macrophylla Benth*). *Journal of European Food Research and Technology* 214: 497–500.

Enujiugha, V.N. (2003). Nutrient changes during the fermentation of African oil bean (*Pentaclethra macrophylla* Benth) seeds. *Pakistan Journal of Nutrition* 2: 320–323.

Enujiugha, V.N., Thani, F.A., Sanni, T.M. and Abigor, R.D. (2004). Lipase activity in dormant seeds of the African oilbean (*Pentaclethra macrophylla Benth*). *Food Chemistry* 88: 405–410.

Enujiugha, V.N. and Akanbi, C.T. (2008). Quality evaluation of canned fermented oil bean seed slices during ambient storage. *African Journal of Food Science* 2: 54–59.

Ezeronye, O.U., Elijah, A.I. and Ojimelukwe, P.C. (2005). Effect of *Saccoglottis gabonensis* and *Alstonia boonei* on the fermentation of fresh palm sap by *Saccharomyces cerevisae*. *Journal of Food Technology* 3(4): 586–591.

FAO, IFAD, UNICEF, WFP and WHO. (2018). The State of Food Security and Nutrition in the World 2018. Building climate resilience for food security and nutrition. Rome, FAO.

Gubag, R., Omoloso, D.A. and Owen, J.D. (1996). Sapal: A traditional fermented taro (*Colocassia esculenta* (L) Short corm and coconut cream mixture from Papua New Guinea. *International Journal of Food Microbiology* 28: 361–367.

Guyot, J.P. (2010). Fermented cereal products. In *Fermented Foods and Beverages of the World*, eds. J.P. Tamang and K. Kailasapathy (New York, NY: CRC Press, Taylor and Francis Group), 247–261. doi:10.1201/ebk1420094954-c8.

Hama, F., Savadogo, A., Ouattara, C.A. and Traore, A.S. (2009). Biochemical, microbial and processing study of Dègué a fermented food (from pearl millet dough) from Burkina Faso. *Pakistan Journal of Nutrition* 8(6): 759–764. doi:10.3923/pjn.2009.759.764.

Hammes, W.P., Brandt, M.J., Francis, K.L., Rosenheim, J., Seitter, M.F. and Vogelmann, S.A. (2005). Microbial ecology of cereal fermentations. *Trends in Food Science and Technology* 16(1–3): 4–11. doi:10.1016/j.tifs.2004.02.010.

Ifesan, B.O.T., Egbewole, O.O. and Ifesan, B.T. (2014). Effect of fermentation on nutritional composition of selected commonly consumed green leafy vegetables in Nigeria. *International Journal of Applied Science and Biotechnology* 2(3): 291–297.

Igwe, E.C., Ojimelukwe, P.C. and Onwuka, G.I. (2011). Effects of processing conditions on the quality of kindirmo made from whole cow milk and cow–soymilk mixtures. *International Journal of Current Research.* 3(11): 11–18.

Iwuoha, C. and Eke, O. (1996). Nigerian indigenous fermented foods: Their traditional process operation, inherent problems, improvements and current status. *Food Research International* 29 (5–6): 527–540.

Jensen, L.J., Kuhn, M., Stark, M., Chaffron, S., Creevey, C., Muller, J., Doerks, T., Philippe J.P., Roth, A., Simonovic, M., Bork, P. and von Mering, C. (2009). STRING 8—a global view on proteins and their functional interactions in 630 organisms. *Nucleic Acids Research* 37: D412–D441.

Jyothi, A.N., Wilson, B., Moorthy, S.N. and George, M. (2005). Physiochemical properties of the starchy flour extracted from sweet potato tubers through lactic acid fermentation. *Journal of the Science of Food and Agriculture* 85: 1558–1563.

Kobawila, S.C., Louembe, D., Keleke, S., Hounhouigan, J. and Gamba, C. (2005). Reduction of the cyanide content during fermentation of cassava roots and leaves to produce bikedi and ntoba mbodi, two food products from Congo. *African Journal of Biotechnology* 4: 689–696.

Koko, A. and Yako, L. (2018). How nutritional composition of commonly consumed vegetable changes under the influence of fermentation. *Medical Biotechnology Journal* 2(2): 59–64. doi:10.22034/mbt.2018.76923.

McKay, D.L., Chen, C.Y., Saltzman, E. and Blumberg, J.B. (2010). *Hibiscus sabdariffa* L. tea (tisane) lowers blood pressure in pre-hypertensive and mildly hypertensive adults. *Journal of Nutrition* 140(2): 298–303.

Njoku, P.C. and Onwu, J.C. (2010). The study of the characteristics and rancidity of three species of *Elaeis guineensis* in south east of Nigeria. *Pakistan Journal of Nutrition* 9(8), 759–761. http://pjbs.org/pjnonline/fin1474.pdf.

Nwaneri, C.B., Ogbulie, J.N. and Chiegboka, N.A. (2017). The Microbiology and Biochemistry of *Treculia Africana* (African Breadfruit). Fermentation. *Nigerian Journal of Microbiology* 31(1): 3666–3670.

Nwosu, C.D. and Ojimelukwe, P.C. (1993). Improvement of the traditional method of ogiri production and identification of the microorganisms associated with the fermentation process. *Plant Foods for Human Nutrition* 43: 267–272.

Oboh, G., Alabi, K.B. and Akindahunsi, A.A. (2008). Fermentation changes the nutritive value, polyphenol distribution and antioxidant properties of *Parkia biglobosa* seeds (African locust beans). *Food Biotechnology* 22: 363–376. doi:10.1080/08905430802463404.

Odibo, F.J.C., Ezeaku, E.O. and Ogbo, F.C. (2008). Biochemical change during the fermentation of Prosopis africana seeds for ogiri-okpei production. *Journal of Industrial Microbiology and Biotechnology* 35: 947e52.

Odunfa, S.A. and Adeyele, S. (1985). Microbiological changes during the traditional production of ogi-baba, a West African fermented sorghum gruel. *Journal of Cereal Science*, 3(2): 173–180. doi:10.1016/S0733-5210(85)80027-8.

Ogbonna, D.N., Sokari, T.G. and Achinewhu, S.C. (2001). Development of an owoh-type product from African yam beans *Sphenostylis stenocarpa*, Hoechst ex A Rich, Harms seeds by solid substrate fermentation. *Plant Foods for Human Nutrition* 56: 183e94.

Ogueke, C.C. and Aririatu, L.E. (2004). Microbial and organoleptic changes associated with ugba stored at ambient temperature. *Nigerian Food Journal* 22: 133–140.

Ogueke, C.C., Nwosu, J.N., Owuamanam, C.I. and Iwuona, J.N. (2010). Ugba, the fermented African Oilbean Seeds: Its production, chemical composition, preservation, safety and health benefits. *Pakistan Journal of Biological Sciences* 13(10): 489–496.

Oguntoyinbo, F.A., Sanni, A.I., Franz, C.M.A.P., Holzapfel, W.H. (2007). *In-vitro* selection and evaluation of Bacillus starter cultures for the production of okpehe, a traditional African fermented condiment. *International Journal of Food Microbiology* 113: 208–218.

Oguntoyinbo, F.A., Huch, M., Cho, G.-S., Schillinger, U., Holzapfel, W.H., Sanni, A. (2010). Diversity of *Bacillus* species isolated from okpehe, a traditional fermented soup condiment from Nigeria. *Journal of Food Protection* 73: 870–878.

Oguntoyinbo, F.A. and Narbad, A. (2012). Molecular characterization of lactic acid bacteria and in situ amylase expression during traditional fermentation of cereal foods. *Food Microbiology* 31 (2): 254–262.

Oguntoyinbo, F.A. (2012). Development of hazard analysis critical control points (HACCP) and enhancement of microbial safety quality during production of fermented legume-based condiments in Nigeria. *Nigerian Food Journal* 30: 59–66.doi: 10.1016/S0189-7241(15)30014-X.

Oguntoyinbo, F.A. (2014). Safety challenges associated with traditional foods of West Africa. *Food Reviews International* 30: 338–358. doi:10.1080/87559129.2014.940086.

Ojimelukwe, P.C., Ojinnaka, M.C. and Okechi, P. (2011a). Physicochemical characteristics of fermenting castor seeds containing lime and NaCl as additive. *African Journal of Food Science*. 5 (14): 754–760.

Ojimelukwe, P.C., Okudu, H.O. and Uche, N.J. (2011b). Effect of pasteurization and preservation treatment on the quality of "kunun zaki" made from Sorghum (*Sorghum biocolor*). *Nigerian Journal of Nutritional Science* 32(2): 42–45.

Ojinnaka, MC and Ojimelukwe, P.C. (2012). Effect of fermentation period on the organic acid and amino acid contents of Ogiri from *Ricunus communis*. *Journal of Food Technology* 10: 140–150.

Ojokoh, A.O., Daramola, M.K. and Oluoti, O.J. (2013). Effect of fermentation on nutrients and antinutrients composition of breadfruit (*Treculia africana*) and cowpea (*Vigna unguiculata*) blend flours. *African Journal of Agricultural Research* 8(27):3566–3570.

Okafor, J.C. (1981). Delimination of a new variety of *Treculia africana* sub sp Africana (Morocceae). *Bulletin du Jardin botanique National de Belgique / Bulletin van de Nationale Plantentuin van België* 51(1/2): 191–199. doi:10.2307/3667741.

Okafor, J.C. (1985). Indigenous fruit production and utilization in Nigeria. *Proceedings of the National Workshop on Fruit Production in Nigeria*, pp. 10–12.

Okonkwo, O.I. (2011). Microbiological analyses and safety evaluation of nono: A fermented milk product consumed in most parts of Northern Nigeria. *International Journal of Dairy Science* 6: 181–189. doi:10.3923/ijds.2011.

Okoro, A.I., Ojimelukwe, P.C., Ekwenye, U.N. and Akaerue, B. (2011). Quality characteristics of indigenous fermented beverage; pito using *Lactobacillus* sake as a starter culture. *Continental Journal of Applied Science* 6: 15–20.

Oladele, O. (2011). Contribution of indigenous vegetables and fruits to poverty alleviation in Oyo State, Nigeria. *Journal of Human Ecology* 34: 1–6.

Olapade, A.A. and Umeonuorah, U.C. (2014). Chemical and sensory evaluation of African breadfruit (*Treculia africana*) seeds processed with Alum and Trona. *Nigerian Food Journal* 52(1): 80–88.

Omafuvbe, B.O., Abiose, S.H., Shonukan, O.O. (2003). Fermentation of soybean (Glycine max) forsoy-daddawaproductionbystarter culturesof Bacillus. *Food Microbiology* 19: 561–566.

Omafuvbe, B.O., Falade, O.S., Osuntogun, B.A., Adewusi, S.R.A. (2004). Chemical and biochemical changes in African locust bean (*Parkia biglobosa*) and melon (*Citrullus vulgaris*) seeds during fermentation to condiments. *Pakistan Journal of Nutrition* 3: 140–145.

Omodara, T.R. and Aderibigbe, E.Y. (2014). Effect of starter culture and different components of "kuru" on nutritional quality of fermented Parkia biglobosa. *International Journal of Applied Microbiology and Biotechnology Research* 2: 73–78.

Omodara, T.R., Adegberigbe, E.Y. (2018). Microbiological, physicochemical and enzymatic changes in fermented African locust bean (*Parkia biglobosa*) seeds using *Bacillus subtilis* and additives. *Sustainable Food Production* 1: 22–29.

Onwuliri, V.A., Attah, I. and Nwankwo, J.O. (2004). Anti-nutritional factors, essential and non-essential fatty acids composition of Ugba (*Pentaclethra macrophylla*) seeds at different stages of processing and fermentation. *Journal of Biological Sciences* 4: 671–675.

Ouoba, L.I., Rechinger, K.B., Barkholt, V., Diawara, B., Traore, A.S. and Jakobsen, M. (2003). Degradation of proteins during the fermentation of African locust bean (*Parkia biglobosa*) by strains of Bacillus subtilis and Bacillus pumilus for production of Soumbala. *Journal of Applied Microbiology*. 94: 396e402.

Ouoba, L.I., Parkouda, C., Diawara, B., Scotti, C. and Varnam, A.H. (2008). Identification of *Bacillus* spp. From Bikalga, fermented seeds of Hibiscus *sabdariffa*: Phenotypic and genotypic characterization. *Journal of Applied Microbiology* 104: 122–131. doi:10.1111/j.1365-2672.2007. 03550.x.

Ouoba, L.I., Nyanga-Koumou, C.A.G, Parkouda, C., Sawadogo, H., Kobawila, S.C., Keleke, S., Diwara, B., Louembe, D., Sutherland, J.P. (2010). Genotypic diversity of lactic acid bacteria isolated from African traditional alkaline-fermented foods. *Journal of Applied Microbiology*, 8(6): 2019–2029.

Ouoba, L.I.I., Parkouda, C., Diawara, B., Scotti, C., and Varnam, A.H. (2007). Identification of Bacillus spp from Bikalga, fermented seeds of Hibiscus sabdariffa: phenotypic and genotypic characterization. *Journal of Applied Microbiology* 104: 122–131.

Panda, S.H. and Ray, R.C. (2007). Lactic Acid Fermentation of β-carotene rich sweet potato (*Ipomoea Batatas* L.) into lacto-juice. *Plant Foods for Human Nutrition* 62: 65–70.

Panda, S.H., Paramanick, M. and Ray, R.C. (2007). Lactic acid fermentation of sweet potato (*Ipomoea batatas* L.) into pickles. *Journal of Food Processing and Preservation* 31: 83–101.

Panda, S.H., Panda, S., Sivakumar, P.S. and Ray, R.C. (2009). Anthocyanin-rich sweet potato lacto-pickle: Production, nutritional and proximate composition. *International Journal of Food Science and Technology* 44: 445–455.

Panda, A.K., Sastry, V.R.B. and Mandal, A.B. (2008). Growth, nutrient utilization and carcass characteristics in broiler chickens fed raw and alkali processed solvent extracted karanj (Pongamia glabra) cake as partial protein supplement. *Journal of Poultry Science* 45: 199–205.

Parkouda, C., Nielsen D.S., Azokpota, P., Ouoba, L.I., Amoa-Awua, W.K., Thorsen, L., Hounhouigan, J.D., Jensen, J.S., Tano-Debrah, K., Diawara, B. and Jakobsen, M. (2009). The microbiology of alkaline-fermentation of indigenous seeds used as food condiments in Africa and Asia. *Critical Reviews in Microbiology* 35:139e56.

Ray, R.C. and Panda, S.H. (2007). Lactic acid fermented fruits and vegetables: an overview. In: Palino, M.V. (ed.). *Food Microbiology Research Trends*. Nova Science Publishers Inc., Hauppauge, New York, USA, pp. 155–188.

Ray, R.C., Panda, S.K., Swain, M.R. and Sivakumar, P.S. (2012). Proximate composition and sensory evaluation of anthocyanin rich purple sweet potato (*Ipomoea batatas* L.) wine. *International Journal of Food Science and Technology* 47(3): 452–458.

Ray, R.C and Ravi, V. (2005). Post-harvest spoilage of sweet potato in tropics and control measures. *Critical Reviews in Food Science and Nutrition* 45:7–8: 623–644.

Sanni, A. (2010). Diversity of *Bacillus* species isolated from okpehe, a traditional fermented soup condiment from Nigeria. *Journal of Food Protection* 73: 870–878.

Sanni, L.O. and Akingbala, J.O. (2000). Effect of drying methods on physicochemical and sensory qualities of fufu. *Drying Technology – An International Journal* 18(1 and 2): 421–431.

Sanni, A.I. and Oguntoyinbo, F.A. (2014). Ntoba Mbodi. In *Handbook of Indigenous Foods Involving Alkaline Fermentation*, eds. P.K. Sarkar and M.J.R. Nout (Boca Raton, FL: CRC Press), 140–143.

Sawadogo-Lingani, H., Lei, V., Diawara, B., Nielsen, D.S., Moller, O.L. and Traore, A.S. (2007). The biodiversity of predominant lactic acid bacteria in dolo and pito wort for the production of sorghum beer. *Journal of Applied Microbiology* 103: 765–777. doi:10.1111/j.1365–2672.2007. 03306.x.

Shivashanakara, K.S., Isobe, S., Al-Haq, M.I., Takenaka, M. and Shiina, T. (2004). Fruit anti-oxidant activity, ascorbic acid, totalphenol, quercetin and carotene of Irwinmango fruits stored a low temperature after high electrical field pre-treatment. *Journal of Agriculture and Food Chemistry* 52: 1281–1286.

Simatende, P., Gadaga, T.H., Nkambule, S.J. and Siwela, M. (2015). Methods of preparation of Swazi traditional fermented foods. *Journal of Ethnic Foods* 2: 119–125. doi:10.1016/j.jef.2015.08.008.

Steinkraus, K.H. (2002). Fermentations in world food processing. *Comprehensive Review in Food Science and Food Safety* 1: 23–32. Available from IFT (ift.org). Posted 20 Nov, 2006.

Subramaniyam, R. and Vimala, R. (2012). Solid state and submerged fermentation for the production of bioactive substances: A comparative study. *International Journal of Science and Nature* 2012(3): 480–486.

Suliman, H.B., Shimmein, A.M. and Shadda, S.A.I. (1987). The pathological and biochemical effects of feeding fermented leaves of Cassia obtusifolia 'Kawal' to broiler chicks. *Avian Pathology* 16(1): 43–49.

Vieira-Dalodé, G., Jespersen, L., Hounhouigan, J., Moller, P.L., Nago, C.M. and Jakobsen, M. (2007). Lactic acid bacteria and yeasts associated with gowé production from sorghum in Bénin. *Journal of Applied Microbiology* 103(2): 342–349. doi:10.1111/j.1365-2672.2006.03252.x.

Vieira-Dalode, G., Madode, Y.E., Hounhouigan, J., Jespersen, L. and Jakobsen, M. (2008). Use of starter cultures of lactic acid bacteria and yeasts as inoculum enrichment for the production of gowe, a sour beverage from Benin. *African Journal of Microbiology Research* 2: 179–186.

Younus, S., Masud, T. and Aziz, T. (2002). Quality evaluation of market yogurt/Dahi. *Pakistan Journal of Nutrition* 1(5): 226–230.

Zamora, F. (2009). Biochemistry of alcoholic fermentation. In: Moreno, M.V.A. and Polo, M.C. (eds). *Wine Chemistry and Biochemistry*, Springer Science, New York, pp. 3–26. Open Access Library.

# 3 Biotechnologies/Fermentation Technologies for Large-Scale Industrial Enzyme Production for the Food and Beverage Industry

*Olusola Oyewole, Sarafadeen Kareem, and Tolulope Adeleye*
Federal University of Agriculture

**CONTENTS**



DOI: 10.1201/9781003178378-3

## 3.1 INTRODUCTION

The food and beverage industry is involved in the processing of raw food materials. Processing of food is carried out for various reasons including the provision of convenient preservation systems, improving the safety of consumed foods, improving the shelf life of food materials, and providing increasing palatability and more food varieties for the consumers.

Several food processing techniques are available which can be categorized as traditional food processing techniques and modern food processing techniques.

Fermentation technology is one of the traditional methods of food processing. Studies have revealed that many of the organisms involved in fermentation technologies achieve their purposes by secreting enzymes into the food system.

The use of enzymes in the food and beverage industry can be traced to 6000 BC when they were employed in the production of beer, cheese, wine making, and bread baking (Poulsen and Buchholz 2003). Enzymes have been found useful in the food industry because of their roles as natural catalysts.

Enzymes are biocatalysts synthesized by living cells. They are proteins that speed up the rate of a chemical reaction. An enzyme converts a specific set of reactants (substrates) into specific products. They perform some significant reactions in food which provide convenience, improve shelf life, offer safety, and increase food palatability. Enzymes are environmentally friendly and produce less residues when compared to chemical catalysts.

Enzymes have been found to be superior to chemical catalysts for their safety, specificity, and stability in the food environment. Some of these enzymes are secreted by microorganisms that are involved in the traditional processes, while some modern technologies require the use of purified and externally added enzymes for them to carry out their catalytical roles in food production. The need for the use of enzymes in food processing is growing every day and new technologies for the production of more efficient enzymes for the production of food and beverages is growing daily.

## 3.2 COMMON ENZYMES USED IN THE FOOD INDUSTRY

The common enzymes used in the food industry include the following.

### 3.2.1 Amylases

Alpha-amylases (EC 3.2.1.1) are enzymes that catalyze the degradation of α-1,4-glucosidic linkage of starch and related compounds to produce simple oligosaccharides. They randomly cleave α-1,4 linkages between adjacent glucose subunits in polysaccharides resulting in the release of short-chain oligomers and α-limit dextrin. α-Amylases operate within the pH range of 2–12 and they are highly stable at high temperatures, and these qualities make them an important starch-degrading enzyme used in the food industry.

Amylases are extensively used in baking, brewing, cake-making, fruit juice and starch syrup-making industries, and preparation of digestive aids. They can be supplemented into bread dough to convert the complex flour starch into simpler dextrins, followed by fermentation using baker's yeast. The mixing of α-amylase to the dough results in accelerating the rate of fermentation and the reduction of the viscosity of dough, resulting in improvements in the volume and texture of the product. The use of amylases in bread production generates additional sugar in the dough to improve the taste, crust color, and toasting qualities of the bread. Amylases are also used for anti-stalling in bread industries to improve the softness retention and shelf life of baked food products.

Amylases also find use in the distilled alcoholic beverages to degrade starch into sugars before fermentation and also to minimize or remove turbidity developed due to starch.

Okafor et al. (2007) reported that amylases are also used to clear the turbidity and complexity of beer or fruit juices. They are also used to catalyze the hydrolysis of unmalted barley and other starchy adjuncts to facilitate the cost reduction of beer brewing (Okafor et al. 2007).

In the food industry, amylases have extensive applications including production of glucose and maltose syrup, clarification of fruit juice for longer shelf life, and saccharification of starch (Christopher and Kumbalwar 2015).

### 3.2.2 Pectinase

Pectinase, an enzyme found in plant cell wall, converts pectin into pectic acid by catalyzing the de-esterification of the methoxyl group of pectin. Pectinases are a class of enzymes that catalyze the disintegration of cell wall cementing materials called pectin.

Pectinases are used in the fruit juice industry and winemaking for clarification and removal of turbidity in the finished product. It also intensifies the colors in the fruit extract while aiding in stabilization, clarification, and filtration (Servili et al. 1992). Pectinase enzymes are one of the most important enzymes in fruit juice industry which help to obtain well clarified and more stable juices with higher yields of production. These are also used for protein enrichment of baby food, oil extraction, canning of orange, and sugar extraction process from the date fruits.

Pectin enzymes are frequently used in other important processes such as in the preparation of hydrolyzed products of pectin in the refinement of vegetable fibers during starch manufacture and in the curing of coffee (de Lima et al. 2010). Pectinases are also used in fruit ripening.

### 3.2.3 Proteases

Proteases (EC 3.4.21.62) are enzymes that perform proteolysis by hydrolyzing the peptide bonds that link amino acids together in polypeptide chains that forms. Proteases are a class of hydrolases that hydrolyze the large proteins into smaller peptides by cleaving peptide bonds. Proteases serve as a coagulant to the bio-protective enzyme and enhance the shelf life and safety of dairy products.

In the food industry, applications of proteases are ranging from baking industry, to brewing, making of various kinds of oriental foods like soy sauce, meat tenderization, and cheese manufacturing.

Proteases have been in use for detergent manufacturing for over 100 years, and since then they have been inducted into various other commercial processes such as food, pharmaceutical, animal feed, leather, diagnostics, waste management, and silver recovery. In fact, proteases dominate the total enzyme sales with a market share of almost 60% (Sawant and Nagendran 2014). The dominance of proteases in the detergent industry is contributed by alkaline proteases due to their unique ability to remain stable and active in the alkaline pH range (Gupta et al. 2002). Serine proteases especially subtilisin A, neutrase, and trypsin are some of the commercially important proteases.

### 3.2.4 Lipases

Lipases (EC 3.1.1.3) are a class of hydrolytic enzymes and a subclass of esterases. Lipases are hydrolases that catalyze the breakage of carboxyl ester bonds present in triacylglycerols to liberate fatty acids and glycerol. In a living system, these enzymes facilitate the breakdown and mobilization of lipids within the cells The substrates for lipases include natural oils, synthetic triglycerides, and esters of fatty acids.

They catalyze the processes of transesterification, hydrolysis and esterification and have significant application in various industries including food, fats and oil, dairy, pharmaceuticals, and baking industries. Lipases play typical roles in the catalysis of fats into fatty acids and glycerol.

Lipases for biodiesel production can be derived from bacterial sources as well as fungal sources. *Burkholderia cepacia*, *Pseudomonas fluorescens*, and *Pseudomonas cepacia* are some of the bacterial lipases used in biodiesel production.

### 3.2.5 Xylanase

Xylanases (EC 3.2.1.8, 1,4-β-xylanxylanohydrolase) are enzymes that break down xylan which is an integral part of plant polysaccharide. Xylan is a complex polysaccharide made of xylose residue backbone with each subunit linked to each other by a β-1,4-glycosidic bond (Ramalingam and Harris 2010).

### 3.2.6 Lactase

Lactase, otherwise known as β-D-galactohydrolase (EC 3.2.1.23), is a glycoside hydrolase that catalyzes the hydrolysis of lactose (disaccharide) into galactose and glucose (monosaccharides). Lactases are mainly found in young animal gut, plants, fungi, yeasts, and bacteria (Holsinger et al. 1997; Almeida and Pastore 2001).

Lactose is the sugar which is found in milk. Humans produce this enzyme only as infants, and this ability lessens as they grow older which leads to lactose intolerance (Kies 2014). Primary lactase deficiency is the most common cause of lactose intolerance and is caused by an inherited genetic defect that runs in families. Primary lactase deficiency develops when your lactase production decreases as your diet becomes less reliant on milk and dairy products. The use of lactase in dairy formulations can be used to ameliorate the lactose deficiency challenges. Hence, lactase is a very important enzyme in the dairy and food industry.

These enzymes have key role in the development of commercial dairy products by improving the solubility and digestibility of milk and milk products (Carminati et al. 2010).

Milk products such as milk candy, condensed milk, and frozen concentrated milk, yogurt and ice creams with low lactose content could be an ideal choice for the consumers with lactose intolerant syndromes. These enzymes produce oligosaccharides with the finest biodegradability of whey second to lactose hydrolysis (Milichová and Rosenberg 2006).

### 3.2.7 Cellulases

Cellulases catalyze the hydrolysis of β-1,4 linkages in cellulose chains, releasing oligosaccharides, cellobiose, and glucose (Dillon 2004). In food industries, cellulases are used for extraction of components from green tea, soy protein, essential oils, aromatic products, and sweet potato starch. These are also used for production of orange vinegar and in extraction and clarification of citrus fruit juices (Orberg and Englehardt 1981).

### 3.2.8 Invertase

Invertase, technically known as β-fructofuranosidase (EC.3.2.1.26), is a glycoprotein which catalyzes the hydrolysis of sucrose into glucose (dextrose) and fructose. The invertase enzyme has a broad range of commercial and industrial applications including confectionery. Invertase is used to produce invert sugar, which was earlier done by acid hydrolysis.

## 3.3 SOURCES OF ENZYMES USED IN THE FOOD INDUSTRY

There are different sources of enzymes – animal, plant, or microbial sources – used in enzyme-based fermentation technologies for the food and beverage industry. The enzymes may be obtained from the whole cells, parts of cells, or cell-free extract of plants, animals, or microorganisms. Microbial enzymes are preferred to animal and plant enzymes due to the following reasons. Firstly, they offer economical advantage because they can be produced on large scales within limited time and space in addition to easy extraction and purification processes. The quantity produced depends largely on the size of fermenter, type of microbial strain, and growth conditions. Secondly, microbial enzymes offer the following technical advantages: (1) they are capable of synthesizing a large variety of enzymes, (2) they can tolerate a wide range of environmental conditions, (3) they can be genetically modified to meet industrial demands, and (4) they have short and rapid generation times.

The extraction of enzymes may follow some complex procedures that may involve cell disruption and solvent extraction and adsorption to a carrier.

### 3.3.1 Amylases

Many microbial species can produce α-amylase especially those from the genus *Bacillus* (e.g., *Bacillus licheniformis*, *Bacillus stearothermophilus*, and *Bacillus amyloliquefaciens*), as well as some plant cells.

### 3.3.2 Pectinases

De Gregorio et al. (2002) showed that many microbial species like *Bacillus*, *Erwinia*, *Kluyveromyces*, *Aspergillus*, *Rhizopus*, *Trichoderma*, *Pseudomonas*, *Penicillium*, and *Fusarium* are high-quality sources of pectinases. Among the pectinase producing fungi, *Aspergillus niger*, *Aspergillus carbonarius*, and *Lentinus edodes* are the most preferred fungi in industries as they secrete approximately 90% of enzymes (Blandino et al. 2001).

### 3.3.3 Proteases

Proteases are commonly present in a wide diversity of animals and plants, but microbes are considered as preferred sources for the production of proteases owing to its technical and economic advantages. Some of the most common protease-producing microorganisms are *Aspergillus niger, A. oryzae, B. amyloliquefaciens, B. stearothermophilus, Mucor miehei, M. pusillus*, etc. Most of the proteases employed in the industry are produced by genetically modified strains of *Bacillus* and *Aspergillus* (Pillai et al. 2011; Radha et al. 2011).

### 3.3.4 Lipases

Microbial lipases are synthesized by fungal, yeast, and bacterial species. The microorganisms *Penicillium restrictum, Candida rugosa, Candida antarctica, Pseudomonas alcaligenes, Pseudomonas mendocina*, and *Burkholderia cepacia* account for most of the industrial microbial lipase producers among others (Jaeger and Reetz 1998). The fungi including *Rhizopus, Geotrichum, Rhizomucor, Aspergillus, Candida*, and *Penicillium* have been reported as lipase producers in making several commercial food products.

### 3.3.5 Xylanase

Xylanases are produced by several bacterial and fungal species. Some insects, crustaceans, and seeds of plants have also been reported to produce xylanase. Filamentous fungi that synthesize this enzyme are of particular interest because they secrete the enzyme into the media in large quantities in comparison to bacteria (Knob et al. 2013).

### 3.3.6 Lactase

Many microorganisms such as bacteria like lactic acid bacteria, *Escherichia coli, Lactobacillus oleracea*, and *Streptococcus* sp.; fungi like *Aspergillus oryzae, Aspergillus niger*, and *Aspergillus sphaericus*; and yeasts like *Kluyveromyces fragilis* and *Kluyveromyces lactis* are preferred sources of lactase for commercial applications.

### 3.3.7 Cellulases

Cellulases are produced by microorganisms either aerobic or anaerobic, mesophilic or thermophilic fungi, bacteria, protozoans, plants, and animals. The significant microbes producing cellulases are fungi (*Trichoderma, Penicillium, Aspergillus, Fusarium*, etc.) and bacteria (*Acidothermus, Bacillus, Staphylococcus, Streptomyces, Xanthomonas, Clostridium, Erwinia, Eubacterium, Ruminococcus*, etc.) (Moreira and Siqueira 2006; Zhang et al. 2011).

### 3.3.8 Invertase

Invertase is potentially produced by the microbes, like filamentous fungi and several bacterial species with its enormous application in various industries

**TABLE 3.1**
**Some Common Enzymes and Their Sources**

| Enzyme | Source | Specific Sources | References |
|---|---|---|---|
| **α-Amylase** | Animals | Hog or pig pancreas | Uhlig (1998) |
| | Fungi | *Aspergillus* spp. | Kvesitadze et al. (1978) |
| | Bacteria | *Bacillus* sp. | Srivastava and Baruah (1986) |
| **β-Amylases** | Plants | Sweet potato | Laurière et al. (1992) |
| | Fungi | *Aspergillus* sp. | Abe et al. (1988) |
| Cellulase | Fungi | *Aspergillus niger* | Kang et al. (2004) |
| Pectinase | Fungi | *Saccharomyces cerevisiae* | Blanco et al. (1994) |
| | Bacteria | *Bacillus* sp. | Kelly and Fogaty (1978) |
| Papain | Plants | *Carica papaya* | Uhlig (1998) |
| Chymosin/Renin | Animals | Calf abomasums | Cheeseman (1981) |
| Catalase | Animals | Liver of *Bos taurus* | Uhlig (1998) |
| | Fungi | *Aspergillus niger* | Fiedurek and Gromada (2000) |
| Glucose oxidase | Fungi | *Penicillium notatum* | Keilin and Hartree (1948) |
| Lipase | Plants | Pig pancreas | Uhlig (1988) |
| | Bacteria | *Bacillus subtilis* | Ruiz et al. (2005) |

(Cardoso et al. 1998). *Aspergillus casiellus* is considered as one of the potential sources for the production of invertase. For food industries, invertase is usually obtained from baker's yeast (*Saccharomyces cerevisiae*) (Uma et al. 2010) (Table 3.1).

## 3.4 COMMON APPLICATION OF ENZYME-BASED FERMENTATION TECHNOLOGIES FOR THE FOOD AND BEVERAGE INDUSTRY

Enzymes have found wide uses in the food industry. These had been widely reviewed by Choi et al. (2015) and Christopher and Kumbalwar (2015). In these various food processing, enzymes have found use in the following areas:

- Improving nutritional quality and bioavailability
- Recovery of new ingredients
- Food modification
- Improving sensory quality – flavor development
- Partially replacing some of the chemical additives
- Increasing yields
- Beverage clarification
- Texture and appearance modification – bakery aids
- Size Reduction – meat tenderization
- Milk coagulating

## 3.5 APPLICATION OF ENZYMES IN THE FOOD INDUSTRY

Enzymes have been found very useful in various food industries, and their roles cover various unit operations from collection of raw food materials to its processing, preserving, packaging, and transporting.

Enzymes have special characteristics to increase total food production and improve its quality like flavor, color, texture, aroma, nutritive values, etc. (Neidleman 1984). Enzymes also have the ability to transform and improve the nutritional as well as functional properties of food materials, hence having a wide range of applications in food processing industries. Due to specificity, enzyme-catalyzed processes generate a lesser number of side products (waste), which in turn give higher-quality products with less generation of pollutants. The enzyme works under very mild conditions which saves the valuable constituents of food materials during processing.

### 3.5.1 Application of Enzymes in Baking

Wheat flour is the major raw material used in the baking industry, and the common products include bread, biscuits, cakes, and various other confectionaries. In the baking industry, enzymes have been used for bread softening and volume and flour adjustment using amylase, obtained from *Aspergillus* sp. and *Bacillus* sp. *Xylanase* and *Lipase* from *Aspergillus niger* have been found useful for *dough conditioning and stability. Glucosidase* from *Aspergillus niger* and *Penicillium chrysogenum* has been found useful for dough strengthening. α-Amylase from *Aspergillus* spp. and *Bacillus* spp. is used for flour adjustment and bread softness. Xylanase from *Aspergillus* is used for dough softening, while lipase from *Aspergillus* spp. is used for dough stability and conditioning. The strength of the dough can be laminated through the use of transglutaminase from *Streptomyces* spp. Enzymes help to increase the production rate of bread and also maintain its qualities such as color, flour enhancement, and volume. In the baking industry, α-amylase catalyzes the hydrolysis of α-1,4-glycosidic linkages of starch polysaccharides to yield dextrins, oligosaccharides, maltose, and D-glucose, thus providing excellent yields and also used for starch modification. Lipases are used to break down the fat matters of various food products as well as flour improvement in baking. Pentosanase is used to break down pentinosans/hemicellulose matter, and finds good use in flour treatment for baked products, instant noodles, pasta, etc. Oxidoreductase is used for dough strengthening and bread whitening. Protease is added for the hydrolysis of gluten, casein, and various animal and vegetable proteins added in baked products.

### 3.5.2 Application of Enzymes in Dairy Industry

Milk is the major raw material in the dairy industry. Dairy products include cheese, milk, and fermented milk products like yogurt, cheese, buttermilk, butter, cream, soured cream, frozen desserts (ice cream, cakes), whey, and milk powders.

Enzymes have been useful for enhancement of yield as well as organoleptic properties like aroma, color, and flavor of dairy products.

Common enzymes used in the dairy industry include lipases, lysozyme, lactase, proteases, aminopeptidase, catalase, lactoperoxidase, esterases, and transglutaminase.

In the dairy industry, challenges of lactose intolerance have been reduced through the application of lactase (β-galactosidase) from *Escherichia coli* and *Kluyveromyces* sp. In addition, acid protease from *Aspergillus* sp. and neutral protease from *Bacillus subtilis* and *A. oryzae* have been found useful for milk clotting and infant formulas (low allergenic), and flavor improvement in cheese production from milk. Lipase from *Aspergillus niger* and *A. oryzae* has helped in cheese flavor development for faster cheese ripening and customized flavor development. Catalase from *Aspergillus niger* and aminopeptidase from *Lactobacillus* sp. have been found useful for faster cheese ripening. Lipase from *Aspergillus* spp. is used to fasten cheese ripening. Lactase from *Escherichia coli* and *Kluyveromyces* spp. is used for lactose reduction in milk and whey, while catalase from *Aspergillus niger* is used for cheese processing.

### 3.5.3 Application of Enzymes in the Brewing Industry

Wheat, barley, maize, rice, sorghum, and cereal grains are the major raw materials used in the brewing industry. Usually, the grain is subjected to an enzymatic process that converts the grains to malt. The production of beer from barley is the oldest example of the enzyme application in brewing industry.

The processing involves the malting of grain, mashing of grist, and fermentation to produce alcohol in the presence of yeasts. During brewing, some enzymes are produced when barley is undergoing germination (malting) which is only capable to break down the plant materials (polysaccharides) (Bamforth 2009; Ray et al. 2016; Aastrup and Erdal 1980). Malting process is difficult to control and also expensive. Enzymes like α-amylase, β-glucanase, and protease are added to unmalted barley to convert the complex polysaccharides into simpler ones resulting in cost reduction and elimination of complicated malting process. Enzymes can also be added to the beer after fermentation to induce faster maturation. They can reduce viscous substances such as polysaccharides (xylans and glucans) and act as filtration improver. They can be used in the production of light beer to remove carbohydrates and induce chill proof. In brewing, proteases are used for malt production and the improvement of yeast growth. Amyloglucosidase helps in improving the glucose contents, while pentosanase and xylanase help in hydrolyzing pentosans of malt, barley, and wheat, as well as aid in extraction and filtration of beer.

Good juice production will not be easily produced without the aid of some enzymes. Pectinases from *Aspergillus oryzae* and *Penicillium funiculosum* have been found useful for depectinization during mashing. α-Amylase from *Aspergillus* sp. and *Bacillus* sp. aids in starch hydrolysis for good juice extraction. β-Glucanase from *Bacillus subtilis* and *Aspergillus* sp. is used to reduce haze formation during mashing.

In cocoa processing to make chocolate, amylase has been used for syrup production. In the animal feed industry, phytase from *Aspergillus niger* and xylanase from *Aspergillus niger* and *Bacillus* species have been added to aid hydrolysis for easy digestion of the hay. β-Glucanase from *Aspergillus niger* has also been found useful as a digestive aid.

### 3.5.4 Application of Enzymes in Beverage Industry

The major activities of the beverage industry include the production of alcoholic beverages such as wine, spirits, and some other alcoholic liquors and the production of nonalcoholic beverages such as malted drinks, tea, coffee, syrups, soft drinks, fruit juices, and packaged water. In the beverage industry, enzymes are used to improve the yield of juice, liquefy the entire fruit, improve the color and aroma, clarify the juice, and break down all insoluble carbohydrates (such as pectins, hemicelluloses, and starch). In the distilled alcoholic beverages, microbial amylase is added to hydrolyze starch to sugars prior to fermentation and to minimize the turbidities generally arising due to starch (Okafor 2007).

Use of microbial enzymes during extraction of plant material improved the color, aroma, and yield and makes product clearer by digesting the cell wall (Kårlund et al. 2014). Pectinases are added to break down the cell wall of the plant materials to aid extraction and filtration of juice. The yield of the juice can also be aided by adding cellulases and amylases (Kumar 2015; Garg et al. 2016). The juice quality and stability are maintained by proper use of enzymes which mainly digested proteins, pectin, cellulose, and starch of fruits and vegetables and facilitate shortening of processing time, improve yields, and enhance sensory characteristics (Garg et al. 2016; Li et al. 2012).

In the coffee industry, enzymes are used for the liquefaction of mucilage of coffee cherry. Cellulases, hemicellulases, galactomannanase, and pectinases from microbial sources such as *Leuconostoc mesenteroides*, *Saccharomyces marscianus*, *Flavobacterium* spp., *Fusarium* spp. are used.

In the tea industry, enzymes such as cellulases, glucanases, pectinases, and tannase are used to break down the cell wall of tea leaf.

In African traditional fermented foods, enzymes are used to detoxify and break down complex starch to sugars. Usually, multiple enzymes produced by mixed cultures of various types of microorganisms are involved.

### 3.5.5 Application of Enzymes in the Starch Industry

The starch industry is another industry where enzymes have been found useful. The objectives of this industry are to rapidly extract the starch materials before the onset of fermentation in the starch slurry. α-Amylases are used for starch extraction. Enzymes are also used for the production of various types of syrups from starch as shown in Figure 3.1.

## 3.6 TECHNOLOGIES FOR LARGE-SCALE INDUSTRIAL ENZYME PRODUCTION FOR THE FOOD AND BEVERAGE INDUSTRY

The technologies for large-scale enzyme production are not complex. However, the process is delicate as the enzymes require specific temperatures and pH conditions for them to remain active. Most enzymes are denatured and become useless if not kept within the range of their stability. Figure 3.2 shows the steps involved in the large-scale industrial enzyme production. The processes involved include the following.

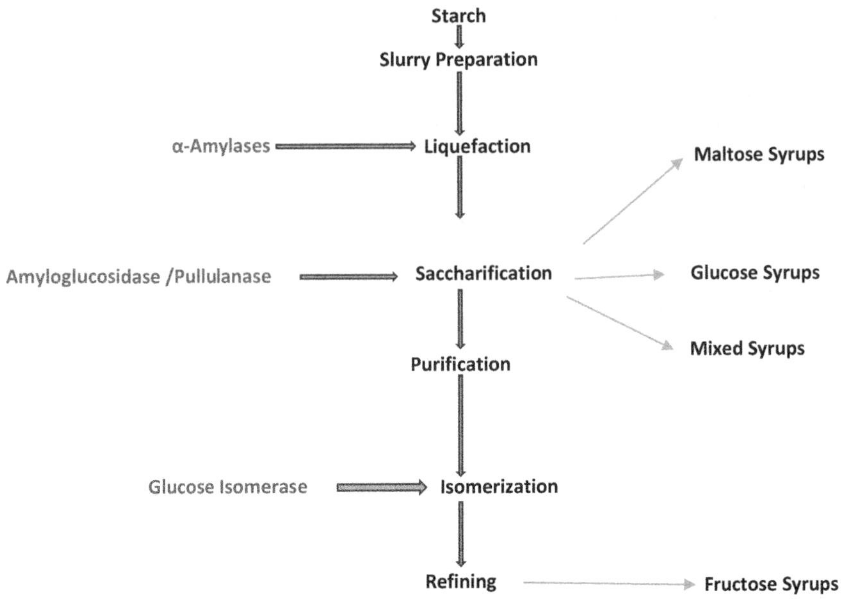


**FIGURE 3.1** Production of various syrups from starch.

### 3.6.1 Cultivation of Microorganisms and Media for Enzyme Production

The enzyme production process consists of fermentation of a suitable microbial culture in growth media, which contain appropriate nutrients for the growth of the enzyme-producing microorganisms, and incubating it under appropriate conditions for the enzyme secretion. The fermentation process is carried by two methods: solid-state fermentation and submerged fermentation (Rana and Bhat 2005). The fermentation process is very specific and its main aim is to obtain high-yield, high-quality products in minimum time.

The solid-state fermentation process involves the secretion of some viable cells of the enzyme-producing microorganisms on some suitable substrates with appropriate nutrients before incubation in a conducive environment, free of contamination, usually in an incubator.

**Advantages of Solid-State Fermentation**

- The medium is easy, simply obtainable, and cheap.
- Substrates would require less pretreatment compared to liquid media.
- Contaminations are restricted since the wet content is low.
- Forced aeration is commonly easier.
- Simplified and decreased downstream method and waste disposal.
- Simple fermentation instrumentation is often used.

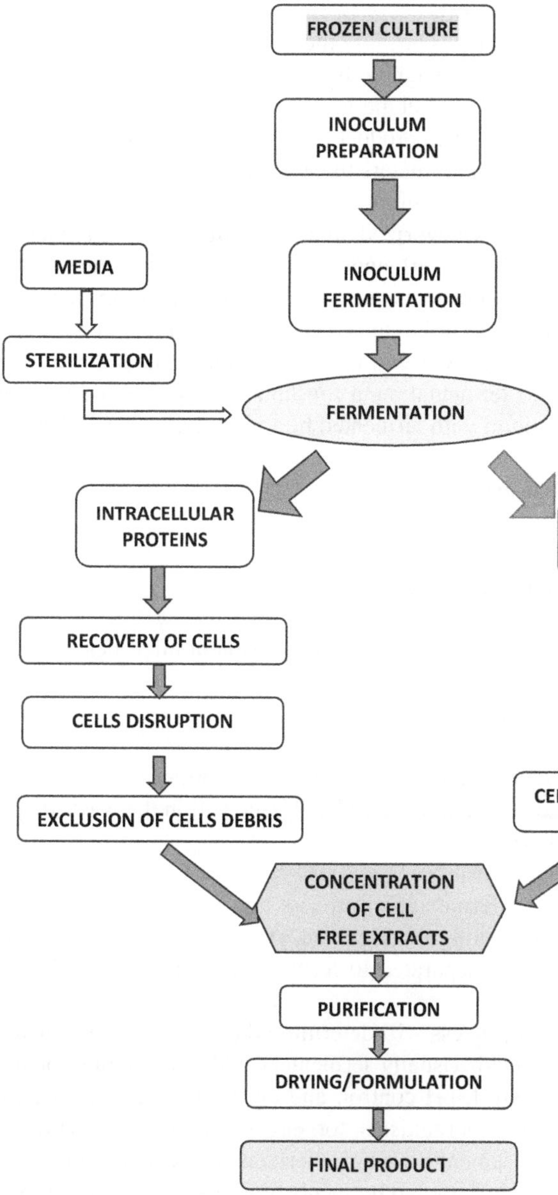

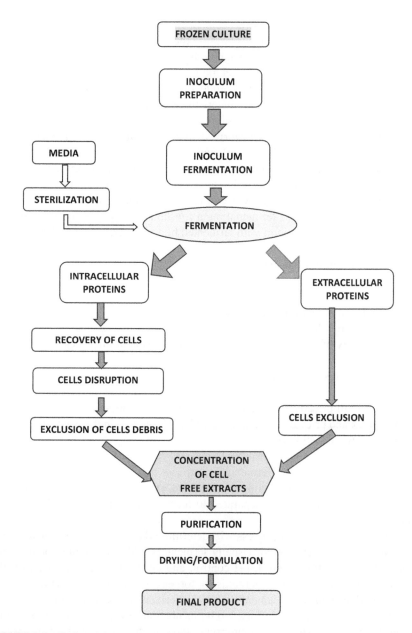

**FIGURE 3.2** Industrial enzyme production.

## Disadvantages of Solid-State Fermentation

- Low moisture level could prohibit the growth of microorganisms.
- Removing of metabolic heat could be a downside in massive-scale solid-state fermentations.
- Difficulties in observing the method parameters.

The **submerged fermentation** process is employed in commercial large-scale production of many enzymes. This is because the process of solid-state fermentation and their downstream processing is slightly difficult compared to the submerged fermentation process (Aunstrup 1979). Submerged fermentation of enzyme production consists of inoculant preparation from production culture and transferal of inoculants into a fermenter, where physical parameters like temperature, pH, agitation, and dissolved oxygen are controlled for maximum enzyme production.

Submerged fermentation is categorized into three different fermentation processes: batch fermentation, fed batch fermentation, and continuous fermentation. With batch fermentation, all growth nutrients are added at the start of fermentation, and fed batch fermentation is very similar to batch fermentation, the only difference being that inoculant culture is fed with extra production nutrients in the fermentation process, while in continuous fermentation, a constant state is obtained by supplying a new fresh production medium with fermented broth harvested from a fermenter.

**Advantages of Submerged Fermentation**

- Easy monitoring of process parameters
- Even distribution of nutrients and microorganisms
- Ability to regulate growth conditions
- Availability of high water content for the expansion of microbes

**Disadvantages of Submerged Fermentation**

- Use of high-priced media and expensive instrumentation
- Complex and pricy downstream method and issue within the waste disposal
- High power consumption

After a certain period of the fermentation process, the fermented broth contains a mixture of cell biomass, remaining nutrients, enzymes, and other metabolites, and now pure enzymes need to be separated to form fermented broth by downstream processing.

The type of fermentation process will determine the type of fermenter in which the organism will be cultivated. Usually fermenters with feed points for nutrients application, temperature control, pH control, and agitation shafts are used for the large-scale cultivation of microorganisms for enzyme production. However, for enzymes produced through filamentous fungi, a fermenter box containing the growth substrates for the organisms are used. The incubation of the fermenter box can be at room temperature. However, appropriate nutrients for the growth of the organisms need to be provided in the growth medium, which should also be at the necessary pH for the production of the enzymes. Defined media are always used for the cultivation of the enzyme-producing microorganisms, and the media should include sources of carbohydrate, minerals, nitrogen, and some vitamins. Extra carbohydrate is usually supplied as starch, sometimes refined but often simply as ground cereal grains. Soybean meal and ammonium salts are frequently used sources of additional nitrogen. Most of these materials will vary in quality and composition from batch to batch causing changes in enzyme productivity. Because the quantity and quality of the enzyme production depend on the media composition, most commercial enzyme producers keep the media composition as their trade secrets.

### 3.6.2 Enzyme Harvesting and Recovery

Following growth, the enzymes need to be recovered. The recovery process depends on the type of enzymes. There are microorganisms that secrete their enzymes into the growth medium and the enzymes can be extracted from the growth medium through many techniques that utilize the differences between the enzymes properties and the growth medium. Some enzymes are secreted intracellularly, in which case the microbial cells will need to be harvested and the cells broken or fractionated for the recovery of the enzymes. The initial steps of the downstream processing are the separation of supernatant containing enzyme solution from the biomass, through either centrifugation or filtration.

After separation, the enzyme solution should be concentrated by means of an ultrafiltration, diafiltration, or evaporation method.

### 3.6.3 Enzyme Purification

Extracellular enzyme extracts from fermentation media or broth generally contain a large quantity of partly fermented agro residues, mycelia, spores, cell debris, and some impurities which can cause technical difficulties in isolation of target enzyme in the purest form and may have adverse effect on the catalytic activity (Kareem et al. 2003). For instance, an undesirable protein known as transglucosidase has been reported as an impurity in gluco-amylase production which catalyzes the formation of unfermentable glucose polymers like isomaltose and oligosaccharides and ultimately results into lower glucose yield (Pandey 1995).

The enzymes thus produced are not pure as they contains various impurities and may contain other enzymes secreted by the cells different from the desired ones. The purification of enzymes is necessary to increase enzyme activity. A commercially viable purification method is column chromatography (Linder et al. 2004). Various technologies have been developed for the purification of enzymes. Purification and separation of enzymes are generally based on solubility, size, polarity, and binding affinity. The various methods of purification have been well articulated at https://www.creative-enzymes.com/service/enzyme-purification_307.html. The production scale, timeline, and properties of the enzymes should all be considered when choosing the proper separation method. Some of the conventional purification techniques include the following.

#### 3.6.3.1 Differential Centrifugation

When the enzymes from cell sources are harvested, they need to be fractionated into components before purification. The first step usually involves homogenization of cells, which disrupts the cell wall to release the enzyme into the homogenate, along with other components. Depending on the cell type, the homogenization could be easy as in the case of mammalian tissue without rigid cell wall, or it may need

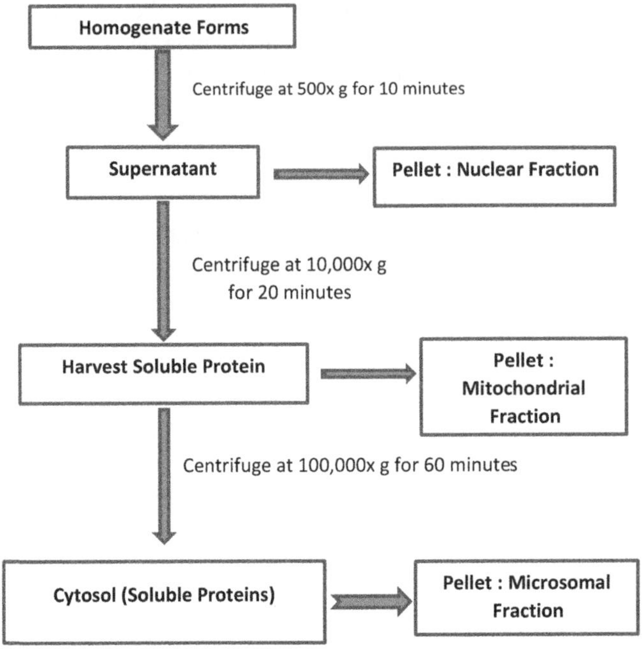


**FIGURE 3.3** Differential centrifugation.

harsher conditions such as abrasion, freezing, and high pressure due to the rigid cell wall of the plant tissue. Sometimes, additional hydrolytic enzymes or detergents are added for better extraction. The mixture is then fractionated by centrifugation, yielding a dense pellet of heavy material at the bottom of the centrifuge tube and a lighter supernatant above (Figure 3.3). The supernatant is again centrifuged at a greater force to yield yet another pellet and supernatant. The procedure, called differential centrifugation, yields several fractions of decreasing density, each still containing hundreds of different proteins, which are subsequently assayed for the activity being purified. Usually, one fraction will be enriched for such activity, and it then serves as the source of material to which more discriminating purification techniques are applied. The choice of temperature, pH, buffering salt, buffer strength, ionic strength, osmolarity, additives (EDTA, SDS, nonionic detergents, etc.), and homogenization technique is important for the success of purification.

#### 3.6.3.2 Solubility-Based Separation, Differential Solubility, and pH

The principle of the type of separation is that enzyme solubility changes drastically when the pH, ionic strength, or dielectric constant changes. For example, most proteins are less soluble at high salt concentrations, an effect called salting out. The salt concentration at which a protein precipitates differs from one protein to another. Hence, salting out can be used to fractionate proteins. Salting out is also useful for concentrating dilute solutions of proteins, including active fractions obtained from other purification steps. Addition of water-miscible organic solvents such as ethanol or acetone will change the dielectric constant of the solvent and therefore precipitate the desired enzyme. Neutral water-soluble polymers can also be used for the same purpose instead of organic solvents. However, the risks of losing enzyme activity during precipitation and further separation of the added salt or polymer need to be considered.

### 3.6.3.3 Size- or Mass-Based Method

Because enzymes are relatively large molecules, separation based on the size or mass of molecules favors purification of enzymes, especially the ones with high molecular weight. Dialysis is a commonly used method, where semipermeable membranes are used to remove salts, small organic molecules, and peptides (Figure 3.4). The process usually needs a large volume of dialysate, the fluid outside the dialysis dag, and a period of hours or days to reach the equilibrium. Countercurrent dialysis cartages can also be used, in which the solution to be dialyzed flow in one direction, and the dialysate in the opposite direction outside of the membrane. Similarly, ultrafiltration membranes, which are made from cellulose acetate or other porous materials, can be used to purify and concentrate an enzyme larger than certain molecular weight. The molecular weight is called the molecular weight cutoff and is available in a large range from different membranes. The ultrafiltration process is usually carried out in a cartridge loaded with the enzyme to be purified. Centrifugal force or vacuum is applied to accelerate the process. Both dialysis and ultrafiltration are quick but somewhat vague on distinguishing the molecular weight, whereas size exclusion chromatography gives fine fractionation from the raw mixture, allowing separation of the desired enzyme from not only small molecules but also other enzymes and proteins. Size exclusion chromatography, also known as gel filtration chromatography, relies on polymer beads with defined pore sizes that let particles smaller than a certain size into the bead, thus retarding their egress from a column. In general, the smaller the molecule, the slower it comes out of the column. Size exclusion resins are relatively "stiff" and can be used in high-pressure columns at higher flow rates, which shortens the separation time. Other factors including the pore size, protein shape, column volumes, and ionic strength of the eluent could also change the result of purification.

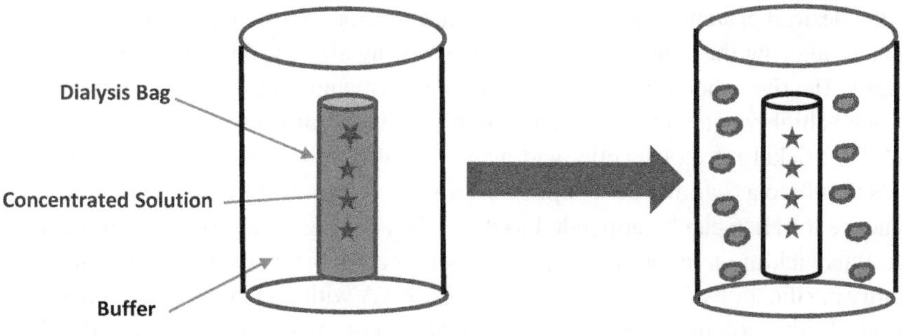


**FIGURE 3.4** The scheme of dialysis from start of dialysis to equilibrium. Enzyme molecules (stars) are retained in the dialysis bag and separated from other smaller molecules (circular).

#### 3.6.3.4 Polarity-Based Separation

Like other proteins, enzymes can be separated on the basis of polarity, more specifically, their net charge, charge density, and hydrophobic interactions. In ion exchange chromatography, a column of beads containing negatively or positively charged functional groups are used to separate enzymes. The cationic enzymes can be separated on anionic columns, and anionic enzymes on cationic column.

Electrophoresis is a procedure that uses an electrical field to cause permeation of ions through a solid or semisolid matrix or surface resulting in separations on constituents on the basis of charge density. The most commonly used methods with an SDS-PAGE matrix are quite well standardized and do not differ much between labs. The distance a protein migrates on SDS-PAGE is inversely proportional to the log of its molecular radius, which is roughly proportional to molecular weight. Similarly, a matrix with gradient pH can be used in isoelectric focusing separation. A protein moves under the influence of an electrical field and stops upon reaching the pH which is the pI for the protein (net charge = 0). The matrix used can be liquid or a gel poured into either a cylindrical shape, or a flat plate.

Hydrophobic interaction chromatography (HIC) employs hydrophobic interactions to distinguish different enzymes, which are adsorbed on matrices such as octyl- or phenyl-sepharose. A gradient of decreasing ionic strength, or possibly increasing nonpolar solvent concentration can be used to elute the proteins, giving fractions that usually contain relatively high-purity enzymes. High-pressure liquid chromatography (HPLC) uses the same principle of separation of HIC, which is filled with more finely divided and tuned materials and thus allows more choices of eluents and results in better separation. Note that HPLC could be based on polarity, affinity, or both.

#### 3.6.3.5 Affinity- or Ligand-Based Purification

Affinity chromatography is another powerful and generally applicable means of purifying enzymes. This technique takes advantage of the high affinity of many enzymes for specific chemical groups. In general, affinity chromatography can be effectively used to isolate a protein that recognizes a certain group by (1) covalently attaching this group or a derivative of it to a column, (2) adding a mixture of proteins to this column, which is then washed with buffer to remove unbound proteins, and (3) eluting the desired protein by adding a high concentration of a soluble form of the affinity group or altering the conditions to decrease binding affinity. Affinity chromatography is most effective when the interaction of the enzyme and the molecule that is used as the bait is highly specific. A special example of ligand-affinity chromatography is the Ni-NTA (nickel-nitrolotriacetic acid-agaraose) affinity chromatography. This ligand binds tightly to a 6-amino acid peptide consisting only of histidines (His6). The cDNA sequence for His6 can be appended to the cDNA coding for a given recombinant protein, thus yielding a recombinant protein which contains a His-TAG. This allows the affinity purification of such a protein using Ni-NTA without having to design a special ligand-affinity column. Other forms of affinity chromatography include dye-ligand chromatography, immune adsorption chromatography, and covalent chromatography.

After purification, the enzymes need to be concentrated, and sometimes lyophilized to give the pure, stable form distributed as the product or added into the final formulation. The following analysis and quality certification is necessary to confirm the enzyme is the desired one, with reasonable concentration, stability, and activity. The enrichment and certification requires experienced researchers to maintain the enzyme quality and choose the correct characterization method.

#### 3.6.3.6 Size Exclusion Chromatography/Gel Permeation Chromatography

Chromatography can be used to separate protein in solution or denaturing conditions by using porous gels. The various chromatographic methods have been highlighted at https://en.wikipedia.org/wiki/Protein_purification. The principle of the size exclusion chromatography is that smaller molecules have to traverse a larger volume in a porous matrix. Consequentially, proteins of a certain range in size will require a variable volume of eluent (solvent) before being collected at the other end of the column of gel.

In the context of protein purification, the eluent is usually pooled in different test tubes. All test tubes containing no measurable trace of the protein to purify are discarded. The remaining solution is thus made of the protein to purify and any other similarly sized proteins.

HIC media is amphiphilic, with both hydrophobic and hydrophilic regions, allowing for separation of proteins based on their surface hydrophobicity. Target proteins and their product aggregate species tend to have different hydrophobic properties and removing them via HIC further purifies the protein of interest.[7] Additionally, the environment used typically employs less harsh denaturing conditions than other chromatography techniques, thus helping to preserve the protein of interest in its native and functional state. In pure water, the interactions between the resin and the hydrophobic regions of protein would be very weak, but this interaction is enhanced by applying a protein sample to HIC resin in high-ionic-strength buffer. The ionic strength of the buffer is then reduced to elute proteins in order for decreasing hydrophobicity (Kennedy 1990).

#### 3.6.3.7 Ion Exchange Chromatography

Ion exchange chromatography separates compounds according to the nature and degree of their ionic charge. The column to be used is selected according to its type and strength of charge. Anion exchange resins have a positive charge and are used to retain and separate negatively charged compounds (anions), while cation exchange resins have a negative charge and are used to separate positively charged molecules (cations).

Before the separation begins, a buffer is pumped through the column to equilibrate the opposing charged ions. Upon injection of the sample, solute molecules will exchange with the buffer ions as each competes for the binding sites on the resin. The length of retention for each solute depends upon the strength of its charge. The most weakly charged compounds will elute first, followed by those with successively stronger charges. Because of the nature of the separating mechanism, pH, buffer type, buffer concentration, and temperature all play important roles in controlling the separation.

Ion exchange chromatography is a very powerful tool for use in protein purification and is frequently used in both analytical and preparative separations.

#### 3.6.3.8 Affinity Chromatography

Affinity chromatography is a separation technique based upon molecular conformation, which frequently utilizes application specific resins. These resins have ligands attached to their surfaces which are specific for the compounds to be separated. Most frequently, these ligands function in a fashion similar to that of antibody–antigen interactions. This "lock and key" fit between the ligand and its target compound makes it highly specific, frequently generating a single peak, while all else in the sample is un-retained. Many membrane proteins are glycoproteins and can be purified by lectin affinity chromatography. Detergent-solubilized proteins can be allowed to bind to a chromatography resin that has been modified to have a covalently attached lectin. Proteins that do not bind to the lectin are washed away, and then specifically bound glycoproteins can be eluted by adding a high concentration of a sugar that competes with the bound glycoproteins at the lectin binding site. Some lectins have high-affinity binding to oligosaccharides of glycoproteins that is hard to compete with sugars, and bound glycoproteins need to be released by denaturing the lectin.

#### 3.6.3.9 High-Performance Liquid Chromatography

High-performance or high-pressure liquid chromatography is a form of chromatography applying high pressure to drive the solutes through the column faster. This means that the diffusion is limited and the resolution is improved. The most common form is "reversed phase" HPLC, where the column material is hydrophobic. The proteins are eluted by a gradient of increasing amounts of an organic solvent, such as acetonitrile. The proteins elute according to their hydrophobicity. After purification by HPLC, the protein is in a solution that only contains volatile compounds, and can easily be lyophilized (Regnier 1983). HPLC purification frequently results in denaturation of the purified proteins and is thus not applicable to proteins that do not spontaneously refold.

### 3.6.4 Enzyme Packaging

Depending upon the application of the enzymes, it can be formulated in the form of powder, granules, or liquid. Enzymes should be stable and be easily released at the site of action, and there should be no dust formation, which can cause allergic reactions.

## 3.7 DEVELOPMENTS WITH ENZYME-BASED FERMENTATION TECHNOLOGIES FOR THE FOOD AND BEVERAGE INDUSTRY

There has been recent development in the biotechnologies/fermentation technologies for large-scale industrial enzyme production for the food and beverage industry. Advances in biotechnology, particularly in genetic and protein engineering, and genetics have provided the basis for the efficient development of enzymes with improved properties for established applications and novel, tailor-made enzymes for completely new applications where enzymes were not previously used. Sukanchan (2018) has provided a good report on these two developments, and most of the technical information contained in this report has been sourced from these researchers.

Two of these advancements are covered in the following sections.

### 3.7.1 Genetic Engineering

A most exciting development is the application genetic engineering techniques to enzyme technology. There are a number of properties that can be improved or altered by genetic engineering including the yield and kinetics of the enzyme, the ease of downstream processing, and various safety aspects. Enzymes from dangerous or unapproved microorganisms and from slow growing or limited plant or animal tissue may be cloned into safe high-production microorganisms. Today, enzymes are now being redesigned to fit more appropriately into industrial processes, for example, making glucose isomerase less susceptible to inhibition by the $Ca^{2+}$ present in the starch saccharification processing stream.

Through genetic engineering, the amount of enzyme produced by a microorganism may be increased by increasing the number of gene copies that code for it. This principle has been used to increase the activity of penicillin-G-amidase in *Escherichia coli*. The cellular DNA from a producing strain is selectively cleaved by the restriction endonuclease HindIII. This hydrolyzes the DNA at relatively rare sites containing the 5′-AAGCTT-3′ base sequence to give identical "staggered" ends. Another extremely promising area of genetic engineering is protein engineering. New enzyme structures may be designed and produced in order to improve on existing enzymes or create new activities.

The main target of genetic engineering in enzyme technology through development of an industrially viable microbial strain is summarized below:

a. To increase product concentration: High product yield may be considered obtainable by any of these basic procedures by: (1) raising the factor or gene dose; (2) breaking down the factor regulation (e.g., catabolite de-repression, metabolite resistance); or (3) altering membrane permeability to enhance product export.
b. To improve process parameters: For economic and technical benefits, improvement of process conditions could be achieved through the following procedures: (1) decrease fermentation time; (2) be ready to metabolize cheap substrates; (3) prevention of undesirable by-products, for example, pigments or substance with chemicals associated with the main product; (4) reduce oxygen needs; (5) decrease foaming; (6) tolerant to high concentrations of carbon or nitrogen sources; (7) resistant to phage …
c. To evolve novel product: Significant changes in the genetic composition of microorganisms will lead to the biosynthesis of new metabolites.

### 3.7.2 Enzyme Immobilization

Use of enzyme in food processing has some limitations. For example, some enzymes have short shelf life, some are easily denatured and become unstable as a result of the high temperatures of the fermenting process, and recovery and re-utilization has been difficult with many of them. In order to overcome these challenges, the process of enzyme immobilization has been developed.

Immobilization of enzymes can be defined as the confinement of an enzyme (biocatalyst) in a distinct phase, separated from the bulk phase but allowing it to exchange with the latter.

Enzyme immobilization can be carried out by various techniques; generally, they are classified by two methods, chemical and physical. In physical methods, various weak interactions between enzymes and supported materials are included, while in chemical methods enzymes and supporting materials are linked by covalent bonds.

Loss in the activity of enzymes is minimized by the use of an appropriate immobilization technique, which can avoid changes in the chemical structure of an enzyme and a reactive group present on the active site of the enzyme. There are several common methods of enzyme immobilization, such as adsorption, encapsulation, covalent-coupling, cross-linking, and entrapment (Cantone et al. 2013).

#### 3.7.2.1 Adsorption

The adsorption of an enzyme on an insoluble matrix is a simple and old technique that has broad applications to other immobilization methods (Nisha et al. 2012). The enzyme is attached to the outside of an inert material. This method includes the attachment of an enzyme through surface binding on glass, alginate beads, or a matrix, in suitable environments of ionic strength and pH. Weak interactions are implied on the adsorption of an enzyme on the supporting material such as hydrogen bonding, hydrophobic interactions, ionic bond, and van der Waals' forces. These are very weak interactions, but they are present in large numbers that provide adequate binding strength. The selection of adsorbent mainly depends on lowering the leakage of the immobilized enzyme. The major disadvantages of the adsorption method are that the enzyme becomes detached from the supporting material by slight changes in the surrounding environment such as temperature, substrate concentration, and ionic strength.

#### 3.7.2.2 Covalent Binding

The formation of covalent bonds is responsible for additional stable interactions between enzyme and support, and these are normally formed during chemical reactions by functional groups (e.g., imidazole, indolyl, phenolic hydroxyl, hydroxyl, amino, and thiol) that are present on the surface of the enzyme. Enzymes can be immobilized covalently by two methods: (1) directly binding the enzyme on the surface of supporting material utilizing the functional group or (2) a spacer arm is used for the attachment of the enzyme on the supporting material. One major advantage of this method is that the binding site does not cover the enzyme's active site, so the activity of the enzyme is only affected by immobility.

#### 3.7.2.3 Entrapment

In entrapment methods, enzymes are trapped in natural polymeric beads by two methods: (1) by thermos-reverse polymerization, where agar, agarose, and gelatin are heated to melt, and when it cools down, the enzyme is applied for entrapment, and (2) by ionotropic gelation, where calcium alginate or carrageenan is extracted from a gel with ions like $Ca^{2+}$ (Datta et al. 2013; Keerti et al. 2014). In addition, numerous synthetic polymers like polyvinylalcohol hydrogel, polyacrylamide, tetramethoxysilane (TMOS), and propyl trimethoxysilane (PTMS) have also been used for the immobilization of enzymes (Deshpande et al. 1987; Grosová et al. 2008). Entrapment includes simple adsorption in which an enzyme is attached to a solid support via ionic interactions, hydrogen bonds, and van der Waals' forces. These have a permeable membrane that permits the substrate and products to cross but keeps the enzyme within the membrane. The entrapment method for the immobilization of enzymes is inexpensive, where the enzyme works very fast under mild environmental conditions.

#### 3.7.2.4 Cross-Linking

This technique of enzyme immobilization is also called copolymerization, where enzymes are attached to each other in multiple points without use of any support or carrier (Datta et al. 2013). These types of enzyme aggregates are called cross-linked enzyme crystals (CLECs). Though it involves covalent bonds, enzymes immobilized by this method often undergo conformational changes, and hence a loss of enzyme activity occurs. This technique can be used in combination with others to minimize enzyme leakage.

#### 3.7.2.5 Encapsulation

Encapsulation is a technique where enzymes are enclosed in a semipermeable membrane that is made by nitrocellulose or nylon. This technique is cheap and simple, and holds a large quantity of enzymes. Its disadvantage is the limitation of pore size of the semipermeable membrane where size limits the crossing of substrate or product. This technique is widely used in beverage industries.

#### 3.7.2.6 Ionic Bonding

In this method, enzymes are reversibly attached to the supporting material by forming a salt bridge linkage. This method is simple, but it is very difficult to determine the conditions in which an enzyme will be strongly bound and fully active. The enzyme can be recovered by modifying the temperature and ionic strength conditions.

Polyethylenimine is widely used and patented for binding a variety of enzymes. This method is similar to interactions of protein and ligands (Guisán et al. 1997; Nisha et al. 2012).

#### 3.7.2.7 Affinity Bonding

In the affinity method, an enzyme is immobilized onto a porous martial by a non-covalent or covalent tag protein. This method is generally used for the purification of proteins and immobilization of several enzymes where polymer-coated controlled porosity glass (CPG) attached to a support material, EziG™, binds protein affinity tags, which has affinity with His-TAG enzymes. There are several affinity interactions that are used for the immobilization of enzymes such as avidin–biotin interactions, cellulose-binding domain, chitin-binding domain, and calmodulin as an affinity tag. The main advantage of the technique is that enzymes are not exposed to the chemical conditions during immobilization (Roy et al. 2005; Nisha et al. 2012).

**Materials Used for Immobilization**

Among the natural polymers that have been widely used as immobilization matrices are calcium alginate, K-carrageenan, gelatin, sepharose, pectin, polyacrylamide gel, and starch, while inorganic materials such as ceramics, silica, celite, glass, zeolites, and activated charcoal have also been considerably exploited in immobilization technology (Shafei and Allam 2010). Recently, some natural polymers from *Detarium microcarpum* and *Irvingia gabonensis* have been reported to compete favorably with conventional entrapment agents (Kareem et al. 2014).

**TABLE 3.2**
**Factors Influencing Performance of Immobilized Enzymes (Cao 2006)**

| Factors | Implications of Immobilization |
|---|---|
| Hydrophobic partition | Enhancement of reaction rate of hydrophobic substrate |
| Microenvironment of carrier | Hydrophobic nature stabilizes enzyme |
| Multipoint attachment of carrier | Enhancement of enzyme thermal stability |
| Spacer or arm of various types of immobilized enzymes | Prevents enzyme deactivation |
| Diffusion constraints | Enzyme activity decreases and stability increases |
| Presence of substrates or inhibitors | Higher activity retention |
| Physical post-treatments | Improvement of enzyme performance |
| Different binding mode | Activity and stability can be affected |
| Physical structure of the carrier such as pore size | Activity retention was often pore size dependent |
| Physical nature of the carrier | Carriers with large pore size mitigate diffusion limitation, leading to higher activity retention |

**Advantages of Immobilized Enzymes**

The advantages of immobilized enzymes are as follows:

(1) Continuous and repeated use of immobilized enzymes, (2) less labor intensive, (3) the process is cost effective, (4) minimum interval, (5) low contamination level, (6) product stability, (7) improved process management, and (8) high enzyme:substrate complex relation (Table 3.2).

## 3.8 CONCLUSION

Enzyme-based fermentation is one of the important food processing technologies in the food and beverage industry. Enzymes have various applications in the bakery products, dairy products, starch processing, and fruit juice production. Advances in biotechnology, particularly in genetic and protein engineering, and genetics have provided the basis for the efficient development of enzymes with improved properties for established applications and novel, tailor-made enzymes for completely new applications where enzymes were not previously used. New technologies such as immobilization and protein cross-linking are being employed to enhance the efficiency of enzyme-based fermentation technologies.

## REFERENCES

Aastrup S., Erdal K. (1980) Quantitative determination of endosperm modification and its relationship to the content of 1,3:1,4-β-glucans during malting of barley. *Carlsb Res Commun* 45(5):369–379.

Abe J.I., Bergmann F.W., Obata K., Hizukuri S. (1988) Production of the raw-starch digesting amylaseof *Aspergillus* sp. K-27. *Appl Microbiol Biotechnol* 27(5–6): 447–450.

Almeida M.D., Pastore G.M. (2001) Galactooligossacarídeos–Produção e efeitos benéficos. *Boletim da Sociedade Brasileira de Ciência e Tecnologia de Alimentos* 35(1/2):12–19.

Aunstrup K. (1979) Production isolation and economics of extracellular enzyme. *Appl Biochem Bioeng* 2:27–70.

Bamforth C.W. (2009) Current perspectives on the role of enzymes in brewing. *J Cereal Sci* 50(3):353–357.

Blanco P., Sieiro C., Diaz A., Villa T.G. (1994) Production and partial characterization of an endopolygalacturonasefrom *Saccharomyces cerevisiae. Can J Microbiol* 40(11):974–977.

Blandino A., Dravillas K., Cantero D., et al. (2001) Utilisation of whole wheat flour for the production of extracellular pectinases by some fungal strains. *Process Biochem* 37(5):497–503.

Cantone S., Ferrario V., Corici L., et al. (2013) Efficient immobilisation of industrial biocatalysts: Criteria and constraints for the selection of organic polymeric carriers and immobilization methods. *Chem Soc Rev* 42(15):6262–6276.

Cardoso M.H., Jackix M.N., Menezes H.C. (1998) Effect of association of pectinase, invertase and glucose isomerase on the quality of banana juice. *J Food Sci Technol* 18(3):275–282.

Carminati D., Giraffa G., Quiberoni A., Binetti A., Suárez V., Reinheimer J. (2010) Advances and trends in starter cultures for dairy fermentations. In: Biotechnology of lactic acid bacteria: Novel applications, p. 177.

Cao L. (2006) Immobilized enzymes: past, present and prospects. In: Carrier-bound immobilized enzymes: Principles, application and design. Wiley-VCH Verlag GmbH & Co. KGaA, Weinheim.

Cheeseman G.C. (1981) Rennet and cheesemaking. In: *Enzymes and Food Processing*, eds. Birch, G.G., Blakebrough, N. and Parker, K.J., pp. 195–211. Springer, Netherlands.

Choi J.-M., Han S.-S., Kim H.-S. (2015) Industrial applications of enzyme biocatalysis: Current status and future aspects. *Biotechnol Adv* 33(7):1443–1454.

Christopher N., Kumbalwar M. (2015) Enzymes used in food industry a systematic review. *Int J Innov Res Sci Eng Technol* 4(10):9830–9836.

Datta S., Christena L.R., Rajaram Y.R.S. (2013) Enzyme immobilization: An overview on techniques and support materials. *3 Biotech* 3(1):1–9.

De Gregorio A., Mandalari G., Arena N., et al. (2002) SCP and crude pectinase production by slurrystate fermentation of lemon pulps. *Bioresour Technol* 83(2):89–94.

De Lima D.A.R., Da Silva T.M., Maller A., et al. (2010) Purification and partial characterization of an exo-polygalacturonase from Paecilomyces variotii liquid cultures. *Appl Biochem Biotechnol* 160(5):1496–1507.

Deshpande A., D'souza S.F., Nadkarni G.B. (1987) Coimmobilization of D-amino acid oxidase and catalase by entrapment of Trigonopsis variabilis in radiation polymerised polyacrylamide beads. *J Biosci* 11(1):137–144.

Dillon (2004) Aldo. Celulases. In: *Enzimas como agentes biotecnológicos*, eds. Said S., Pietro R.C.L., pp. 243–270. Ribeirão Preto, Legis Summa

Fiedurek J., Gromada A. (2000) Production of catalase and glucose oxidase by *Aspergillus* unconventional oxygenation of culture. *J Appl Microbiol* 89(1): 85–89.

Garg G., Singh A., Kaur A., et al. (2016) Microbial pectinases: An ecofriendly tool of nature for industries. *3 Biotech* 6(1):47.

Grosová Z., Rosenberg M., Rebroš M., et al. (2008) Entrapment of β-galactosidase in polyvinylalcohol hydrogel. *Biotechnol Lett* 30(4):763–767.

Guisán J.M., Penzol G., Armisen P., et al. (1997) Immobilization of enzymes acting on macromolecular substrates. In: *Immobilization of Enzymes and Cells*, ed. Bickerstaff G.F., pp. 261–275. Humana Press, Totowa.

Gupta R., Beg Q., Lorenz P. (2002) Bacterial alkaline proteases: Molecular approaches and industrial applications. *Appl Microbiol Biotechnol* 59(1):15–32.

Holsinger V., Rajkowski K., Stabel J. (1997) Milk pasteurisation and safety: A brief history and update. *Rev Sci Tech Oie* 16(2):441–466.

Jaeger K.-E., Reetz M.T. (1998) Microbial lipases form versatile tools for biotechnology. *Trends Biotechnol* 16(9):396–403.

Kareem S.O., Adio O.Q, Osho M.B. (2014) Immobilization of Aspergillus niger F7-02 lipase in polysaccharide hydrogel beads of Irvingia gabonensis matrix. *Enzyme Res* 2014:967056.

Kang S.W., Park Y.S., Lee J.S., Hong S.I., Kim S.W. (2004) Production of cellulases and hemicellulases by *Aspergillus niger* KK2 from lignocellulosic biomass. *Bioresource Technology* 91(2): 153–156.

Kårlund A., Moor U., Sandell M., et al. (2014) The impact of harvesting, storage and processing factors on health-promoting phytochemicals in berries and fruits. *PRO* 2(3):596–624.

Keerti G.A., Kumar V., et al. (2014) Kinetic characterization and effect of immobilized thermostable β-glucosidase in alginate gel beads on sugarcane juice. *ISRN Biochem* 2014:178498. doi:10.1155/2014/178498.

Keilin, D., Hartree, E.F. (1948). Properties of glucose oxidase (notatin): Addendum. Sedimentation and diffusion of glucose oxidase (notatin). *Biochem J* 42(2):221.

Kelly C.T., Fogarty W.M. (1978) Production and properties of polygalacturonate lyase by an microorganism *Bacillus* sp. RK9. *Can J Microbiol* 24(10):1164–1172.

Kennedy R.M. (1990) Hydrophobic chromatography. *Methods Enzymol* 182:339–343. doi:10.1016/0076-6879(90)82029-2. ISBN 9780121820831. PMID 2314246.

Kies A.K. (2014) Authorised EU health claims related to the management of lactose intolerance: Reduced lactose content, dietary lactase supplements and live yoghurt cultures. In: *Foods, Nutrients and Food Ingredients with Authorised Eu Health Claims*, ed. Sadler M.J., 1st edn. Woodhead Publishing, Sawston.

Knob A., Beitel S.M., Fortkamp D., et al. (2013) Production, purification, and characterization of a major Penicillium glabrum xylanase using Brewer's spent grain as substrate. *BioMed Res Int* 2013:8.

Kumar S. (2015) Role of enzymes in fruit juice processing and its quality enhancement. *Adv Appl Sci Res* 6(6):114–124.

Kvesitadze G.I., Svanidze R.S., Buachidze T., Bendianishvili M.B. (1978) Acid-stable and acid-unstable alpha-amylases of the mold fungi *Aspergillus*. *Biokhimiia* 43(9):1688–1694.

Laurière C., Doyen C., Thévenot C., Daussant J. (1992) β-Amylases in cereals: A study of the maize β-amylase system. *Plant Physiol* 100(2): 887–893.

Li S., Yang X., Yang S., et al. (2012) Technology prospecting on enzymes: Application, marketing and engineering. *Comput Struct Biotechnol J* 2:e201209017.

Linder M.B., Qiao M., Laumen F., Selber K., Hyytia T., Nakari-Setala T., Penttila M.E. (2004) Efficient purification of recombinant proteins using hydrophobins as tags in surfactant-based two-phase systems. *Biochemistry* 43(37):11873–11882.

Milichová Z., Rosenberg M. (2006) Current trends of β-galactosidase application in food techonology. *J Food Nutr Res* 45(2):47–54.

Moreira F.M.S., Siqueira J.O. (2006) *Microbiologia e bioquímica do solo*. Universidade Federal de Lavras, Lavras, p. 729.

Neidleman S.L. (1984) Applications of biocatalysis to biotechnology. *Biotechnol Genet Eng Rev* 1(1):1–38.

Nisha S., Karthick A.S., Gobi N. (2012) A review on methods, application and properties of immobilized enzyme. *Chem Sci Rev Lett* 1(3):148–155.

Okafor U., Emezue N., Okochi V. et al. (2007) Xylanase production by Penicillium chrysogenum (PCL501) fermented on cellulosic wastes. *Afr J Biochem Res* 1(4):48–53.

Orberg L., Englehardt W. (1981) Apparatus for locating therein a pipe union. Google Patents

Poulsen P.B., Buchholz H.K. (2003) History of enzymology with emphasis on food production. In: *Handbook of Food Enzymology*, eds. Whitaker, J.R., Voragen, A.G.J., Wong, D.W.S., pp. 1–20. Marcel, Dekker, New York.

Radha S., Nithya V., Himakiran R., et al. (2011) Production and optimization of acid protease by Aspergillus spp under submerged fermentation. *Arch Appl Sci Res* 3:155–163.

Ramalingam C., Harris A.D. (2010) Xylanases and its application in food industry: A review. *J Exp Sci* 1(7). https://updatepublishing.com/journal/index.php/jes/article/view/1737.

Rana N.K., Bhat T.K. (2005) Effect of fermentation system on the production and properties of tannase of Aspergillus niger var tieghem MTCC 2425. *J Gen Appl Microbiol* 51:203–212.

Ray L., Pramanik S., Bera D. (2016) Enzymes – an existing and promising tool of food processing industry. *Recent Pat Biotechnol* 10(1):58–71.

Regnier F.E. (1983) High-performance liquid chromatography of biopolymers. *Science* 222 (4621): 245–252. Bibcode:1983Sci...222..245R. doi:10.1126/science.6353575. PMID 6353575.

Roy I., Sardar M., Gupta M.N. (2005) Cross-linked alginate–guar gum beads as fluidized bed affinity media for purification of jacalin. *Biochem Eng J* 23(3):193–198.

Ruiz C., Pastor F.I., Diaz P. (2005) Isolation of lipid- and polysaccharide degrading microorganisms from subtropical forest soil, and analysis of lipolytic strain *Bacillus* sp. CR-179. *Lett Appl Microbiol* 40:218–227.

Sawant R., Nagendran S. (2014) Protease: An enzyme with multiple industrial applications. *World J Pharm Sci* 3:568–579.

Shafei M.S., Allam R.F. (2010) Production and immobilization of partially purified lipase from *Penicillium chrysogenum*. *Malays J Microbiol* 6(2):196–202.

Servili M., Begliomini A.L., Montedoro G., et al. (1992) Utilisation of a yeast pectinase in olive oil extraction and red wine making processes. *J Sci Food Agric* 58(2):253–260.

Srivastava R.A.K., Baruah J.N. (1986) Culture conditions for production of thermostable amylase by *Bacillus stearothermophilus*. *Appl Environ Microbiol* 52(1):179–184.

Uhlig H. (1998). *Industrial Enzymes and Their Applications*. John Wiley & Sons, New York.

Uma C., Gomathi D, Muthulakshmi C., et al. (2010) Production, purification and characterization of invertase by Aspergillus flavus using fruit peel waste as substrate. *Adv Biol Res* 4(1):31–36.

Zhang K., Ren N.-Q., Cao G.-L., et al. (2011) Biohydrogen production behavior of moderately thermophile Thermoanaerobacterium thermosaccharolyticum W16 under different gas-phase conditions. *Int J Hydrog Ener* 36(21):14041–14048.

# 4 Recent Advances in Dairy Industries

## *Monitoring, Standard, Quality, and Process*

*Charles Oluwaseun Adetunji,*
*Olugbemi Tope Olaniyan, and*
*Kingsley Eghonghon Ukhurebor*
Edo State University Uzairue

*Julius Kola Oloke*
Ladoke Akintola University of Technology

*Olusola Olawale Olaleye*
Nigerian Institute for Trypanosomiasis Research

*Benjamin Ewa Ubi*
Ebonyi State University
Tottori University

*Daniel Ingo Hefft*
University Centre Reaseheath

**CONTENTS**



DOI: 10.1201/9781003178378-4

## 4.1 INTRODUCTION

Adopting strategies that could enhance and improve the processes involved in the milk process optimization has been shown to be essential for rapid development in the dairy industry. The quick decline of milk products has led dairy processors to optimize various parameters that are involved in the effective production of good products, with an effective planning strategy as well as production schedules. The application of business models has been recognized to perform a vital purpose in the workforce that could help in the decrease or prevention of resource wastage, and time, mistakes, unnecessary costs, and elimination and prevention of unnecessary errors while aiming at the processes that are involved in the improvement of adequate production in terms of quality and quantity (Ahern, 2017).

There have been several changes that have been reported worldwide, especially for the dairy sector. Burke et al. (2018) indicated that the global diary industry is currently undergoing developmental transformation. Dairy food products include milk, cheese, milk powders, butter, yogurt, and ice cream. In Asia, particularly China, the request for milk products is slow, due to trade sanctions with Russia and some other European Union (EU) countries resulting into excess supply and reduction in prices (Douphrate et al., 2013). Nevertheless, the dairy sector is anticipated to expand and grow yearly at a rate of 1.8 % in the next 10 years, about 177 million tons of powdered milk by the year 2025 due to urbanization, rising incomes, and the emergence of new markets (FAO-UN, 2018). In Europe, farmers committed about 16,597 tons of skimmed milk powder to the interventions stock at 1.698€ in 2017. Studies have revealed that the world's milk production is principally from cow, followed by buffalo. Thus, the leading producers are Asia (30 %), then the EU (28 %), North and Central America (18 %), South America (9 %), other European countries (9 %), Africa (5 %), and Oceania (5 %) (Burke et al., 2018). The Food and Agricultural Organization of the United Nations (FAO-UN) stated that the price of the dairy products has risen drastically which indicated that the price of milk products increased up to 26 % below its peak from February 2014 to 2017 (FAO-UN, 2018). This high level of price increase might be linked to increase in the level of growing incomes and emerging markets. Furthermore, altering consumer demand patterns are upsetting food production. The majority of these dairy food products are required to meet standards for healthy foods, in terms of wellness and health, social impact, taste, convenience, and lower prices (Adetunji and Anani 2021a,b,c; Adetunji et al. 2021; Dwivedi et al., 2021; Olaniyan and Adetunji 2021, Mishra et al., 2020; Thangadurai et al., 2021; Anani et al., 2020; Islam et al 2021). Therefore, there is the need for producers to adopt innovations in delivering novel products and stratagems optimization techniques that will allow and maintain proper protection and value of dairy products (Deloitte, 2015). Thus, this chapter details comprehensive information on the use of the standardization of dairy and fermented foods by focusing on evolving integrated quality management systems. Special highlights will also be provided on the components of an integrated quality management system and risk management in dairy and fermented food chain, respectively, particularly milk dairy production process.

## 4.2 RECENT ADVANCES IN STANDARDIZATION

Milk and dairy products have been identified as major source of essential minerals, protein, and several vitamins. Some enzymes derived from dairy products have been recognized to possess the capability to disintegrate hydrogen peroxide and lysozyme that could break the cell wall of bacterial cells, thus supporting the antimicrobial activity in cow milk. Other important enzymes include glycosidases, nucleases, proteases, and protease activators. Moreover, milk has been recognized to possess nine essential amino acids which are required by humans (US-FDA, 2015). There is a need for standardization because it has been reported that the presence of contamination in these products normally affects their acceptability in most traded agricultural commodities. Typical examples of such contamination include the presence of pathogens that may develop mycotoxin, particularly aflatoxin M1 (FAO-UN 2017a; Safe food, 2017). The presence of aflatoxin M1 has been recognized as the most investigated mycotoxin available in milk, and it has been recognized as a threat to food safety (Flores-Flores et al., 2015). The incident of climate change also has been established to produce 14.5 % of greenhouse gas emissions that are liberated from livestock most especially from beef, while the production of milk has been reported to be affected by this situation also. Climate change could also lead to enhancements in the development of animal diseases. This has also resulted to increases in the level of drug resistance as a result of high persistence in the usage of synthetic drugs. Moreover, the application of chemical pesticides also affects the production of safer food because these pesticides could enter into the food production and distribution chain (Ukhurebor et al., 2020). The upsurge in the level of rainfall might also lead to intensification in the dissemination of synthetic residues that could affect the atmospheric features (Ukhurebor and Umukoro, 2018), normal ecosystems, and food chains as well as go into the food chain through cow's milk (EPA, 2014). Moreover, it has been discovered that perchlorate could decrease most especially in the thyroid gland. The standardization of dairy and fermented foods entails an integrated quality management system, special highlights were also provided on the components of an integrated quality management system and risk management in dairy and fermented food chain, respectively. Milk quality entails a blend of characteristic features such as chemical, bacteriological, aesthetic, and physical qualities that enhance the appropriateness of the milk product, which includes density, color, flavor, pH value, and freezing point (Merwan et al., 2018).

## 4.3 HISTORY OF DAIRY

It has been reported that the industrial revolution benefited tremendously from agricultural practice, due to several key inventions made between 1850 and 1900 particularly in the dairy industry such as the cream separator, developed by G. P. de Laval, and the milk condensation by Gail Borden to expand the dairy industry. Subsequently, pasteurization, filling machines and capping bottles, and refrigeration brought huge expansion in the dairy industry. Advancement in feed storage, nutrition of cattle, feeding practices, and handling methods are noted to improve feed effectiveness and milk production capability in the dairy sector, thereby decreasing operational cost linked with feeding on dairy farms globally. Also, new management and housing methods assisted the science of dairy farming such as computers, automation of milking, and information management (Jones, 2006; McGuffey and Shirley, 2011).

## 4.4 RISK MANAGEMENT IN DAIRY AND FERMENTED FOOD CHAIN

Kok and Hutkins (2018) revealed that the ingesting of fermented dairy products is linked with the decrease in disease conditions such as metabolic syndrome, type 2 diabetes, gastrointestinal tract colonization resistance and defense system plus cardiovascular disease and fat metabolism. Cissé et al. (2019) revealed that fermented foods perform vital functions in human nutrition and many of the fermented products are consumed all over the world. Hence, they decided to investigate the physiochemical properties and microbial composition of the fermented milk taken in Burkina Faso. From their findings, the fermented milk contained high amounts of microbial load such as yeasts, molds, total coliforms, *Staphylococcus aureus*, thermotolerant coliforms, *Escherichia coli*, *Bacillus* sp., lactic acid bacteria, and total microbial flora but devoid of *Shigella* sp. and *Salmonella* sp. Even though these two microorganisms are absent in the fermented milk product consumed in Burkina Faso, it was revealed that the fermented milk products showed high variability in microbial load, mineral concentration, and physicochemical quality. Thus, the occurrence of certain types of bacteria strains such as *S. aureus*, *E. coli*, *Bacillus* sp., and lactic acid bacteria coliforms poses high level of menace for the milk quality and consumer's health. The authors therefore suggested a comprehensive training program for the producers of most of the fermented products on the issue of sanitation and hygiene to promote fermented milk and dairy products containing high amounts of organoleptic and other nutritional qualities.

Van Asselt et al. (2017) demonstrated the importance of monitoring the dairy supply chain that is focused on important food safety hazards; hence, their study investigated the microbial, physical, and chemical risks of contamination in the dairy supply chain. From their reports, it was indicated that significantly high levels of microbiological hazards are present in dairy products. Many species of bacteria were found in human milk, semi-soft and soft cheeses, and powdered infant formula products. Some toxic chemicals like dioxins and dioxin-like molecules, aflatoxin as well as extracts of veterinary drugs were found. Also, their findings showed the presence of physical hazards such as plastic particles, glass, and metal.

Nicola et al. (2014) revealed that *Listeria monocytogenes* is known to cause two types of disease conditions in pregnant women: noninvasive gastrointestinal listeriosis and invasive listeriosis. Generally, the mother can recover from these diseases, but the developing fetus can have serious detrimental health outcomes; hence, the authors decided to monitor the presence of *L. monocytogenes* in dairy products obtained from different livestock. They revealed that risk assessment is an important process to protect the public from many pathogenic diseases. Their finding revealed that *L. monocytogenes* is not only found in dairy products but also present in the soil, environment, water, and vegetation. Furthermore, it was demonstrated that milk supported the growth of *L. monocytogenes*, which may become contaminated through equipment, animals, or environment through the fecal waste of animals. Moreover, prolonged as well as inadequate storage of these dairy products might be linked to the presence of *L. monocytogenes* and the growth of other microbes.

Carrasco et al. (2011) showed that rigorous efforts are continuously being made to provide quality and safe food in line with modern trends and practice in the use of technology. Omics in form of proteomics and genomics have been adapted to food production and to monitor the activity of different microbial strains within the food products and the environment. Biosensors are now being used to detect pathogens with virulent genes to reduce the hazardous effect and promote food safety. Hazard Analysis and Critical Control Point (HACCP) systems guarantee quantitative risk evaluation and continuous expansion of food standards, formulations, new foods and specifications in line with international standards to promote trade. Many different food safety management system structures are being promoted to encourage food enterprises on managing food risk.

Chari and Ngcamu (2017) investigated the effects of disaster menaces on Zimbabwe's enactment of dairy supply chains based on the fact that Zimbabwe has been found to be susceptible to risks such as cyclones, crop pests, animal diseases, floods, and drought. From their findings, it was discovered that disaster risk significantly impacted negatively on the routine of the dairy food chains like job losses, reduced milk productivity, food insecurity, and overall decreased growth in dairy enterprises. It was suggested that there should be a deliberate effort by relevant stakeholders to increase investment in reducing disaster risk to promote and enhance sustainable performance in the dairy food chain of Zimbabwe.

Akabanda et al. (2013) revealed that contaminated dairy food products accounted for more than 90% of all dairy-linked infections and disease conditions in man. This report prompted an investigation by the authors to find out milk handling practices, bacterial infection, and selected milk-borne zoonotic pathogens from some particular regions in Tanzania. From the results, it was discovered that more that 57% of dairy farmers practiced unhygienic milking and more than 92% of farmers are untrained and did not meet the international standard in the dairy supply chain. Most of the collected milk products are already contaminated with coliforms, and *B. abortus* was found to significantly contribute to the high rate of microbial infection among the populace.

Kouamé-Sina et al. (2012) revealed that pathogenic microbes are found in animal sources of foods such as milk and dairy products causing food-borne diseases. The authors investigated the danger of consuming contaminated milk with different pathogenic bacteria in Abidjan, Côte d'Ivoire. Their results showed that a significant number of consumers ingest contaminated milk, which does not meet up with international standards and is clearly linked to production under dirty environment that is contaminated with pathogenic bacteria. They concluded by suggesting that to decrease the consumption of contaminated dairy products and milk, awareness on well-being and sanitization practice in dairy food chain should be promoted.

Ezeonu and Ezeonu (2017) revealed that a lot of health-promoting benefits have been shown to be derived from fermented food products such as reduced lactose level for lactose intolerant individuals, enhanced function of the gastrointestinal tract with the supply of beneficial microorganisms to the human flora, as well as protecting the shelf-life of perishable food products by increasing the amount of micro- and macronutrients. The authors stressed that certain microorganisms such as *Clostridium botulinum* could produce by-products like aldehydes and biogenic amines when in unfavorable conditions causing detrimental effects to humans. Bishop (2004) revealed the role of International Dairy Federation in dairy-related national interest groups such as food monitoring agencies, dairy farmers, dairy suppliers, dairy processing industry, academics, and governments to promote safe and quality dairy food products. This body has been involved in the regulation and production of guidelines in form of tests, manuals in the control of dairy commodities relating to safety and microbial hygiene and other chemical contaminants and equally achieving an integrated food supply chain management approach. Paul (2004) showed that with effective quality management systems in place, the dairy industry can maintain and ensure that quality products get to the consumers. The author, however, stressed that some of the contaminants cannot be regulated by the dairy producers from the plant as they are toxic residues transferred to the dairy products before getting to the dairy plants.

Bosch et al. (2018) revealed that food-borne viruses are the most prominent pathogens in food safety priority during the risk assessment utilizing the Delphi method. This necessitates a control measure across the food supply chain. Van Asselt et al. (2016) revealed that physical (glass, plastic particles, or metal) and chemical contaminants or hazards (dioxins, aflatoxin) are introduced to dairy products along the food supply chain. The authors revealed that in the Netherlands, most dairy producers are licensed with KKM-SYSTEM as a form of quality assurance method that ensures quality production. They noted that food safety globally can be influenced by trade and economy, climate change, and technology. Cheng et al. (2017) revealed the challenges of food management systems in ensuring safe livestock production due to climate change and demographic factors. Studies have shown that Mali is becoming one of the largest producers of dairy foods; hence, the need for institutional systems to ensure animal safety and good products, risk assessment, and management systems at the farm level is necessary.

Choi et al. (2016) demonstrated that food-borne diseases associated with cheese consumption are becoming pronounced. Many microbial risk management assessments have been conducted. These microbial menace evaluations have noted that danger becomes amplified in cheese with increased moisture content, particularly raw milk cheese, but could be suppressed by preharvest and postharvest preventive measures. Nduko et al. (2017) revealed that the dairy food products of Kenya serve as a source of income generation and food security program with probiotic and therapeutic roles; but recent reports have indicated that they are now a source of pathogenic contaminants such as biotoxins causing cancer, infections, and death. Most of the plants for production of dairy products are done under very unhygienic environments with contaminated and poor-quality materials. Schaper et al. (2009) discussed the risk management and perception stratagems of dairy farmers in selected European countries. The study revealed that the most reasonable threats that dairy farmers presently perceive are different market menaces and policy and production risks. Ozlem and Zehra (2012) demonstrated how people across the globe are constantly being exposed to physical and chemical contaminants through dairy products; hence, proper quality and safety practices for food sufficiency must be adopted by preventing hazards and enhanced processes. The authors identified critical control points in the dairy sector, and all possible hazards in dairy products.

## 4.5 RECENT ADVANCES IN MONITORING AND PROCESSING

### 4.5.1 Processes Involved from Milking to Milk Powder

The raw milk is taken to the dairy factory, which is then pasteurized and separated into cream and skim milk utilizing a centrifugal cream separator. The next step is to preheat at a temperature range of 75–120 °C, held for certain amount of time (pasteurization 72 °C for 15 s). Generally, the heating is to denature the proteins, destroy bacteria, generate natural antioxidants, inactivate enzymes, and cause heat stability in the milk. The next step is to concentrate the preheated milk in the evaporator by boiling under a vacuum at temperatures below 72 °C in a falling film to remove about 85 % of the water in the milk. Subsequently, the concentrated milk is spray-dried through atomizing into fine droplets. Lastly, the milk powders are properly packaged into very stable products compared to the fresh milk and protected from moisture, light, heat, and oxygen to maintain quality and improved shelf-life (Pearce, 2000).

## 4.6 RECENT ADVANCES IN QUALITY CONTROL

### 4.6.1 Important Quality Standards in Dairy Milk Production

During milk production process, many quality control checks are applied in the dairy factory such as reception of antibiotics-, insecticides-, and additives-free milk, pass dye reduction test, filtration process to remove all debris, and cleaning the filters daily. Chilling, storage, and preheating at temperatures <4 °C and 35 °C are carried out to kill all bacteria and inactivate enzymes. Standardization, homogenization, and final heating (HACCP system) are done to meet the desired temperature duration for pasteurization, with cooling and filling/packing (aseptic against dust, microbes, and insects) (Merwan et al., 2018).

Tania (2019) revealed that despite the high rate of consumption of dairy products, the sector still faces a lot of challenges in terms of contamination. Mantovani and Frazzoli (2010) analyzed the different ways of food contamination such as environment, unauthorized additives, and during food production processes by cross-contamination. The authors highlighted detailed critical risk assessment issues such as the possible pathways of feed contamination, characterization of toxicological hazards, and the carry-over of parent molecules or metabolites to foods of animal origin together with pinpointing of conditions that may need risk management assessments. Benkerroum (2013) showed that in North Africa, food technology plays a vital role in food security and economic growth. Thus, there should be a risk assessment to ascertain the safety level of what the public is exposed to. Therefore, dearth of scientific reports on the occurrence of dangers in this area of foods put more challenge onto the need in conducting sound risk assessment profiling. Hence, it was suggested that all stakeholders should help in the conduct of risk assessments and take the necessary measures to reduce risks as required.

Mwangi et al. (2016) demonstrated the role of Suusa, a naturally fermented milk product, which is usually obtained from raw camel milk and is utilized by communities in Kenya. The authors determined the handling practices in line with the safety and microbiological quality of Suusa. From the results obtained, it was discovered that the microbial load in the milk sample was high. Thus, it was suggested that sanitized practices in raw camel milk and Suusa production should be improved upon. Cissé et al. (2018) showed that livestock products generated the third biggest economic value in Burkina Faso; thus, the need for assessment of socioeconomic characteristics in line with sanitary risks along the food chain for dairy products is imminent. The authors discovered that unhygienic practices resulted into significant sanitary risks for animals and consumers. Angelo et al. (2008) demonstrated that ochratoxin A is produced by black *Aspergilli* sp. in wine food supply chain. Studies have shown that climatic conditions facilitate ochratoxin A accumulation in berries and grapes. Proper insecticidal or fungicidal treatments can reduce ochratoxin A contamination. Gadaga et al. (2013) demonstrated that conventional methods of making fermented beverages and foods in Lesotho. The findings showed that many of the conventional aspects of the preparative methods, like utilization of earthenware pots, are taken over by modern equipment and suggested that there is need for studies on the biochemical and microbiological characteristics of these fermented beverages and foods including dairy in Lesotho.

Meijuan et al. (2015) revealed that food safety is a serious issue that could cause a detrimental effects as a result of the presence of heavy metals when transmitted from food to human. The authors thus evaluated the risk of heavy metal contamination on milk and dairy products in China. The authors discovered that all the elements measured in the food products were within the physiological limits as compared with the reference point. However, the heavy metal indicated high amount; thus, continuous consumption will lead to adverse effects. Bilska and Kolozyn-Krajewska (2019) revealed that indiscriminate utilization of food may constitute risk to the environment. Therefore, the authors developed a risk management system for dairy products. Three areas of food loss in dairy were focused upon: handing over for disposal, feed, and reprocessing. The level of threat was discovered to be high, hence reselling of product, and the redistribution has been suggested. Frias et al. (2017) showed that fermented foods display an exciting opportunity and promising role as target to multitude pathogenic conditions directly by probiotic effect of microorganisms or indirect effects by biogenic role of metabolites. The authors showed that despite all the beneficial roles, safety is still a very important aspect owing to the peril of food-borne contaminants to consumers.

## 4.7 IMPLEMENTING AN INTEGRATED QUALITY MANAGEMENT SYSTEM IN FOOD PRODUCTION AND CONTROL

An "Integrated Management System (IMS)" assimilates all of an establishment's systems and procedures into one comprehensive framework, allowing the establishment to work as a single component with combined purposes and objectives.

Popov et al. (2018) revealed the advantages of utilizing IMSs for agro-industrial sectors and enterprises in ensuring safety and quality of products. The authors showed that national and international standards with ISO 9001, 22000 series are adapted to test the quality of various products. Different procedural protocols are followed such as staff requirement, quality assurance, documentation, equipment and environments, production arrangement, inspection, and quality control measures. In addition, risk management and hazards were also identified and determined. Other important areas for application of IMSs after production as also described by the authors which can serve as business ventures include testing and regulation of finished products, marketing, financial, certification, availability of raw materials, repairs and maintenance, legal framework, personnel, documentation, storage, and industrial safety. Therefore, based on the analysis and research carried out, proper integration system was instituted and documented for continuous improvement. The authors revealed that after rigorous analysis of the data generated from the irregularities and inconsistencies, it was estimated by 2014 that these inconsistencies and irregularities decreased significantly compared with 2013. Hence, it was concluded that IMSs of quality and safety control of products and services is considered to be effective.

Van Heerden (2013) carried out an investigation to analyze the advantages of IMSs based on so many pessimistic and optimistic opinions on how many industries have different results in line with outcome or output from implementing IMSs based on international quality control systems. Suttiprasit (2012) revealed that the survival and development of large food industry is dependent on many factors such as efficient availability of energy, raw materials, and water for sustainable development. Therefore, safety and quality control management are essential to guarantee that there is continuous support and growth of the industry and the communities. The author also stressed the importance of various certifications to improve on standard and hygiene. Giving example of how Thailand has utilized the Innovative Integration Management System to improve and develop the food production companies over the decades in reducing cost and improving efficiency, the author showed that integrated quality management system in food sector will allow every critical unit in the organization to integrate and maintain harmonious relationship with others.

Knaflewska and Pospiech (2007) described how the EU free-trade interaction has enabled health and food safety of consumers and improving standards through legislations. The authors revealed how legislation has enhanced food industries across the EU and individual sectors implementing the integrated quality management system.

Qingxin (2018) revealed how in China many dairy and food-related industries suffered from contamination which affected over 294,000 babies across China resulting into kidney stones in 2007. Hence, a quality management system with food research was introduced by the government and relevant stakeholders to confirm proper value and safety of products. The author proposed a mixed method of approach like quantitative and qualitative research after critically examining the impact of implementing Chinese quality management systems in terms of external factors, barriers, quality standards, supply chain, regulations, and legislation. Using pilot interviews to investigate the concept, management, and quality control of four different companies, it was discovered that quality management certifications and legislation on safety and quality of dairy products have been implemented, but there is a need for improvement such as training of employees, quality control, and proper implementation of quality management system.

Sadikoglu and Olcay (2014) evaluated the impacts of Total Quality Management (TQM) performance on different routine due to the discrepancies noticed in association between recital and complete quality management practices in Turkish firms. The authors discovered that performance output was influenced by various TQM practices. Lack of employee commitment and awareness of workers, availability of resources, inappropriate structural organization seem to be the major barriers, and the necessity of continuous implementation of TQM practices was recommended to enhance performance.

Baert et al. (2005) described how consumers must be protected from food contamination and infections; hence, the government and the producers must ensure proper safety of food and products before reaching the market. Adequate provision must be made to protect the food when passing through the production and supply chain in terms of following proper legislations, quality, and standards. The authors pointed out the laid down guidelines in ensuring that normal food quality and safety through legal regulation are complied with by different companies. They emphasized that the ultimate goal of a producer is to ensure good distribution of the products; therefore, certain level of compliance must be adequately followed to satisfy and maintain the need of the consumers. HACCP guides for Good Sanitary Practices for transportation of products is equally important which might be linked to the influence of chemical, physical, and biological factors on transported food products. Bhutto (2004) suggested that contractor industry plays a key role in UK economic sector where it reduces unemployment rate but the level of compliance with safety and quality was very disappointing, hence the need to effectively engage in implementing management systems to fulfill the legal obligations and attention to societal concerns in terms of health and safety. IMSs have been shown to influence the nuclear, manufacturing, and chemical industries, but lack of proper guidance on the implementation is witnessed across many UK organizations. The author pointed out that most of the top contractors including the dairy and other industrial sectors are beginning to implement basic to advanced models in IMSs.

Salima et al. (2017) suggested ways to tackle agricultural businesses in relations to well-being and quality of food products and improvements on the quality management system utilizing strategic analysis to implement and develop tools for management or optimizing resources utilized in agricultural businesses particularly in Kazakhstan due to economy in transition. Other factors responsible for poor quality and low production experienced by most companies were also highlighted and ways to overcome such were suggested to include the implementation of proper quality management systems. Laux et al. (2015) revealed how novel quality management practices by the agricultural industry can meet up with the production gap experienced by the population in order to enhance efficiency in the food supply chain and security. The authors showed the influence of executing a quality management system on a large scale for grain elevator company. After implementation, it was discovered that grain quality improved with increased monetary value after utilizing quality metrics to enhance their blending, storage, and inventory strategy and management. Barendsz (1998) revealed that based on international trade competitiveness, food safety is a major factor through the adoption of TQM to improve products and services to meet consumer demand and satisfaction. The author revealed that the implementation needs integrated methods, while designing a management system requires adequate information of the agricultural food industry. They concluded by suggesting the importance of various certifications in enhancing standardization and improvements in chain monitoring, assessment, and formation in food industries.

Pozo et al. (2018) examined the quality management system together with the food safety management system and other various factors influencing them globally in order to facilitate decision-making by relevant stakeholders. The authors showed that after critical analysis, prioritization and selection of 14 crucial factors for successful outcome in terms of food safety and quality management systems were observed. Sumaedi and Yarmen (2015) revealed that ISO 9001 Quality Management System was adopted by several multinational companies, but the effectiveness is still questionable. Thus, the authors evaluated the effectiveness on food industries and how the existing quality management system can be enhanced. Also, Krieger and Schiefer (2006) described the process involved in maintaining a quality management system using legislative approach that are applicable in supply food chain. The authors developed a decision support model to solve challenges based on the existent environmental, quality, and occupational health qualities. Maria and Adina (2011) also highlighted the association between quality management and hazard management together with the integrated methods. They emphasized on the need to enhance effectiveness by the combination of both management systems such as integrated quality risk management.

Nordenskjöld (2012) examined the enactment of food quality processes in different companies in Sweden with the challenges encountered during certification. The author analyzed the role of quality management in enhancing food hygiene and quality in the production and supply chain. It was observed that consumer demands were met, problems were effectively handled, and the economy was impacted positively upon implementation. Different effective control measures were put in place to supervise quality such as HACCP, ISO 22000 and International Featured Standards (IFS), British Retail Consortium (BRC), and Good Manufacturing Practices. The author concluded that the implementation of a standard and quality is enhanced if every stakeholder in the organization focuses attention on the same vision with corresponding commitment to perform. Oloo (2010) highlighted the impact of relevant stakeholders in verifying the importance of food quality and safety together with the operational legislation in Kenya. The author discovered many weak links in the production, supply chain, and further suggested appropriate recommendations to improve the processes. The Department of Public Health in Kenya has the sole regulation to ensure multilevel monitoring of food security and safety. Even though several legal and regulatory international guidelines exist to guarantee proper quality and safety are followed, majority of processed food in Kenya markets are poor in quality. The author, therefore, suggested that adequate monitoring of the informal sector must be followed as a major player in the domestic market to promote hygiene-based demand in a poor economy.

Mensah and Julien (2011) revealed factors that impact the present state of food quality and safety in the food manufacturing sector of UK. It was discovered that quite a number of regulations have been implemented to deal with risk, improved safety, and quality bearing in mind the cost implications. Also, there was no significant influence of enterprise size on the profits, challenges, and drivers to agree with food safety, quality, and security regulation. Yusof (1998) evaluated the implementation of quality management among different industries, especially the small- and medium-scale enterprises. It was discovered that most of the enterprises are in a state of dilemma as to implement certain tools and techniques; hence, implementation plans and mechanisms are designed to identify the associated gaps. It was also revealed that a linear association existed among quality management practices and performance levels especially with accreditation to ISO 9000 for short-term substantial benefits. Filipović et al. (2008) highlighted the different quality managements seen in the food sector with the purpose of gaining consumer trust and meeting up with distribution. It is revealed that implementation of food management systems places a company on a competitive advantage in a global market. Gurudasani and Sheth (2009) also opined that the competitive advantage of a company is a reflection of the adopted policies and strategies to standardization and quality to the operational market space to which the company belongs.

Armstrong (1999) demonstrated that outstanding improvement in food quality and safety could be because of implementation of integrated food hygiene and a safety management system during production and distribution. The author went further to suggest a novel approach called hygieneomics to be introduced to various management systems. Broadly, there are four main generational states to food quality and safety such as rules setting, certifying compliance, personnel commitment, and action. Naresh et al. (2006) wrote on the quality management system and how it influences safety control measures in seeking for hazardous areas that can pose danger to the consumers. The authors explained that the main goal of food safety management is to identify and address hazards that can cause deterioration and barriers to international market. It is known that safety measures and regulators in several countries have adopted HACCP as a quality control system to prevent substandard products from getting into the market. Akhmetova et al. (2019) revealed several projects done on the creation of integrated quality management system on the foundation of international ISO standards of series 9000 and the principles of HACCP. They showed that this process helped to identify allocation of the processes functioning in the HACCP and quality management system.

Eurostat (2009) described guidelines on how to implement quality assurance frameworks for many international organizations to enhance quality assessment, utilizing various factors such as costs and benefits, evaluation, quality tools, enactment experiences, and monitoring. Table 4.1 shows the description of the integrated system adaptable to food and dairy industries at the basic and advanced levels.

**TABLE 4.1**
**Description of the Integrated System Adaptable to Food and Dairy Industries at the Basic and Advanced Levels**

| SN | Integrated Management System | Industry | References |
|---|---|---|---|
| 1. | Staff requirement and training | Food | Pozo et al. (2018) |
| 2. | Quality assurance | Food | Eurostat (2009) |
| 3. | Documentation | Agro | Popov et al. (2018) |
| 4. | Equipment | Food and pharmaceuticals | Yusof (1998), |
| 5. | Environment | Food | Nordenskjöld (2012) |
| 6. | Production arrangement | Food | Baert et al. (2005) |
| 7. | Inspection and quality Control measures | Fish | Van Heerden (2013) |
| 8. | Risk management and hazards | Food | Maria and Adina (2011) |
| 9. | Testing and regulation of finished products | Food and grain | Laux et al. (2015) |
| 10. | Marketing and financial | Food | Krieger and Schiefer (2006) |
| 11. | Certification | Food and dairy | Qingxin (2018), Suttiprasit (2012), Nordenskjöld (2012) |
| 12. | Availability of raw materials and energy | Food | Suttiprasit (2012) |
| 13. | Legal framework and legislations | Food | Knaflewska and Pospiech (2007) |
| 14. | Personnel | Food | Oloo (2010) |
| 15. | Storage and industrial safety | Food | Baert et al. (2005) |
| 16. | Qualitative and quantitative research | Food | Qingxin (2018) |

## 4.8 CONCLUSION AND FUTURE PERSPECTIVES

This chapter has provided comprehensive details on the recent advances in dairy food industries by focusing on evolving integrated quality management systems such as monitoring, standardization, quality control, and processing. Special highlights were also provided on the components of an integrated quality management system as well as risk management in dairy and fermented food chains, respectively. There is a need to enforce adequate quality (most especially in the numerous stages of milk processing chain) right from the farm to the point where it is processed into milk powder. There is a need to enforce necessary optimization techniques and some standards and regulations as well as ISO standards. There are several advances that need to be developed to ensure that contemporary and optimized methodologies are sanctioned. Moreover, most regulatory bodies are putting in place novel standards from certified interlaboratory trainings, leveling the encroachment in instrumentation, and for at-line and in-line manufacture investigation for enhanced obviousness and control of manufacturing procedures. The finished product most especially from the dairy products must be innocuous and conform to regulatory requirements. Therefore, there is a need to ensure standards in the level of production of safer foods most especially from quality dairy products.

## REFERENCES

Adetunji, C.O. and Anani, O.A. (2021a). Recent advances in the application of genetically engineered microorganisms for microbial rejuvenation of contaminated environment. *Microbial Rejuvenation of Polluted Environment. Microorganisms for Sustainability* (Adetunji, C.O., Panpatte, D.G., and Jhala Y.K. eds), vol 27. Springer, Singapore. https://doi.org/10.1007/978-981-15-7459-7_14

Adetunji, C.O., Anani, O.A., and Panpatte, D. (2021). Mechanism of actions involved in sustainable ecorestoration of petroleum hydrocarbons polluted soil by the beneficial microorganism. *Microbial Rejuvenation of Polluted Environment. Microorganisms for Sustainability* (Panpatte, D.G. and Jhala Y.K. eds), vol 26. Springer, Singapore. https://doi.org/10.1007/978-981-15-7455-9_8

Adetunji, C.O. and Anani, O.A. (2021b). Utilization of microbial biofilm for the biotransformation and bioremediation of heavily polluted environment. *Microbial Rejuvenation of Polluted Environment. Microorganisms for Sustainability* (Panpatte, D.G. and Jhala Y.K. eds), vol 25. Springer, Singapore. https://doi.org/10.1007/978-981-15-7447-4_9

Anani, O.A. and Adetunji, C.O. (2021c). Bioremediation of polythene and plastics using beneficial microorganisms. *Microbial Rejuvenation of Polluted Environment. Microorganisms for Sustainability* (Adetunji, C.O., Panpatte, D.G. and Jhala, Y.K. eds), vol 27. Springer, Singapore. https://doi.org/10.1007/978-981-15-7459-7_13

Anani O.A., Mishra R.R., Mishra P., Enuneku A.A., Anani G.A., and Adetunji C.O. (2020) Effects of Toxicant from Pesticides on Food Security: Current Developments. In: Mishra P., Mishra R.R., Adetunji C.O. (eds) Innovations in Food Technology. Springer, Singapore. https://doi.org/10.1007/978-981-15-6121-4_22

Ahern, J. (2017). Applying lean techniques to dairy farming. *Acountance Plus* 03 (Sept): 26–27.

Akhmetova, S.O., Baibolova, L.K., and Serikkyzy, M.S. (2019). *Entrepreneurship and Sustainability* 6(4):1807–1822. doi:10.9770/jesi.2019.6.4(19), Issues ISSN 2345-0282 (online) http://jssidoi.org/jesi/.

Angelo, V., Perrone, G., Cozzi, G. and Solfrizzo, M. (2008). Managing ochratoxin A risk in the grape-wine food chain. *Food Additives and Contaminants* 25(2): 193–202.

Armstrong, G.D. (1999). Towards integrated hygiene and food safety management systems: The Hygieneomic approach. *International Journal of Food Microbiology* 50 19–24.

Akabanda, F., Owusu-Kwarteng, J., Tano-Debrah, K., Glover, R.L., Nielsen, D.S., Jespersen, L. (2013). Taxonomic and molecular characterization of lactic acid bacteria and yeasts in nunu, a Ghanaian fermented milk product. *Food Microbiology* 34(2):277–283. doi: 10.1016/j.fm.2012.09.025. Epub 2012 Nov 16. PMID: 23541194.

Baert, K., Devlieghere, F., Jacxsens, L. and Debevere, J. (2005). Quality management systems. In: *The Food Industry in Safety in the Agri-Food Chain*. Wageningen Academic Publishers, pp. 877–879.

Barendsz, A.W. (1998). Food safety and total quality management. *Food Control* 9(2–3):163–170.

Benkerroum, N. (2013). Traditional fermented foods of North African Countries: Technology and food safety challenges with regard to microbiological risks. *Comprehensive Reviews in Food Science and Food Safety* 12. doi:10.1111/j.1541-4337.2012.00215.x.

Bhutto, K.H. (2004). The application of integrated management systems (IMS) by contracting organisations. Doctoral, Sheffield Hallam University (United Kingdom). http://shura.shu.ac.uk/id/eprint/19353.

Bilska, B. and Kołozyn-Krajewska, D. (2019). Risk management of dairy product lossesasa tool to improve the environment and food rescue. *Foods* 8:481. doi:10.3390/foods8100481.

Bishop, J.R. (2004). The role and contribution of IDF in promoting dairy safety and hygiene in emerging and developed countries. The food chain approach from an international perspective. A farm-to-table approach for emerging and developed dairy countries. IDF/FAO International Symposium on Dairy Safety and Hygiene. Cape Town, South Africa, 3–6.

Bosch, A., Gkogka, E., Le Guyader, F.S., Le., F.S., Loisy-Hamon, F., Lee, A., van Lieshout, L., Marthi, B., Myrmel, M., Samson, A., Shultz, S.A.C., Winkler, A., Zuber, S. and Phister, T. (2018). Foodborne viruses: Detection, risk assessment, and control options in food processing. *International Journal of Food Microbiology* 285: 110–128.

Burke, N., Zacharski, K.A., Southern, M., Hogan, P., Ryan, M.P. and Adley, C.C. (2018). The Dairy industry: Process, monitoring, standards, and quality. *Descriptive Food Science* (Diaz, A.V. and Garcia-Gimeno, R.S. ed.) Chapter 1: 1–25. http://dx.doi.org/10.5772/intechopen.80398.1,

Carrasco, E., Valero, A., Pérez-Rodríguez, F., García-Gimeno, R.M. and Zurera, G. (2011). Food Safety Risk Management. Source: In Tech. 78–102. Book Chapter. doi: 10.5772/17757.

Chari, F. and Ngcamu, B.S. (2017). An assessment of the impact of disaster risks on dairy supply chain performance in Zimbabwe. *Cogent Engineering*, 4: 1–14. doi:10.1080/23311916.2017.1409389.

Cheng, R., Alberto Mantovani, A. and Frazzoli, C. (2017). Analysis of food safety and security challenges in emerging African food producing areas through a one health lens: The dairy chains in mali. *Journal of Food Protection* 80 (1): 57–67 doi:10.4315/0362-028X.JFP-15-561.

Choi, K.H., Lee, H., Lee, S., Kim, S. and Yoon, Y. (2016). Cheese microbial risk assessments — A review. Asian Australas. *Journal of Animal Science* 29 (3): 307–314. doi:10.5713/ajas.15.0332.

Cissé, H., Muandze-Nzambe, J.U., Somda, N.S., Sawadogo, A., Drabo, S.M., Tapsoba, F., Zongo, C., Traoré, Y. and Savadogo, A. (2019). Assessment of safety and quality of fermented milk of camels, cows, and goats sold and consumed in five localities of Burkina Faso. *Veterinary World* 12(2): 295–304.

Cissé, H., Sawadogo, A., Kagambèga, B., Zongo, C., Traoré, Y. and Savadogo, A. (2018). Milk production and sanitary risk along the food chain in five cities in Burkina Faso. *Urban Science* 2:57. doi:10.3390/urbansci2030057.

DeLoitte (2015). Capitalizing on the Shifting Consumer Food Value Equation [Internet]. 2015. Deloitte Development LLC. Available from: https://www2.deloitte.com/content/dam/Deloitte/us/Documents/consumer-business/us-fmi-gma-report.pdf.

Douphrate, D.I., Hagevoort, G.R., Nonnenmann M.W., Kolstrup, C.L, Reynolds, S.J., Jakob, M. and Kinsel, M. (2013). The dairy industry: A brief description of production practices, trends, and farm characteristics around the world. *Journal of Agromedicine*, 18(3): 187–197. doi:10.1080/1059924X.2013.796901.

Dwivedi, N., Dwivedi, S., and Adetunji, C.O. (2021). Efficacy of microorganisms in the removal of toxic materials from industrial effluents. *Microbial Rejuvenation of Polluted Environment. Microorganisms for Sustainability* (Adetunji, C.O., Panpatte, D.G., and Jhala, Y.K. eds), vol 27. Springer, Singapore. https://doi.org/10.1007/978-981-15-7459-7_15

EPA (2014). US Environmental Protection Agency. Technical Fact Sheet—Perchlorate [Internet]. Available from: https://www.epa.gov/sites/production/files/2014-03/documents/ ffrrofactsheet_contaminant_perchlorate_january2014_final.pdf.

Ezeonu, C.S. and Ezeonu, N.C. (2017). Biological risks associated with fermented diary products, fruits, vegetables and meat: A critical review. *Advances in Biotechnology and Microbiology* 2(1): 555–577. doi:10.19080/AIBM.2017.02.555577 009.

Eurostat (2009). Guidelines for the implementation of quality assurance frameworks for international and supranational organizations compiling statistics. November. Committee for the Coordination of Statistical Activities. Fourteenth Session. Bangkok.

FAO-UN (2017a). Food and Drug Organisation of the United Nations [Internet]. Milk and Milk Products: Price and Trade Update Price Index. Available from: http://www.fao.org/3/a-i8326e.pdf.

FAO-UN (2018). Gateway to Dairy Production and Products [Internet]. Available from: http://www.fao.org/dairy-production-products/en/.

FAO-UN (2017b). Livestock Solutions for Climate Change [Internet]. Food and Agriculture Organisation of the United Nations. Available from: http://www.fao.org/3/a-i8098e.pdf.

Flores-Flores, M.E., Lizarraga, E., López de Cerain, A., and González-Peñas, E. (2015). Presence of mycotoxins in animal milk: A review. *Food Control* 53: 163–176.

Filipović, I., Njari, B., Kozačinski, L., Cvrtila Fleck, Z., Mioković, B., Zdolec, N. and Dobranić, V. (2008). Quality management systems in the food industry. Conference Paper. Vol. X. studeni - prosinac br. 465–468.

Frias, J., Martinez-Villaluenga, C. and Peñas, E. (2017). Fermented Foods in Health and Disease Prevention. Fermented Foods in Health and Disease Prevention. doi:10.1016/B978-0-12-802309-9.00001-7.

Gadaga, H., Lehohla, M. and Ntuli, V. (2013). Traditional fermented foods of Lesotho. *Journal of Microbiology, Biotechnology and Food Sciences* 2(6): 2387–2391.

Gurudasani, R., and Sheth, M. (2009). Food safety knowledge and attitude of consumers of various food service establishments. *Journal of Food Safety* 29: 364–380.

Islam, S., Thangadurai, D., Adetunji, C.O., Nwankwo, W., Kadiri, O., Makinde, S., Michael, O.S., Anani, O.A., and Adetunji, J.B. (2021). Nanomaterials and nanocoatings for alternative antimicrobial therapy. *Handbook of Nanomaterials and Nanocomposites for Energy and Environmental Applications* (Kharissova, O.V., Martínez L.M.T., and Kharisov B.I. eds). Springer, Cham. https://doi.org/10.1007/978-3-030-11155-7_3-1

Kok, C.R. and Hutkins, R. (2018). Yogurt and other fermented foods as sources of health-promoting bacteria. *Nutrition Reviews* 76 (S1):4–15. doi:10.1093/nutrit/nuy056.

Kouamé-Sina, S., Makita, K., Costard, S., Grace, D, Dadié, A., Dje, M. and Bonfoh, B. (2012). Hazard identification and exposure assessment for bacterial risk assessment of informally marketed milk in Abidjan, Côte d'Ivoire. *Food and Nutrition Bulletin* 33(4): 223–234. The United Nations University.

Krieger, S. and Schiefer, G. (2006). Quality systems in the agri-food industry – implementation, cost, benefit and strategies. Poster paper prepared for presentation at the International Association of Agricultural Economists Conference, Gold Coast, Australia, August 12–18.

Knaflewska, F. and Pospiech, E. (2007). Quality assurance systems in food industry and health security of food. *Acta Scientiarum Polonorum, Technol Aliment* 6(2), 75–85.

Laux, M.C., Gretchen, A., Mosher, G.A. and Charles, R., Hurburgh Jr., C.R. (2015). Application of quality management systems to grain handling: An inventory management case study. *Agricultural and Biosystems Engineering* 31:313–321. doi:10.13031/aea.31.10860.

Mensah, L.D. and Julien, D. (2011). Implementation of food safety management systems in the UK. *Food Control* 22:1216–1225. doi:10.1016/j.foodcont.2011.01.021.

Mantovani, A. and Frazzoli, C., (2010). Risk assessment of toxic contaminants in animal feed. CAB reviews: Perspectives in agriculture. *Veterinary Science, Nutrition and Natural Resources*. 5, No. 046.

Mwangi, L.W., Matofari, J.W., Muliro, P.S. and Bebe, B.O. (2016). Hygienic assessment of spontaneously fermented raw camel milk (suusa) along the informal value chain in Kenya *International Journal of Food Contamination* 3:1–9. doi:10.1186/s40550-016-0040-8.

Maria, P. and Adina, D. (2011). Considerations on Integrating Risk and Quality Management. Annals of "Dunarea de Jos" University of Galati Fascicle I. Economics and Applied Informatics Years XVII – no1/2011, 49–54.

McGuffey, R.K. and Shirley, J.E. (2011). *History of Dairy Farming*. Elsevier, Amsterdam, pp. 1–11.

Meijuan, Y., Yonglin, L., Achal, V., Qing-Long, F. and Lanhai, L. (2015). Health risk assessment of al and heavy metals in milk products for different age groups in China. *Polish Journal of Environmental Studies* 24 (6): 2707–2714. doi:10.15244/pjoes/58964.

Merwan, A., Amza, N. and Tamiru, M. (2018). Review on milk and milk product safety, quality assurance and control. *International Journal of Livestock Production* 9(4): 67–78. doi:10.5897/IJLP2017.0403.

Mishra, P., Mishra, R. R., and Adetunji, C. O. (eds.). (2020). *Innovations in Food Technology: Current Perspectives and Future Goals*. Springer, Singapore. https://doi.org/10.1007/978-981-15-6121-4.

Naresh, L., Merchant, S.U. and Dhuldhoya, N.C. (2006). Food Safety Using Haccp Quality Management System. Food Promotion Chronicle.

Nicola, K., Lake, R. and Cressey, P. (2014). Risk Profile: Listeria monocytogenes in Raw Milk. MPI Technical Paper. Institute of Environmental Science and Research Limited Christchurch Science Centre Location. No: 2014/16.

Nduko, J.M., Matofari, J.W., Nandi, Z.O., and Sichangi, M.B. (2017). Spontaneously fermented kenyan milk products: A review of the current state and future perspectives. *African Journal of Food Science* 11(1): 1–11.

Nordenskjöld, J. (2012). Implementation of a quality management system in food production. The Faculty of Natural Resources and Agricultural Sciences Uppsala Bio Centrum Department of Microbiology. Sverigeslantbruksuniversitet Swedish University of Agricultural Sciences.

Oloo, J.E.O. (2010). Food Safety and Quality Management in Kenya: An Overview of the Roles Played By Various Stakeholders. Volume 10 No. 11. African journal of food agriculture nutrition and development, pp. 4379–4397.

Olaniyan O.T. and Adetunji C.O. (2021). Biochemical role of beneficial microorganisms: an overview on recent development in environmental and agro science. *Microbial Rejuvenation of Polluted Environment. Microorganisms for Sustainability* (Adetunji C.O., Panpatte D.G., and Jhala Y.K. eds), vol 27. Springer, Singapore. https://doi.org/10.1007/978-981-15-7459-7_2

Ozlem, Y.K. and Zehra, G. (2012). Quality management systems in dairy industry. In *Proceedings of the 2012 International Conference on Industrial Engineering and Operations Management Istanbul*, Turkey, pp. 523–531.

Paul, J.M. (2004). Tools for integrated chain management in food safety. The food chain approach from an international perspective, pp. 33–35. In *IDF/FAO International Symposium on Dairy Safety and Hygiene*, Cape Town, South Africa.

Pearce, K.N. (2000). III-Dairy-C-Milk Powder, Food Science Section, New Zealand Dairy Research Institute, pp. 1–5.

Popov, V.G., Kadochnikova, G.D., Pozniakovsky, V.M., Ermolaeva, E.O., and Surkov, I.V. (2018). Development and implementation of management systems - a state priority in providing quality and safety of AIC products. *RJPBCS* 9(6):66–77.

Pozo, H., Barcelos, A.F. and Akabane, G.K. (2018). Critical factors of success for quality and food safety management: Classification and prioprization. *Universal Journal of Industrial and Business Management* 6(2): 30–41. http://www.hrpub.org. doi:10.13189/ujibm.2018.060202.

Suttiprasit, P. (2012). Innovative integrated management system (IIMS) for sustainable food industry. *Nang Yan Business Journal* 1(1):137–142.

Qingxin, C. (2018). Study of the Impact of Quality Management System on Chinese Baby Dairy Product Industry. A thesis submitted in partial fulfilment for the requirements for the degree of Doctor of Philosophy at the University of Central Lancashire.

Sadikoglu, E. and Olcay, H. (2014). The effects of total quality management practices on performance and the reasons of and the barriers to TQM practices in Turkey. *Hindawi Publishing Corporation Advances in Decision Sciences* 2014:1–17. Article ID 537605. doi:10.1155/2014/537605.

Safe food (2017). Impact of Climate Change on Dairy Productoin [Internet]. Safefood, Little Island, Ireland. Available from: http://www.safefood.eu/Publications/Research-reports/.

Salima, M., Umbetaliev, N., Aitzhanova, A. and Bogomolov, A. (2017). The quality management system improvement for the enhancement of production competitiveness. *Revista ESPACIOS* 38 (42):29.

Schaper, C., Lassen, B. and Theuvsen, L. (2009). Risk Management in Milk Production: A Study in Five European Countries Paper prepared for presentation at the 113th EAAE Seminar "A resilient European food industry and food chain in a challenging world", Chania, Crete, Greece.

Sumaedi, S. and Yarmen, M. (2015). International symposium on food and agro-biodiversity (ISFA2014) the effectiveness of ISO 9001 implementation in food manufacturing companies: A proposed measurement instrument. *Procedia Food Science* 3:436–444.

Tania, A. (2019). Risk management in dairy contracting: A study in France, Australia, and the United States. *PM World Journal Risk Management in Dairy Contracting* 8(7): 1–19.

Thangadurai, D., Dabire, S.S., Sangeetha, J., Said Al-Tawaha, A.R.M., Adetunji, C.O., Islam, S. Shettar, A.K., David, M., Hospet, R., and Adetunji, J.B. (2021). Greener composites from plant fibers: preparation, structure, and properties. *Handbook of Nanomaterials and Nanocomposites for Energy and Environmental Applications* (Kharissova O.V., Martínez L.M.T., Kharisov B.I. eds). Springer, Cham. https://doi.org/10.1007/978-3-030-11155-7_21-1

Ukhurebor, K.E., Aigbe, U.O., Olayinka, A.S., Nwankwo, W. and Emegha, J.O. (2020). Climatic change and pesticides usage: A brief review of their implicative relationship. *Assumption University eJournal of Interdisciplinary Research* 5(1): 44–49.

Ukhurebor, K.E., Umukoro, O.J. (2018). Influence of meteorological variables on UHF radio signal: Recent findings for EBS, Benin City, South-South, Nigeria. *IOP Conference Series: Earth & Environment Science* 173:012017.

US-FDA (2015). Grade "A" Pasteurized Milk. Revision U.S. Department of Health and Human Services, Public Health Service, Food and Drug Administration [Internet]. Available from: https://www.fda.gov/downloads/Food/GuidanceRegulation/GuidanceDocumentsRegulatoryInformation/Milk/UCM513508.pdf.

van Asselt, E.D., Marvin, H.J.P., Boon, P.E., Swanenburg, M., Zeilmaker, M., Mengelers, M.J.B., and van der Fels-Klerx, H.J. (2016). Chemical and physical hazards in the Dairy Chain. Wageningen, RIKILT Wageningen UR (University and Research centre), RIKILT report. 003, pp. 46.

van Asselt, E.D., van der Fels-Klerx, H.J., Marvin, H.J.P., van Bokhorst-vande Veen, H. and Nierop Groot, M. (2017). Overview of food safety hazards in the European dairy supply chain. *Comprehensive Reviews in Food Science and Food Safety* 16: 59–75. doi:10.1111/1541-4337.12245.

Van Heerden, M.J. (2013). The Effect of an Integrated Quality Management System on a Leading Fish Product Manufacturer: A Pre-Implementation Viability Study. Thesis: University of South Africa.

Yusof, S and Aspinwall, E. (1999). Critical success factors for total quality management implementation in small and medium enterprises. *Total Quality Management* 10(4–5): 803–809. doi:10.1080/0954412997839.

# 5 Soy-Based Food Products Consumed in Africa

## *A Panacea to Mitigate Food Insecurity and Health Challenges*

*Olugbemi Tope Olaniyan and Charles Oluwaseun Adetunji*
Edo State University Uzairue

*Juliana Bunmi Adetunji*
Osun State University

*Julius Kola Oloke*
Ladoke Akintola University of Technology

*Olusola Olawale Olaleye*
Nigerian Institute for Trypanosomiasis Research

*Daniel Ingo Hefft*
University Centre Reaseheath

*Benjamin Ewa Ubi*
Ebonyi State University
Tottori University

**CONTENTS**



DOI: 10.1201/9781003178378-5

## 5.1 INTRODUCTION

Soybean (*Glycine max* [L.] Merrill) has been identified as a significant crop with several crucial agronomic features such as protein and oil contents, productivity, weight per plant, height, days to maturity, and branches that have been discovered in China since 2200 BC (Kolapo, 2011). Soybean is a legume that is normally cultivated, especially in the temperate, subtropical, tropical climate region. Moreover, soybean has now been utilized in most parts of the globe forfood, medicine, and animal feeds (CGIAR, 2005).

It has been stated by FAO in the year 2005 that about 95.2 million ha of land are being utilized for the planting of soybean globally while the total planting was approximately 212.6 million tones. The first three countries that produce the highest quantity of soybean were Argentina (14 million ha), Brazil (23 million ha), and the USA (29 million ha), while the average of 1.16 million ha was utilized for the soybean cultivation in the whole of Africa with an average production of 1.26 million tons (Kolapo, 2011). The soybean production in Africa has been stagnated over the years, and this has been attributed to utilization of poor-performance varieties, poor use of rhizosphere, biofertilizers, and inoculants particularly among the four leading producers of soybeans, namely, Nigeria, South Africa, Uganda, and Zambia (Khojeley et al., 2018; Zachary et al., 2020). Food insecurity, environmental pollution and malnutrition have been major challenges in Africa, which has been observed to increase daily (Adetunji and Anani, 2021a,b,c; Adetunji, et al., 2021; Dwivedi, et al., 2021; Olaniyan and Adetunji, 2021, Mishra et al., 2020; Thangadurai et al., 2021; Anani et al., 2020; Islam et al., 2021). It has been reported that the available natural resources that could be utilized to eradicate these aforementioned challenges are not enough; therefore, there is need to search for a sustainable food production that could help to resolve all these challenges (Hadad and Martorell, 2002; FAO, 2016). In addition, the high population explosion in most African cities coupled with bad governance and economic system have increased the number of malnourished children which is expected to further rise exponentially by the year 2030 (Rosegrant and Meijer, 2002; Benson and Shekar, 2006; UNICEF, 2019). Therefore, there is a great demand to search all the available biological resources available in Africa that could be used to manage all these problems, most especially malnutrition and hunger (United Nations, 2015).

Furthermore, it has been identified that Africa's edaphic and climatic factors, especially the tropical soybean varieties could be grown in about half of the African cultivation land (Kolapo, 2011). The proper knowledge about the food values and the nutritional features of soybeans could also help in resolving all the aforementioned problems. The introduction of genetically modified soybeans that are tolerant to abiotic stress such as drought and biotic factors such as diseases and pest could play a significant role toward enhancing food production. Taking all the nutritional and medicinal features that could be derived in soybean like vitamins, phytosterols, minerals, isoflavones, and high protein contents; this will go a long way in reducing some of the problem of food insecurity and malnutrition in Africa. Therefore, this chapter intends to provide detailed information on the nutritional and health benefits of consuming soy-based foods as a typical type of nutritional food and give better insight on how the pharmacological active compounds present in these soy-based foods exhibit their biological activities for maximum benefits of mankind.

## 5.2 NUTRITIONAL AND HEALTH BENEFITS OF CONSUMING SOY-BASED FOODS

Rodríguez-García et al. (2019) revealed that there are great biological and health benefits when plant-based diets are consumed as part of nutritional supplements. This has alluded to the fact that many essential nutritional components are present in the plants such as phytochemicals and polyphenol representing important nutrients for maintaining physiological function, enhancing growth and development of nervous tissues. Recent scientific interest in increasing awareness on the health importance of macronutrient in plant-based diets has increased tremendously due to the fact that several studies like Forni et al. (2019) and Adegbola et al. (2017) have reported the role of antioxidant and anti-inflammatory properties of plants macronutrients in ameliorating some metabolic disorders and prevention of disease conditions in the body. Studies have shown that polyphenols can be broadly subdivided into several groups such as lignans, stilbenes, phenolic acids, and flavonoids among others. Across the globe, renewed interest in flavonoids and lignans (described as steroid-like chemical compound—phytoestrogens) has increased drastically. Various health benefits have been ascribed to the consumption of flavonoids and lignans from plant-based diets such as reduced cardiometabolic disorders, boosting fertility rate, and improvement in neurological functions, and hence, these bioactive molecules have been suggested to play a fundamental role in promoting good health status (Rajalakshmy et al., 2011; Olaniyan et al., 2018).

Desmawati and Sulastri (2019) revealed that the biodiversity of phytoestrogens across Asian and Africa regions is huge, and its existence in soybean has been confirmed with similar properties and chemical structure like the endogenously produced estradiol with diverse health benefits. Recently, Messina (2016) revealed that the research into the potential health benefits of macromolecules like high quality proteins and fat derived from soybean dated back as far as 25 years. Their roles in reducing cardiometabolic disorders, cancer, and reproductive dysfunctions have been confirmed recently to be due to the presence of isoflavones. It is generally known that isoflavones can modulate estrogen receptors and equally act as phytoestrogens, and therefore concerns have been raised about its negative effects. Though these concerns have been generated from animal studies, European Food Safety Authority has alluded to the fact that isoflavones are safe and do not pose any kind of danger when consumed in humans. Our aim in this review is to uncover each of the major research areas involving soybeans focusing primarily on the basic research and its biomedical uses.

Stephen et al. (2010) have demonstrated that consumption of soy food is associated with low occurrence of diseases, and for several decades, scientists have carried out different studies to identify some major bioactive compounds in soybeans. It has been recently gathered that isoflavones and soy proteins are the major active biomolecules responsible for the biological effects of soybeans even though there are other phytochemicals present (Ahsan et al., 2018). Stephen et al. (2010) have also proposed that soy proteins or isoflavones without the presence of other active constituents are devoid of certain important pharmacological benefits derived from consumption of the entire soybeans. Hence, it was suggested that other phytochemicals present in soybean complement the biological effects of isoflavones and soy proteins in the body, such as maintaining blood pressure and hormonal balance, regulating reproductive and endocrine functions.

Fasiha et al. (2018) reported that soy proteins are naturally found in soybeans which have been recognized to regulate many biological activities in humans such as lipolysis and estrogenic ability. It serves as an alternate means of amino acids for individuals who cannot eat protein-based diet but along with other benefits. Soybeans are extremely versatile beans that can be processed into flour, oil, and meal. Majorly, about 25 g/day of soybean protein is generally known to minimize low-density lipoprotein and total cholesterol. Soybeans have different active substances such as saponins, isoflavones, trypsin inhibitors, fibers, and phytic acid. Research has highlighted some concerns about the possible negative impact of soybeans, due to the presence of anti-nutrients and allergenic agents which prevent micronutrient absorption resulting in hypothyroidism, hormonal imbalance, and deficiency in a few essential amino acids and considered as an incomplete protein source (Aneta and Dasha, 2019). In the last 5 years, serious attention and scientific report on the beneficial role of soybeans have not been seen when compared with other types of protein including their effects on low-density lipoprotein cholesterol, but recently soy protein has been discovered to reduce cardiovascular effects of LDL cholesterol level even in combination with other types of protein (Soetan and Oyewole, 2009).

Sacks et al. (2006) revealed that tremendous attention have been attributed to soy protein and isoflavones by the American Heart Association Scientific Advisory Council on isoflavones and soy protein which has resulted in positive impact in the treatment of cardiovascular disorders and improvement in cardiometabolic functions as well as other risk factors. Reports have also shown that isoflavones are less effective when compared with soy protein. Moreover, other active constituents present in soybean play a crucial role in the lowering of excessive fat and carbohydrate in human being. Sacks et al. (2006) hence concluded that further research on the impacts of high protein diets on cardiometabolic function should be conducted to ascertain the severity of excessive soy protein intake on cardiovascular status.

Lokuruka (2011) showed that due to the increased consumption of soybeans worldwide and its pharmacological properties, increased bioprocessing of soybeans may alter sensual and physical attractiveness thereby reducing its potential value. During bioprocessing of soybeans, heating above 100 °C might alter and denature some trypsin inhibitory enzymes, distort globulins, changing serine, threonine, cysteine, and lysine to dehydroprotein intermediary, hence reacting with lysine. This can later cause the formation of lysinolanine, loss of methionine and tryptophan, increase soy protein bioavailability, and alter hemagglutinins, thus changing the nutritive value. In this process, acylation reaction is activated through alkali to generate toxic substance lysinoalanine (Finley et al., 2006). Studies have shown that lysinoalanine can cause severe gastrointestinal disorders, loss of hair growth, renal dysfunctions, and severe enlargement of organs in rodents, but in humans, much of these have not been seen (Poulsen et al., 2013). Other detrimental effects seen include conversion of cysteine to dehydroalanine and arginine changing to ornithine, urea, citrulline, and ammonia. Maillard reaction is a process between proteins and carbohydrates, specifically, between the neophilic amino group of amino acids and the carbonyl group of sugars. Racemic mixtures of D and L isoforms are formed when there is isomerization of amino acids due to high temperature above 200 °C causing reduction in nutritive value and toxicity of soybeans. Free radical generation causes oxidative stress, membrane damage by lipid peroxidation, destruction of lipid soluble vitamins, and reduction in tissue essential lipid contents. Studies have shown that the combination of oxidized protein and lipid generates products that cause atherosclerotic plaques in the blood vessels (Fisher et al., 2012; Yang et al., 2017). Most of the beneficial vitamins like vitamins E and A and carotenoids are destroyed in the process. Also, it has been confirmed that subjecting isoflavones in soybeans to toasting and roasting does not reduce its nutritive value, but fermentation may cause bioavailability through enzyme interactions which might alter molecules making them more available to the human microbiome. Hence to reduce drastic changes in the nutritive value of soybeans, it has been suggested that fermentation, reduced washing, and lowered heating below 100 °C are recommended during bioprocessing.

Adelakun et al. (2005) showed that soybean flour has tremendous benefits in augmenting the nutritional status of individuals who are deficient in daily dietary needs. Based on the rich diversity of soybeans, wheat and corn flour products which are low in nutrients are often fortified to give them some sort of nutritious value. Individuals can then consume soybeans, wheat, and corn flour–fortified products to enhance their nutritional profile. Notwithstanding, additional efforts should be garnered and refocused toward advocating for utilization and consumption of fortified food products and addressing the technical issue in the bioprocessing methods. Recently, Ciabotti et al. (2016) reported the isoflavone contents in a study using soybeans containing different genes to alter the phenotypic coat colors (brown, black, and different shades of yellow), and their nutritional status. The authors revealed that significant differences were observed between the existing genes in terms of nutritional values and showed that albumin and glutelin are the main proteins compared with other leguminous plants. They concluded that isoflavones content in soybean of the yellow seed coat expressed the highest content, and therefore can serve as a potential source of ingredients for food production and consumption.

Martina et al. (2002) evaluated soy flour using enzymatic hydrolysis plus about three different proteases to produce soy protein hydrolysate with protease flavor enzyme showing the highest degree of hydrolysis. The authors also measured the foaming stability and gelation and concluded that flavourzyme improved the foaming of soy flour proteins with good gelation activity due to breakdown of β-conglycinin and glycinin. Hence, the free amino acids during the hydrolysis were histidine, tyrosine, arginine, leucine, and phenylalanine. Miroljub et al. (2005) showed that diverse beneficial advantages are seen when soy protein products are utilized for food formulations such as flavor, low cost, nutrient, and functionality in the body. It is generally known that different bioprocessing is employed in the production of soybean protein products such as chemical, physical, and enzymatic treatments with restricted proteolysis showing the best protective and safety level. The authors demonstrated that subjecting soybean protein to different treatment such as heating, hydrolysis, and pH adjustment could hamper the physiochemical properties causing disintegration of the structure, formation of aggregates via electrostatic force, and denaturing and dissociation of the interchange bond (Veličković et al. 1994).

Nguyen (2015) revealed that soybean protein is very important for vegans and vegetarians because of its rich nutritive value in minerals, vitamins, fibers, and proteins. The authors also demonstrated that soybean-based foods possess beneficial physiological bioactive components or peptides with anti-inflammatory and antioxidant functions. They concluded that in the process of producing bioactive peptides from soybean, adequate precaution must be taken to ensure that its nutrition and safety is maintained for optimal human benefit. In another study by Preeti et al. (2008), the author revealed that soybean protein has remained the potential ingredient of many food products since the FDA's approval in 1998. Products containing beneficial ingredients like soybean proteins have appealing flavor and taste to many consumers, therefore high acceptability is given to soybean protein–fortified dairy products. The authors revealed that soy fortification helps to increase viscosity and coating of the mouth with higher acceptability, hence it was suggested that knowledge about optimization and design will help in formulating acceptable standard for soy fortified products.

Hossein (2011) revealed that in recent years consumers have become very conscious of the health implications of dietary products, and hence great attention has been invested in soybean protein products that have potential health benefits. Epidemiological data has revealed that there is increased risk of developing chronic diseases due to different lifestyles, and thus consumption of soybean protein may offer a great deal of health benefit to individuals by lowering the incidence and prevalence. Even though the particular active constituent in soybean responsible for these health effects has not been identified, scientists have continued to study the phytochemistry and their biological effects.

Chatterjee et al. (2018) revealed the immense benefits linked with the consumption of soybean protein products. The authors highlighted that physiological benefits have been derived from it such as reducing the risk of obesity, insulin-resistance/type II diabetes, and immune disorders. These effects were attributed to the bioactive molecules derived from soybeans, which has led to the approval of its consumption in Canada and the USA. The available information on soybean peptide revealed that soymorphins and lunasin are the major bioactive components with multiple effects, and therefore studies on their pharmacological and biological effects and mechanisms of action have started to accumulate recently to utilize them for drug and food formulations (Chatterjee et al., 2018).

Bo and Doo (2012) worked on the physiology of lactic acid bacterium in yogurt-fermenting culture by adding glasswort composed of soy fruit (bean). The authors discovered that the fiber of the soy yogurt was greatly influenced but no significant improvement was seen in malic and citric acid, fat, and protein level. They also noticed a great increase in mineral and number of viable lactic acid bacterial, total phenolic and polyphenol compounds after adding glasswort. Fasiha et al. (2018) also reported the usefulness of soybean protein and other beans as an alternative source of dietary protein for those vegetarian and vegans. Other derived products include trypsin inhibitors, fibers, saponins, phytic acid, and isoflavones. Bibu (2010) remarked that plant polyphenol is known to cause immense biological response with high similarity with estrogen, and its receptors have been found in soybeans and are generally called phytoestrogens. Epidemiological studies have shown little information on phytoestrogen from plant and animal origin. Very few have shown the beneficial importance of phytoestrogen. Recently, scientists are beginning to study the biology of phytoestrogens utilizing liquid chromatography-mass spectrometry and automated extraction (solid phase).

Peeters et al. (2013) wrote on phytoestrogen as a potential antitumor agent with estrogenic characteristics classified into three main types: coumestans, lignans, and isoflavones (Mazur et al., 1998). The authors analyzed the role of phytoestrogen in soybean against breast cancer from different studies, and it was discovered that soybean phytoestrogen did not produce any significant effect against cancer. Urinary elimination of isoflavones was observed in some of the groups analyzed, and enterolactone's role in breast cancer produced a mild prevention of breast cancer in the study evaluated, and hence, it was concluded that soybean phytoestrogen does not have a significant preventive role on cancer.

**TABLE 5.1**
**Summary of Nutritional Composition of Soybean**

| SN | Components | Physiological Benefits | References |
|---|---|---|---|
| 1 | Glycitein | Antioxidant, anti-inflammatory Estrogenic activity | Song (1998) |
| 2 | Genistein | Antifungal, antibacterial, estrogenic activity | Haidong et al. (2018) |
| 3 | Daidzein | Cardiovascular functions | Sacks et al. (2006) |
| 4 | Glyceollins | Antitumor, immune response | Charles et al. (2006) Anatoliy et al. (2013) |
| 5 | Phytoalexins | Antitumor | Zimmermann et al., (2010) |
| 6 | Vitamins and minerals | Antioxidants, growth of nervous system | Rodríguez-García et al. (2019) |
| 7 | Phytoestrogens | Endocrine effects | Dezmawati and Sulasti (2019) |
| 8 | Protein | Lipolysis, endocrine function, reducing risk of obesity, cardiometabolic function fertility, neurological activity | Fasiha et al. (2018) |
| 9 | Lignans | Antitumor, immune response | Peñalvo et al. (2004) |
| 10 | | | |
| 11 | Saponins | Cardiometabolic function | Moses et al. (2014) |
| 12 | Trypsin inhibitors | Cardiovascular function and fat | Aneta and Dasha (2019) |
| 13 | Fibers | Metabolism | Soetan and Otewole (2009) |
| 14 | Phytic acid | | |
| 15 | Proanthocyanidins | Anti-inflammatory, antioxidant | Jehun et al. (2011) |

## 5.3 HEALTH BENEFITS OF SOY ISOFLAVONES DERIVED FROM SOYBEANS

Wolf (1974) revealed that soybean phytochemicals such as phytoesterogens and other isoflavones have become major crop biomolecules in the Western world. Research has gone to the extent of evaluating the clinical trials on some of its potential benefits in treatment of cardiometabolic disorders, cancer, bone, and reproductive dysfunctions. The author analyzed some of the clinical studies and discovered that soybean protein could significantly reduce lipids effects and showed that the phytoestrogen is a potential therapy for postmenopausal effect of estrogen deficiency and its clinical manifestations (Anderson, 1997). Jehun et al. (2011) also wrote extensively on soybean and its phytochemical constituents like proanthocyanidins, polyphenolic molecules, and small proteins in line with their therapeutic effects as antioxidant, anti-inflammatory, and UV radiation protective effects.

Anatoliy et al. (2013) also studied the role of transgenic and normal soybean on some pathogenic diseases such as soybean chlorotic mottle, soybean crinkle leaf, soybean dwarf, soybean mosaic, soybean severe stunt, and bud blight. The transgenic plant was incorporated with a lot of phytoconstituents and thus the inhibition of isoflavone synthesis resulted in accumulation of isoflavone precursor in the transgenic plant cotyledons. Hence, these findings showed that promoter non-coding DNA sequences of lectin may be activated during germination. Also, the transgenic plant was able to resist the inoculated cotyledon with inducible pathogenic glyceollin using *P. sojae*, and thus the authors revealed that the soybean's innate immune response could be stimulated by increased glyceollin build up during infection.

Barnes (2010) evaluated the pharmacology of soybean isoflavones which has been shown to be accepted in the USA and South America owing to the tremendous beneficial role. They revealed that the absorption process for soybean isoflavones and other metabolites is normal. Isoflavones were observed to bind with estrogen receptors, act as antioxidant, inhibit steroid biosynthesis enzymes, stimulate specific T cell, inhibit tyrosine kinases, influence natural killer cell function, and stimulate peroxisome proliferator regulators. Moses et al. (2014 remarked that in the Western world, the highest environmental risk factor is that of dietary issues. The authors showed the health-promoting benefits of consuming leguminous plants and showed that greater attention is given to soy and soybean products today due to their health-promoting benefits. They showed that soybean is very rich in many active compounds such as isoflavones and saponins which are of significant interest to scientists working in the area of cardiometabolic disorders. There is high blood concentration of these metabolites among individuals residing in countries with reduced incidence of cardiovascular disorders or cancer sufficient to affect various cellular processes like growth factor, protein biosynthesis, malignant cell growth, intracellular proteins, angiogenesis, and differentiation. The authors concluded that leguminous seeds such as soybean with regard to their high level of phytoestrogens are potential agents for cancer-protective food.

Kumaran and Gitarasu (2015) demonstrated the antibacterial effects of the soybean phytoconstituents using liquid micro-culture technique. Their study revealed that soybean active molecules were able to resist potential pathogens. They suggested that the growth of rhizobia must be due to the ability of isoflavonoid to stimulate resistance in rhizosphere of soybean. Peñalvo et al. (2004) revealed that there has been recent commercialization of isoflavone extracted from lignans plant for mainly soybean-based food supplement to promote health. Their studies showed that different types of lignans, such as matairesinol, syringaresinol, secoisolariciresinol, pinoresinol, lariciresinol, and isolariciresinol, have so far been identified and extracted using different spectrometry method showing positive relation between isoflavones level and lignans concentration, and the authors also went further to analyze the quantity of isoflavones and lignans in the soybean and soybean supplements (Carmen et al., 2019).

Jyoti et al. (2015) demonstrated that phytoestrogen in terms of physiochemical properties have similar physiological function with estrogen and estrogen receptor activity which are plant-based nonsteroidal polyphenolic molecules. Studies have shown that three main examples of phytoestrogen are lignans, isoflavones, and coumestans. The authors revealed that among these three, isoflavones are the group with the most researched soybean phytoestrogens with very strong estrogenic properties. It is shown that phytoestrogens may act as antagonists based on the concentration or level, thereby preventing the entering or binding of estrogen on the receptor. Also, genistein a typical example of isoflavones is gradually gaining more recognition and attention owing to the fact that scientists have discovered its important role in the regulation of cardiometabolic functions in the body.

Adrian et al. (1998) carried out a quantitative analysis of the level of isoflavones found in the human milk utilizing HPLC approach and used absorbance analysis to identify and separate the metabolites and the isoflavones (Franke and Custer, 1996). They later administered roasted soybean and used similar methods to evaluate the levels of isoflavones in urine and human milk, resulting in the detection of 1–3 pmol in the analytes tested. Stephen et al. (2010) in their own study used clinical investigation and epidemiological findings to critically examine the relationship between the incidence of reduced levels of cardiometabolic diseases and the consumption level of soybean-based food or supplement. They revealed that over the past few years, substantial efforts have been made to isolate and identify the bioactive ingredient in soybean leading to the discovery of soybean protein and isoflavones as the main constitutes in soybean with important physiological functions. Subsequently, other studies revealed numerous important phytochemicals in soybeans-based food that is yet to be characterized with any known function. It is believed that the physiological function and health-promoting effects of isoflavones and soybean proteins are significantly enhanced by the presence of other phytochemicals present in soybean.

Zimmermann et al. (2010) wrote in their study that many estrogens stimulated pathological effects are prevented by soybean through an important active biomolecule called glyceollin, a known phytoalexin isolated from soybean. The authors explained that there are three isomers of glyceollin (isomers 1–3) with known tumor suppressive effects, especially estrogenic induced cancer formation, but glyceollin 1 is the main isomer responsible for the antiestrogenic function in human estrogen receptor $\alpha$. In their analysis, they discovered that the hormone-receptor complex conformation displayed by glyceollin 1 is unique and different from that of isomers

2 or 3. They concluded that glyceollin 1 is an important component of phytoalexin-enriched food diet with chemoprevention function in tumors. Haidong et al. (2018) carried out a research on the physiological role of soybean isoflavones administered to rats with ovariectomized induced bone degeneration with reduced body weight, estradiol, alkaline phosphatase, serum levels of osteocalcin and minerals. It was discovered that administration of soybean isoflavones reversed these effects via reduced Notch1 proteins in muscle and bone tissue. They concluded that soybean isoflavones administration is a potential therapy for preventing postmenopausal bone degeneration by stimulating the Notch transduction signaling pathway (Liang et al. 2018). Isoflavones from soybean and other plant sources are of interest because of their estrogenic, antifungal, antibacterial, and nutritional properties. Soybean isoflavones possess similar features or characteristics with estrogen or 17-ß estradiol which makes these molecules have ability to bind to estrogen receptors plus other sex hormone–binding proteins.

Further research should be carried out to evaluate the physiological activity of soybean isoflavones estrogenic effects on reproductive functions plus other health effects. Charles et al. (2006) described the important role of glyceollins, a novel class of soybean phytoalexins, with potential chemoprotective antiestrogenic effects in cancer patients. It was discovered that glyceollin-enriched soybean protein has significant physiological regulation on breast cancer biomarkers with reduction in proliferation, progesterone receptor, and trefoil factor 1gene expression as compared with the control group. They concluded that soybean glyceollins are novel natural biomolecules with tremendous potential for estrogen-modulating activity against breast cancer.

Thompson et al. (2007) revealed and analyzed for isoflavones in 21 non-vitamin, non-mineral food supplements consumed in Canada by women to support their previous research on food phytoestrogen. Supplements containing soybean are seen to possess the highest amount of isoflavones and other phytoestrogens with other food supplements having far less amount of isoflavones and total phytoestrogens. Lignans were revealed to be significantly present in all food supplements, and the daily consumption of isoflavones with lignans is suggested to exceed those from soybean or vegetables. Hence, they suggested that in every clinical study, in order to have accurate estimate of the phytoestrogen consumption, supplements intake should be taken into serious consideration.

Meghwal and Sahu (2015) remarked that soybean-based isoflavones have similar structure and physiological characteristics with 17 β-estradiol and their health-promoting effects such as anti-inflammatory and antioxidant have been well characterized in different animal experimental models. The authors showed that isoflavones are effective in the treatment and management of several cardiometabolic conditions. They reduce plasma lipid level, hence minimizing the risk of cardiometabolic remodeling, osteoporosis, and cancer through the inhibition of tyrosine kinase signaling and androgen receptor. The authors thus confirmed the nutraceutical and pharmaceutical uses in biomedical science in many health-related conditions.

Riyanto and Lelyana (2017) wrote extensively on acne vulgaris as one of the most common skin inflammatory conditions. In their study, they investigated the role of soybean isoflavones in acne vulgaris lesions among women and the results revealed that soybean isoflavones possess anti-inflammatory and anti-androgenic effects due to the reduction in total acne vulgaris lesions. Song (1998) demonstrated that soybean biomolecules particularly isoflavones possessed certain health-promoting activities, such as reducing blood lipid levels, cancer prevention, and reducing postmenopausal bone loss. The estrogenic activity of the isoflavone glycitein was evaluated utilizing *in vitro* estrogen receptor binding assay. The authors revealed that glycitein showed a significant role in uterine enlargement in mice compared with genistein. Glycitein was equally compared with the role of genistein and daidzein in terms of the affinity for uterine cytosol estrogen receptor proteins, and they discovered that glycitein possesses higher estrogenic activity comparable to genistein or daidzein. In another study, utilizing the addition of soy germ, daidzein, or soy germ extract to casein reduced the plasma level of lipid in hamsters. Soybean isoflavones and protein significantly reduced the plasma lipid levels (Tongtong et al., 2003).

Pham et al. (2019) wrote on the physiological activity of plant-based molecules, referred to as phytochemicals or biomolecules capable of influencing the activity of protein or genes plus affect the function of diverse receptors, signaling pathways plus transcription factors that play crucial roles in cellular physiology. The authors revealed that great potential benefits may be generated from utilization of phytochemicals in the management and treatment of many pathophysiological conditions. Glyceollins have attracted serious recognition from biomedical scientists in recent times owing to ability to exert numerous physiological activities (Pham et al., 2019). It is believed that further scientific evaluation will help to properly give the precise mechanism of action used by these phytochemicals in their metabolic regulations such as protein kinase, estrogen receptor, and lipid kinase signaling transduction pathways. Payton-Stewart et al. (2009) described glyceollins as soy-based phytoalexins, a potential biomolecule that can be utilized for cancer prevention. The authors went further to evaluate the molecular activity of soy-based phytoalexin (glyceollins) on prostate cancer cell line to elucidate the physiological signaling pathways involved. It was discovered that glyceollins were able to suppress cancer growth particularly by suppressing protein levels, G1/S interphase, and upregulation of cyclin-dependent kinase inhibitor 1 mRNA. Other effects of glyceollin include downregulation of mRNA gene levels for androgen responsiveness, thought to be controlled by estrogenic pathway. Thus, glyceollins displayed ameliorative effects against cancer cells which appear to operate a different mode of action from other soy-derived phytochemicals.

Mulligan et al. (2012) suggested that diets rich in phytoestrogen have been shown to have a therapeutic activity against cardiometabolic disorders. In their study, they evaluated the mean intake levels of soybean biomolecules such as isoflavones, aenterolignans, lignans, and coumestrolin enriched with dairy foods and established an empirical data on the total phytoestrogen, lignin, and isoflavone consumption. Also, Boué et al. (2012) described the role of soybean glyceollins during stress-induced cancer cell growth. Their results revealed significant decrease in glucose level, stimulation of adipocytes increased insulin release, and glucose uptake in dose response manner. On molecular basis, the authors revealed that using PCR (polymerase chain reaction) technique they observed a significant increase in the amount of glucose transporter mRNA gene in the adipocytes, particularly when treated with glyceollins. Thus, they remarked that the effect of glyceollins on glucose uptake in the adipocytes may be linked with upregulation seen in glucose transporters. Hence, glyceollins displayed a potential therapeutic effects, pre-diabetic conditions and type 1 and 2 diabetic treatment option through the stimulation of both insulin-mediated plus basal, insulin-independent and glucose uptake by adipose tissue (Lygin et al., 2013).

Seung-Hyun et al. (2007) demonstrated that including β-cyclodextrin with the complex molecules of isoflavone can improve the bioavailability and solubility of the soybean isoflavones. It was shown that the bioavailability of soybean glycitein and genistein increased significantly in comparison with inclusion complexes of the isoflavone extract alone. These results show that the absorption of inclusion complexes of the isoflavone extract is enhanced by the addition of β-cyclodextrin. Table 5.1 below shows the summary of nutritional composition of soybean.

## 5.4 CONCLUSIONS

Soy-based food is an important food consumed as a staple food in most part of the African countries. Different studies have revealed great biological and health-promoting benefits of soy proteins and isoflavones when plant-based diets are consumed as part of nutritional supplements. This has been credited to the fact that many essential nutritional components present in the plants represent important macronutrients for growth and development, and in the protection against many cardiometabolic diseases including cancer. Hence, it is suggested that many of soybean phytochemicals should be utilized as nutraceutical and pharmaceutical supplement to enhance many biological and physiochemical processes in the body.

## REFERENCES

Adegbola, P., Aderibigbe, I., Hammed, W., and Omotayo, T. (2017). Antioxidant and anti-inflammatory medicinal plants have potential role in the treatment of cardiovascular disease: A review. *American Journal of Cardiovascular Disease* 7(2): 19–32.

Adelakun, O.E., Adejuyitan, J.A., Olajide, J.O., and Alabi, B.K. (2005). Effect of soybean substitution on some physical, compositional and sensory properties of kokoro (A Local Maize Snack). *European Food Research and Technology* 220:79–82.

Adrian, A.F., Laurie, J.C., and Yuichiro, T. (1998). Isoflavones in human breast milk and other biological fluids. *The American Journal of Clinical Nutrition* 68(suppl): 1466S–73S. American Society for Clinical Nutrition.

Adetunji, C.O. and Anani, O.A. (2021a). Recent advances in the application of genetically engineered microorganisms for microbial rejuvenation of contaminated environment. In: Adetunji C.O., Panpatte D.G., Jhala Y.K. (eds) *Microbial Rejuvenation of Polluted Environment. Microorganisms for Sustainability*, vol 27. Springer, Singapore. https://doi.org/10.1007/978-981-15-7459-7_14

Adetunji, C.O., Anani, O.A. and Panpatte, D. (2021). Mechanism of actions involved in sustainable ecorestoration of petroleum hydrocarbons polluted soil by the beneficial microorganism. In: Panpatte D.G., Jhala Y.K. (eds) *Microbial Rejuvenation of Polluted Environment. Microorganisms for Sustainability*, vol 26. Springer, Singapore. https://doi.org/10.1007/978-981-15-7455-9_8

Adetunji, C.O. and Anani, O.A. (2021b). Utilization of microbial biofilm for the biotransformation and bioremediation of heavily polluted environment. In: Panpatte D.G., Jhala Y.K. (eds) *Microbial Rejuvenation of Polluted Environment. Microorganisms for Sustainability*, vol 25. Springer, Singapore. https://doi.org/10.1007/978-981-15-7447-4_9

Anani, O.A. and Adetunji, C.O. (2021c). Bioremediation of polythene and plastics using beneficial microorganisms. In: Adetunji C.O., Panpatte D.G., Jhala Y.K. (eds) *Microbial Rejuvenation of Polluted Environment. Microorganisms for Sustainability*, vol 27. Springer, Singapore. https://doi.org/10.1007/978-981-15-7459-7_13

Anani, O.A., Mishra, R.R., Mishra, P., Enuneku, A.A., Anani, G.A., and Adetunji, C.O. (2020). Effects of toxicant from pesticides on food security: Current developments. In: Mishra P., Mishra R.R., Adetunji C.O. (eds) *Innovations in Food Technology*. Springer, Singapore. https://doi.org/10.1007/978-981-15-6121-4_22

Ahsan, F., Imran, M., Gilani, S.A., Bashir, S., Khan, A.A., Khalil, A.A., Hassan Shah, F., and Mughal, M. H. (2018). Effects of dietary soy and its constituents on human health: A review. *Biomedical Journal of Scientific & Technical Research* 12:2. doi:10.26717/BJSTR.2018.12.002239.

Anderson, J.W. (1997). Phytoestrogen effects in humans relative to risk for cardiovascular disease, breast cancer, osteoporosis, and menopausal symptoms. In: Pavlik, E.J. (eds) *Estrogens, Progestins, and Their Antagonists. Hormones in Health and Disease.* Birkhäuser Boston. https://doi.org/10.1007/978-1-4612-4096-9_2.

Aneta, P. and Dasha, M. (2019). Antinutrients in plant-based foods: A review. *The Open Biotechnology Journal*, 13:68–76. doi:10.2174/1874070701913010068.

Benson, T., and Shekar, M. (2006). Trends and issues in child undernutrition. In: Jamison, D.T., Feachem, R.G., Makgoba, M.W., editors. *Disease and Mortality in Sub-Saharan Africa*, 2nd edition. Washington (DC): The International Bank for Reconstruction and Development / The World Bank. Chapter 8.

Bibu, J.K. (2010). Phytoestrogens in Animal Origin Foods. *Veterinary World*, 3(1):43–45.

Bo, R.K., and Doo, H.P. (2012). Nutritional and functional evaluation of soy yogurt supplemented with Glasswort (*Salicorniaherbacea* L.). *BTAIJ*, 6(6):166–171.

Boué, S.M., Isakova, I.A., Burow, M.E., Cao, H., Bhatnagar, D., Sarver, J.G., Shinde, K.V., Erhardt, P.W., and Heiman, M.L. (2012). Glyceollins, Soy Isoflavone – Phytoalexins -, improves oral glucose disposal by stimulating glucose uptake. *Journal of Agricultural and Food Chemistry* 60:6376–6382. doi:10.1021/jf301057d.

Carmen, R.-G., Sánchez-Quesada, C., Toledo, E., Delgado-Rodríguez, M., and Gaforio, J.J. (2019). Naturally Lignan-rich foods: A dietary tool for health promotion? *Molecules* 24:917; 1–25. doi:10.3390/molecules24050917.

CGIAR (2005). CGIAR: Research and impact areas of research: soybean. www.cgiar.org/impact/research/soybean.html (accesed 14/07/2010).

Chatterjee, C., Gleddie, S., and Xiao, C.-W. (2018). Soybean bioactive peptides and their functional properties. *Nutrients* 10(9):1211. doi:10.3390/nu10091211.

Ciabotti, S., Silva, A.C.B.B., Juhasz, A.C.P., Mendonça, C.D., Tavano, O.L., Mandarino, J.M.G., and Gonçalves, C.A.A. (2016). Chemical composition, protein profile, and isoflavones content in soybean genotypes with different seed coat colors. *International Food Research Journal* 23(2):621–629.

Desmawati, D., and Sulastri, D. (2019). Phytoestrogens and their health effect. *Open Access Macedonian Journal of Medical Sciences* 7(3):495–499. doi:10.3889/oamjms.2019.086.

Dwivedi, N., Dwivedi S., and Adetunji, C.O. (2021). Efficacy of microorganisms in the removal of toxic materials from industrial effluents. In: Adetunji, C.O., Panpatte, D.G., Jhala Y.K. (eds) *Microbial Rejuvenation of Polluted Environment. Microorganisms for Sustainability*, vol 27. Springer, Singapore. https://doi.org/10.1007/978-981-15-7459-7_15

Fasiha, A., Muhammad, I., Syed, A.G., Shahid, B., Amar, A.K., Anees, A.K., Faiz-ul, H.S., and Muhammad, H.M. (2018). Effects of dietary soy and its constituents on human health: A review. *Biomed Journal Science & Tech Research* 12(2). BJSTR. MS.ID.002239. doi:10.26717/ BJSTR.2018.12.002239.

Finley, J.W., Snow, J.T., Johnston, P.H., and Friedman, M. (2006). Inhibition of lysinoalanine formation in food proteins. *Journal of Food Science* 43(2):619–621. doi:10.1111/j.1365-2621.1978.tb02368.x.

Fisher, E.A., Feig, J.E., Hewing, B., Hazen, S.L., and Smith, J.D. (2012). High-density lipoprotein function, dysfunction, and reverse cholesterol transport. *Arteriosclerosis, Thrombosis, and Vascular Biology* 32:2813–2820. [PMC free article] [PubMed].

Food and Agriculture Organization of the United Nations (FAO) (2016). Agriculture and Food Insecurity Risk Management In Africa. Concepts, lessons learned and review guidelines, pp. 1–77.

Forni, C., Facchiano, F., Bartoli, M., Pieretti, S., Facchiano, A., D'Arcangelo, D., Norelli, S., Valle, G., Nisini, R., Beninati, S., Tabolacci, V., and Jadeja, R.N. (2019). Beneficial role of phytochemicals on oxidative stress and age-related diseases. *BioMed Research International* 8748253. doi:10.1155/2019/8748253

Franke, A.A., and Custer, L.J. (1996). Daidzein and genistein concentrations in human milk after soy consumption. *Clinical Chemistry* 42(6): 55–964.

Hadad, L., and Martorell, R. (2002). Feeding the world in the coming decades requires improvements in investment, technology and institutions. *Journal of Nutrition* 132:3435S–3436S.

Haidong, L., Fang, Y., Bo, Y., Zheng, N.Z., Chang, G.T., Xue, H.L., ShuFang, W. (2018). Effect of dietary soy isoflavones on bone loss in ovariectomized rats. *Tropical Journal of Pharmaceutical Research* 17(1): 91–96. doi:10.4314/tjpr.v17i1.14.

Hossein, J. (2011). Soy products as healthy and functional foods. *Middle East Journal of Scientific Research* 7(1):71–80. Project: Fruit juice beverage production.

Islam, S., Thangadurai, D., Adetunji, C., Nwankwo, W., Kadiri, O., Makinde, S., Michael, O.S., Anani, O.A., and Adetunji, J.B. (2021). Nanomaterials and nanocoatings for alternative antimicrobial therapy. In: Kharissova, O.V., Martínez, L.M.T., and Kharisov B.I. (eds) *Handbook of Nanomaterials and Nanocomposites for Energy and Environmental Applications*. Springer, Cham. https://doi.org/10.1007/978-3-030-11155-7_3-1

Jehun, C., Sun-Hwa, K., Kun-Young, P., Byung, P. Y., Nam, D.K., Jee, H.J., Hae, Y.C. (2011). The anti-inflammatory action of fermented soybean products in kidney of high-fat-fed rats. *Journal of Medicinal Food* 14(3):232–239. doi: 10.1089/jmf.2010.1039.

Jyoti, S.S. Agrawal, Shikha Saxena, Archana Sharma. (2015). Phytoestrogen "genistein": Its extraction and isolation from soybean seeds. *International Journal of Pharmacognosy and Phytochemical Research* 7(6): 1121–1126.

Kang, J., Badger, T.M., Ronis, M.J., and Wu, X. (2010). Non-isoflavone phytochemicals in soy and their health effects. *Journal of Agricultural and Food Chemistry* 58(14): 8119–8133. doi:10.1021/jf100901b.

Khojely, D.M., Elrayah, S., Sapey, I.E., and Han, T. (2018). History, current status, and prospects of soybean production and research in sub-Saharan Africa. *The Crop Journal* 6(3):226–235.

Kolapo, A.L. (2011). Soybean: Africa's potential Cinderella food crop. In book: *Soybean - Biochemistry, Chemistry and Physiology*. InTech. doi:10.5772/15527.

Kumaran, T. and Gitarasu T. (2015). Isolation, characterization and antibacterial activity of crude and purified saponin extract from seeds of soyabean (Glycine Max). *International Journal of Pure and Applied Researches* 1:33–36.

Liang, H., Yu, F., Yuan, B., Zhao, Z.N., Tong, C.G., Liu, X.H., and Wu, S.F. (2018). Effect of dietary soy isoflavones on bone loss in ovariectomized rats. *Tropical Journal of Pharmaceutical Research January* 17 (1):91–96.

Lokuruka, M.N. (2011). Effects of processing on soybean nutrients and potential impact on consumer health: An overview. *African Journal of Food, Agriculture, Nutrition and Development* 11 (4) 5000–5017.

Lygin, A.V., Zernova, O.V., Hill, C.B., Kholina, N.A., Widholm, J.M., Hartman, G.L., and Lozovaya, V.V. (2013). Glyceollin is an important component of soybean plant defense against phytophthorasojae and macrophominaphaseolina. *Biochemistry and Cell Biology* 103(10):984–994.

Martina, H., Monika, R., and Jaroslav, Z. (2002). Enzymatic hydrolysis of defatted soy flour by three different proteases and their effect on the functional properties of resulting protein hydrolysates. *Czechoslavakia Journal of Food Science* 2(1):7–14.

Mazur, W.M., Duke, J.A., Wahala, K., Rasku, S., and Adlercreutz, H. (1998). Isoflavonoids and lignans in legumes: Nutritional and health aspects in humans. *Journal of. Nutritional Biochemistry* 9:193–200. Elsevier Science Inc.

Meghwal, M., and Sahu, C.K. (2015). Soy isoflavonoids as nutraceutical for human health: An update. *Journal of Cell Science and Therapy* 6:194. doi:10.4172/2157-7013.1000194.

Messina, M. (2016) Soy and health update: Evaluation of the clinical and epidemiologic literature. *Nutrients* 8 (754): 1–42. doi:10.3390/nu8120754.

Miroljub, B.B., Sladjana, S., and Mirjana, P. (2005). Biolgically active components of soybeans and soy protein products-A review. *Acta Periodica Technologica* 2005(36). DOAJ. doi: 10.2298/APT0536155B.

Mishra, P., Mishra, R. R., & Adetunji, C. O. (Eds.). (2020). *Innovations in Food Technology: Current Perspectives and Future Goals.* Springer, Singapore. https://doi.org/10.1007/978-981-15-6121-4.

Moses, T., Papadopoulou, K. K., and Osbourn, A. (2014). Metabolic and functional diversity of saponins, biosynthetic intermediates and semi-synthetic derivatives. *Critical Reviews in Biochemistry and Molecular Biology 49*(6): 439–462. doi:10.3109/10409238.2014.953628.

Mulligan, A.A., Kuhnle, G.G.C., Lentjes, M.A.H., van Scheltinga, V., Powell, N.A., McTaggart, A., Bhaniani, A., and Khaw, K.-T. (2012). Intakes and sources of isoflavones, lignans, enterolignans, coumestrol and soya-containing foods in the Norfolk arm of the European Prospective Investigation into Cancer and Nutrition (EPIC-Norfolk), from 7d food diaries, using a newly updated database. *Public Health Nutrition* 16(8): 1454–1462.

Nguyen, P.M. (2015). Alcalase hydrolysis of bioactive peptides from soybean. *Internatiional Journal of Pure and Applied Bioscience* 2:19–29.

Olaniyan, O.T, Kunle-Alabi, O.T., and Raji, Y. (2018). Protective effects of methanol extract of Plukenetia conophora seeds and 4H-Pyran-4-One 2,3-Dihydro-3,5-Dihydroxy-6-Methyl on the reproductive function of male Wistar rats treated with cadmium chloride. *JBRA Assisted Reproduction* 22(4). doi: 10.5935/1518-0557.20180048.

Olaniyan, O.T. and Adetunji, C.O. (2021). Biochemical role of beneficial microorganisms: an overview on recent development in environmental and agro science. In: Adetunji, C.O., Panpatte, D.G., Jhala, Y.K. (eds) *Microbial Rejuvenation of Polluted Environment. Microorganisms for Sustainability*, vol 27. Springer, Singapore. https://doi.org/10.1007/978-981-15-7459-7_2

Payton-Stewart, F., Schoene, N.W., Kim, Y.S., Burow, M.E., Cleveland, T.E., Boue, S.M., and Wang, T.T.Y. (2009). Molecular effects of soy phytoalexinglyceollins in human prostate cancer cells LNCaP. *MolCarcinog* 48(9):862–871. doi:10.1002/mc.20532.

Peeters, P.H.M, Keinan-Boker, L., van der Schouw, Y.T., and Grobbee, D.E. (2003). Phytoestrogens and breast cancer risk review of the epidemiological evidence. *Breast Cancer Research and Treatment* 77:171–183.

Peñalvo, J.L., Heinonen, S.M., Nurmi, T., Deyama, T., Nishibe, S., and Adlercreutz, H. (2004). Plant lignans in soy-based health supplements. *Journal of Agricultural and Food Chemistry* 52(13): 4133–4138. doi:10.1021/jf0497509.

Pham, T.H., Lecomte, S., Efstathiou, T., Ferriere F., and Pakdel, F. (2019). An update on the effects of glyceollins on human health: Possible anticancer effects and underlying mechanisms. *Nutrients* 11:79; 1–24. doi:10.3390/nu11010079.

Poulsen, M.W., Hedegaard, R.V., Andersen, J.M., de Courten, B., Bugel, S., Nielsen, J., Skibsted, L.H., Dragsted, L.O. (2013). Advanced glycation endproducts in food and their effects on health. *Food and Chemical Toxicology* 60:10–37.

Preeti, S., Kumar, R., Sabapathy, S.N., and Bawa, A.S. (2008). Functional and edible uses of soy protein products. *Comprehensive Reviews in Food Science and Food Safety* 7(1):14–28. doi:10.1111/j.1541-4337.2007.00025.x.

Rajalakshmy, I., Ramya, P., and Kavimani, S. (2011). Cardioprotective medicinal plants-A review. *International Journal of Pharmaceutical Invention* 1: 24–41.

Riyanto, P., and Lelyana, R. (2017). Soy Isoflavones reduce toll-like receptor-2 levels in acne vulgaris. *Journal of Nanomedicine and Nanotechnology* 8:430. doi:10..4172/21577439.1000430.

Rosegrant, M., and Meijer, S. (2002). Past accomplishments and alternative futures for child malnutrition to 2020. *J. Nutr.* 132:3437S–3440S.

Sacks, F.M., Lichtenstein, A., Van Horn, L., Harris, W., Kris-Etherton, P., and Winston, M., American Heart Association Nutrition Committee (2006). Soy protein, isoflavones, and cardiovascular health: An American Heart Association Science Advisory for professionals from the Nutrition Committee. *Circulation* 113(7):1034–1044. doi:10.1161/CIRCULATIONAHA.106.171052. Epub 2006 Jan 17. PMID: 16418439.

Seung-Hyun, L., Young, H.K., Heui-Jong, Y., Nam-Suk, C., Tae-Hyu, K., Dong-Chool, K., Chan-Bok, C., Yong-Il, H., Ki, H.K. (2007). Enhanced bioavailability of soy isoflavones by complexation with beta-cyclodextrin in rats. *Bioscience, Biotechnology, and Biochemistry* 71(12):2927–2933. doi:10.1271/bbb.70296.

Soetan, K and Oyewole, O. (2009). The need for adequate processing to reduce the anti-nutritional factors in plants used as human foods and animal feeds: A review. *African Journal Food Science* 3(9): 223–32. doi:10.3109/08039489709090718.

Song, T. (1998). Soy isoflavones: Database development, estrogenic activity of glycitein and hypocholesterolemic effect of daidzein. Retrospective Theses and Dissertations. 12528. https://lib.dr.iastate.edu/rtd/12528.

Stephen, B. (2010). The biochemistry, chemistry and physiology of the isoflavones in soybeans and their food products. *Lymphatic Research and Biology* 8(1):89–98. doi:10.1089=lrb.2009.0030.

Thompson, L.U., Boucher, B.A., Cotterchio, M., Kreiger, N., and Z. Liu (2007). Dietary phytoestrogens, including isoflavones, lignans, and coumestrol, in nonvitamin, nonmineral supplements commonly consumed by women in Canada. *Nutrition and Cancer* 59(176):184. doi:10.1080/01635580701420616.

Tongtong, S., Sun-Ok, L., Patricia, A.M., and Suzanne, H. (2003). Soy protein with or without isoflavones, soy germ and soy germ extract, and daidzein lessen plasma cholesterol levels in golden Syrian hamsters. *Experimental Biology and Medicine* 228(9): 1063–1068. doi:10.1177/153537020322800912.

Thangadurai, D., Dabire, S.S., Sangeetha, J., Said Al-Tawaha, A.R.M., Adetunji, C.O., Islam, S., Shettar, A.K., David, M., Hospet, R., Adetunji, J.B. (2021). Greener composites from plant fibers: Preparation, structure, and properties. In: Kharissova O.V., Martínez L.M.T., Kharisov B.I. (eds) *Handbook of Nanomaterials and Nanocomposites for Energy and Environmental Applications*. Springer, Cham. https://doi.org/10.1007/978-3-030-11155-7_21-1

United Nations (2015). Department of Economic and Social Affairs, Population Division. Population 2030: Demographic challenges and opportunities for sustainable development planning (ST/ESA/SER.A/389), 1–54.

United Nations Children's Fund (UNICEF) (2015). World Health Organization, International Bank for Reconstruction and Development/The World Bank. Levels and trends in child malnutrition: key findings of the 2019 Edition of the Joint Child Malnutrition Estimates. Geneva: World Health Organization; 2019 Licence: CC BY-NC-SA 3.0 IGO.

Veličković, D., Vucelić–Radović, B., Simić, D., Barać, M., and Ristić, N. (1994). Characterisation of the change of soybean flour protein composition during thermal inactivation of trypsin inhibitor. *Rev. of Res. Work Fac. Agr. Belgrade* 39:41–48.

Waqas, M.K., Akhtar, N., Mustafa, R., Jamshaid, M., Shoaib Khan, H.M., and Murtaza, G. (2015). Dermatological and cosmeceutical Benefits of *Glycine Max* (Soybean) and its active components. *Acta Poloniae Pharmaceutica - Drug Research* 72 (1):3–11.

Wolf, W.J. (1974). Soybean Proteins: Their production, properties, and food uses. A selected bibliography. *Journal of the American Oil Chemist Society* 51(I):63A–66A.

Wood, C.E., Clarkson, T.B., Appt, S.E., Franke, A.A., Boue, S.M., Burow, M.E., McCoy, T., and Cline, J.M. (2006). Effects of soybean glyceollins and estradiol in postmenopausal female monkeys. *Nutrition and Cancer* 56(1):74–81.

Yang, X., Li, Y., Li, Y., Ren, X., Zhang, X., Hu, D., Gao, Y., Xing, Y., and Shang, H. (2017). Oxidative stress-mediated atherosclerosis: Mechanisms and therapies. *Frontiers in Physiology*, 8: 600. doi:10.3389/fphys.2017.00600.

Zachary, S., William, M.S., and Bo, Z. (2020). Soybean production, versatility, and improvement, legume crops - prospects, production and uses, Mirza Hasanuzzaman. *IntechOpen*. doi: 10.5772/intechopen.91778. Available from: https://www.intechopen.com/chapters/71498.

Zimmermann, M. C, Tilghman, S.L., Boué, S.M., Salvo, V.A., Elliott, S., Williams, K.Y., Skripnikova, E.V., Ashe, H., Payton-Stewart, F., Vanhoy-Rhodes, L., Fonseca, J.P., Corbitt, C., Collins-Burow, B.M., Howell, M.H., Lacey, M., Shih, B.Y., Carter-Wientjes, C., Cleveland, T.E., McLachlan, J.A., Wiese, T.E., Beckman, B.S., and Burow, M.E. (2010). Glyceollin I, a novel antiestrogenic phytoalexin isolated from activated soy. *The Journal of Pharmacology and Experimental Therapeutics* 332(1):35–45.

# 6 Nutritional and Health Benefits of Nutraceutical Beverages Derived from Cocoa and Other Caffeine Products

## *A Comprehensive Review*

*Juliana Bunmi Adetunji*
Osun State University

*Charles Oluwaseun Adetunji, Olugbemi Tope Olaniyan, and Florence U. Masajuwa*
Edo State University Uzairue

*Saher Islam*
University of Veterinary and Animal Sciences

*Devarajan Thangadurai*
Karnatak University

*Olusola Olawale Olaleye*
Nigerian Institute for Trypanosomiasis Research

*Daniel Ingo Hefft*
University Centre Reaseheath

DOI: 10.1201/9781003178378-6

*Wadazani Palnam Dauda*
Federal University Gashua

*Benjamin Ewa Ubi*
Ebonyi State University
Tottori University

## CONTENTS



## 6.1 INTRODUCTION

Caffeine, whose scientific nomenclature is 1,3,7-trimethylxanthine, has been recognized as an important biologically active substance present in most beverages, nutraceutical foods, dietary supplements, and drugs. This pharmacoactive constituent has been recognized in the treatment of newborns suffering from apnea, i.e. short-term stoppage of breathing. Caffeine could be found in fruits, leaves, seeds where it could serve as an attractant for pollution, herbicidal agents, and an insect repellant (Wright et al., 2013, Adetunji and Anani, 2021a,b,c; Adetunji et al., 2021; Dwivedi et al., 2021; Olaniyan and Adetunji, 2021, Mishra et al., 2020; Thangadurai et al., 2021; Anani et al., 2020; Islam et al., 2021). This biologically active constituent has been observed as one of the most highly ingested stimulants globally (Salinardi et al., 2010; Doré et al. 2011). Caffeine goes into the human food chain majorly through plants such as coffee and cocoa beans, kola nuts, tea leaves, and guarana (Sanchez, 2017; Ananya, 2019). A beverage is a type of drink that could be grouped into alcoholic and non-alcoholic. Non-alcoholic beverages include fruit juices, hot beverages, and soft drinks, while the alcoholic beverages contain alcohol in different concentrations.

Caffeine has been identified as one out of the numerous constituents with physiological effects on humans when consumed. It has been proven through history and scientifically that ingestion of 400 mg/day of caffeine will not demonstrate any adverse effect on an adult human being (Heckman et al., 2010). Consumption of coffee and cocoa, as well as caffeine-containing beverages, has been documented to exhibit some pharmacological benefits which include enhancements in mental readiness, fatigue, concentration, and athletic performance (Nawrot et al., 2003). Some other benefits include enhanced glucose tolerance, reduction in the emergence of Parkinson's illness, prevention of cancer, eradication of type II diabetes, and improved glucose tolerance (Floegel et al., 2012).

Moreover, whenever 1,3,7-trimethylxanthine-containing substances or beverages are consumed in higher quantities, there is a tendency for restlessness, headaches, anxiety, and nausea. Also, it has been observed that some side effects like drowsiness, headache, and fatigue might be observed when there is sudden stoppage of caffeine and the symptom might be temporary and mild (Nawrot et al., 2003; Mielgo-Ayuso et al., 2019). Some other authors have also highlighted the capability of caffeine-containing substances to enhance the incidence of cardiovascular disease and hypertension (Mesas et al., 2011). Furthermore, oral intake of about 400 mg/day dose by healthy adults will have no detrimental effect on the cardiovascular health status of adult human being, but the ingestion of concentration below 300 mg/day might have antagonistic influence on pregnancy or the well-being of the reproductive system (Kuczkowski, 2009; Brent et al., 2011).

This review therefore provides more insight into the analytical techniques used for the determination of caffeine-containing products, beneficial and detrimental effect of caffeine, caffeine and maternal health during pregnancy and the antimicrobial action of caffeine.

## 6.2 ANALYTICAL TECHNIQUES FOR THE DETERMINATION OF CAFFEINE

Rudolph et al. (2012) assessed the level of caffeine present in different food item available in Australian market with the aid of caffeine assessment tool. The authors assessed the level of caffeine present in 124 different products containing energy drinks, yogurt, coffee, chocolate, colas, and coffee-based beverages using ultraviolet (UV) detection reversed phase high-performance liquid chromatography (RP-HPLC) following solid-phase extraction. The greatest concentration of caffeine was discovered from coffee prepared from 755 mg/L pads and 659 mg/L steady filtered coffee. The overall amount of caffeine available in the chocolate-based beverages and coffee ranges from 15 mg for a liter of chocolate milk and 448 mg for a liter of canned ice coffee, while from 266 to 340 mg/L for energy drinks, normal, and ice teas ranges between 13 and 183 mg/L, and the values from 33 to 48 mg/kg were observed for coffee-flavored yogurts. Moreover, the amounts of caffeine concentration present in the chocolate bar and chocolate are between 17 and 551 mg/kg in whole milk chocolate and chocolate with coffee filling, respectively. A caffeine examination tool was established and authenticated using a 3-day dietary record containing ($r^2 = 0.817$, $p < 0.01$) with the aid of these analytical data, while the level of caffeine saliva concentrations was equally observed with this result ($r^2 = 0.427$, $p < 0.01$).

Müller et al. (2014) evaluated the availability of bioactive components nicotine, myosmine, and caffeine present in milk/dark chocolate using headspace solid-phase micro-extraction alongside gas chromatography-tandem mass spectrometry (GC-MS). The restrictions of detection evaluated from linearity data were 0.000120 mg/kg for nicotine, 216 mg per kg for caffeine, and 0.000110 mg per kg for myosmine. Also, 30 samples from five different brands with different cocoa content of 30%–99% were evaluated in triplicate. The results obtained showed that nicotine and caffeine were present in all the chocolate samples, while myosmine was not detected in any sample. The value of nicotine ranged from 0.000230 to 0.001590 mg/kg, while caffeine content varied from 420 to 2,780 mg/kg.

Olmos et al. (2009) assessed the level of caffeine contained in all beverages in Argentina's market. The amount of caffeine in beverages like chocolate milk, mates, coffees, energy, and soft drinks was evaluated with HPLC with UV detection. The study was carried out among 471 people who came across different age group ranging from 2 to 93 years of age that consumed various levels of caffeine. The result of the evaluation showed that maximum level of caffeine was detected from short coffee containing (1.38 mg/mL), while the maximum amount detectable in a particular serving was found to contain 95 mg of caffeine per serving. The average caffeine consumption within adults was 288 mg daily, while mate was the highest caffeine contributor. The average caffeine consumption was within ages 10 in children receiving 35 mg daily, while soft drinks were discovered to be the foremost contributor to that consumption. Also, neonates whose ages fell between 11 and 15 years had an average of 120–240 mg daily caffeine consumption, respectively; while mate still formed the significant sources of caffeine from all the samples taken during consumption. On the whole, it was observed that drinking mate is an entrenched habit among Argentinians, and it could be a factor that might be responsible for the elevated daily caffeine consumption.

## 6.3 BENEFICIAL AND DETRIMENTAL EFFECT OF CAFFEINE

Franco et al. (2013) reviewed the influence of methylxanthines available in different beverages like chocolate and cocoa frequently consumed by humans. The authors highlighted various health benefits most especially the physiological effects associated with methylxanthines which have been in existence for close to 30 decades, and they are predominantly interceded by adenosine receptors. The authors observed that theobromine and caffeine were the most active components available in cocoa and their physiological effect has been observed for a long time. Their health-promoting benefits are specially highlighted, making the consumption of certain types of chocolate as nutraceutical and functional foods attractive. Moreover, the significance of adenosine receptor blockade available in the natural compounds like chocolate as well as cacao was also reviewed in detail. In addition, the health benefits and palatability advantages of methylxanthines and theobromine are well known (Baggott et al., 2013; Martínez-Pinilla et al., 2015).

The list of benefits is manifold, but some are listed below:

- decrease in the risk of cancer
- lowering of blood pressure
- enhancements in brain function
- decreases in tooth decay
- decreased gouts
- enhanced airflow in lungs
- prevention of kidney stone formation
- enhanced weight loss
- better sleep facilitation
- improved cholesterol level
- stimulated heart activities
- suppressed LDL oxidative susceptibility
- enhanced HDL-cholesterol concentrations

Moreover, the health benefits of methylxanthines have been reported for their therapeutic effect in the inhibition and alleviation of neurodegenerative diseases,

Alzheimer's disease, and Parkinson's disease, as well as their application in the fabrication of novel and more effective drugs (Oñatibia-Astibia et al., 2017).

Maughan and Griffin (2003) discussed comprehensively on methylxanthine compounds and caffeine present in some beverages. The authors reported that critical consumption of caffeine in large concentration ranging from 250 to 300 mg about 5–8 tea cups or 2–3 coffee cups could stimulate the level of urine output in people.

Wierzejska (2012) presented a compressive review about the significance of caffeine present in the caffeine-containing products such as food supplement and energy drinks. Caffeine has been highlighted as a major source of psychoactive substance globally. The combinatory effect of the bio-constituents of caffeine and dearth of nutrition value could have significant detrimental effect on human health most particularly the menace of cardiovascular disorders, though the influence of caffeine might depend on individual rates of metabolism which also depend on the environmental and endogenic factors. The author also observed that the safe dose (400 mg a day) of caffeine for human consumption by healthy individuals might not be related with any antagonistic influence but also depend on the lifestyle of the concerned individual. However, excessive consumption of caffeine might cause negative consequences including gastrointestinal complaints, insomnia, psychomotor agitation, and headache.

The World Health Organization's via International Classification of Diseases (ICD-10) classified the negative effect of caffeine to be ICD-10. Also, the absorption of caffeine most especially by maternal mothers has reduced because there is a tendency that caffeine and its metabolite could cross the placenta into the uterus, thereby affecting the developing fetus. Therefore, it is advisable that pregnant women, adolescents, and children should minimize their daily caffeine consumption. It has also been observed that caffeine could affect the neurons starting from speedy development of the brain to altering sleep duration, and calcium balance. The mean caffeine consumption in Europe has been observed to range from 280 to 490 mg per day, while Scandinavia has the maximum caffeine consumption, and it is believed that they consume a lot of coffee. The authors claimed in their study that caffeine consumption available in coffee and tea beverages differs.

## 6.4 CAFFEINE AND MATERNAL HEALTH DURING PREGNANCY

Bakker et al. (2010) commented on the impact of caffeine ingestion from tea and coffee on pregnancy. The community-based cohort study involved 7,346 pregnant women in the Netherlands and was carried out within the years 2001 and 2005. Their result was focused on caffeine consumption during the first, second, and third trimesters via the use of a questionnaire. Fetal development was also examined using ultrasound, while the detailed data regarding fetal delivery was collected from the hospital. The result showed no relationship between caffeine consumption, weight of fetus, and circumference of fetus head most especially during pregnancy. Moreover, it was observed that the consumption of caffeine was related to smaller crown-rump length from the first to second trimesters and femur or birth length of third trimester. Moreover, the maternal who consumed > or =6 units/d of caffeine had a tendency to enhance the danger of small-for-gestational-age infants at delivery. Their result showed that the consumption of caffeine with > or =6 units/d after conception is related to a compromised length of the fetus.

The teratogenic potential of caffeine was evaluated in animals by Souza and colleagues in 2016, by exposing the animals to a moderate dose (0.1 g/L) and elevated dose (0.3 g/L) of caffeine from the onset of gestation to the end of gestation. Though it was observed that there was a stillbirth in just one of the animals exposed to the moderate dose, malformation was not observed. However, malformations were noticed in the group exposed to high dose which were linked to fetus growth and cardiovascular alterations like labial malformations, bruises, poor placental formation, macrocephaly, having short limbs, hydrops fetalis, and abnormal development/growth (or absence) of head structures and limbs (Souza et al., 2016). It was also suggested that the mechanism of action employed by caffeine to cause the alterations could be catecholamines increase in pregnancy and down-regulation of adenosine A1 receptors. It is therefore suggested that exposure to high caffeine intake in pregnancy should be discouraged.

Also, Collier et al. (2009) observed that high doses of caffeine have been linked with limb malformations and craniofacial which is then attributed to a decrease in branchial bars and somites counts in connection with forelimb formation impairment during morphogenesis.

Nehig and Debry (1994b) reported that different animals demonstrated different levels of sensitivity when treated with caffeine. The authors discovered that mice administered with 50–75 mg/kg of caffeine had malformations, though the minimal needed dose that can induce malformation is 80 mg/kg/rats. Moreover, it has been observed that administration of 330 mg/kg per day of caffeine to rats could induce teratogenicity in the treated rats. Some of the observed signs of malformation include craniofacial malformations, deformation of the digits and limbs, deferments in ossification of sternum, limbs, jaw, and ectrodactyly. Also, congenital malformation is a major symptom that could be observed when caffeine is administered to rats, most especially when administered a higher dose. It was also reported that the consumption of caffeine might enhance the teratogenic influence of substances like alcohol and tobacco and could react in a combinatory effect with propranolol and ergotamine that could result into materno-fetal vasoconstrictions which might later lead to malformations prompted by ischemia. It was also discovered that some complications might be possible in the human fetus when caffeine is taken with anti-migraine medications, tobacco, and alcohol. The author validated that maternal ingestion of caffeine potentiates the capability to distress the neurons' composition most especially in low-protein diet (LPD) and has the capability to react with zinc fixation in the neurons. Furthermore, exposure of maternal to caffeine could increase anxiety in rat offspring and learning abilities and caused long-term consequences on sleep and movement. However, there is still a need for more clinical trials to validate caffeine effect on human behavior most especially when consumed by pregnant mothers.

In 2003, Nawrot and his research team evaluated the consequence of caffeine consumption in popular beverages like tea, soft drink, and coffee on human health. It was revealed that adequate consumption of 400 mg caffeine-containing drinks daily might not induce any negative effect such as cardiovascular effects, alterations in behavior, influence on calcium balance and bone position, effects on male fertility, enhanced occurrence of cancer, and general toxicity. However, the data obtained indicates that children and women of reproductive age need to take adequate precaution when consuming caffeine. However, it was advised that the maximum level of caffeine that children should ingest should be </=2.5 mg/kg bwt/day and reproductive age women should ingest </=300 mg/day which might be comparable to 4.6 mg/kg bwt/day for 65 kg adults.

Furthermore, Nehlig and Debry (1994b) wrote a detailed review on coffee intake and its influence on lactation, fertility, growth, and reproduction. They documented that caffeine present in the coffee could result in dysfunctions in rats' gestational period which might have a detrimental effect on the birth weight. Interestingly, it depends on the concentration of the caffeine administered most especially when given in a high dose containing 7 cups/day, while there was no adverse effect when given at a moderate level. Furthermore, it was suggested that the hematological parameters of the progeny or rat could be affected. Moreover, the caffeine present in coffee might affect some characteristic behaviors in adult rats such as locomotion, emotivity, long-term consequences on sleep, anxiety, and learning capability, though no report established its effect on new-born rats. This also supports the findings that caffeine or coffee consumption at adequate quantity during the period of lactation and gestation period does not have any adverse effect on maternal health. It was later suggested that the minimal level of caffeine that pregnant mothers should consume should be 300 mg caffeine/day.

Jensen and colleagues in 1998 assessed the effect of caffeine intake and level of fecundity within 430 Danish couples in their first pregnancy. This was carried out by evaluating the combined and independent effect of caffeine obtained from smoking and other various sources, on conception probability. The various data was obtained by filling a questionnaire which was based on different factors which included the consumption of cola beverages, tea, coffee, chocolate, and chocolate bars. The cycle-specific relationship between fecundity and the consumption of caffeine was assessed using a logistic regression model for each cycle irrespective of existence of pregnancy in a Cox discrete model evaluating the fecundity odds-ratio. The results showed that women that do not smoke but ingest 300–700 mg/day caffeine contain a fecundity odds-ratio of 0.88, while women consuming high amount of caffeine had a fecundity odds-ratio of 0.63 when compared to women that do not smoke with less than 300 mg/day caffeine after adjusting the body mass index of the female and alcohol consumption, duration of menstrual cycle, quality of the semen, and female reproductive organs dysfunctions. Moreover, there was no correlation with dose-response discovered among smokers. On the whole, Jensen et al. (1998) suggested that there exists a relationship between smoking and consumption of caffeine biologically, while its dearth of influence within the smokers might be linked to its high rate of metabolism in the human body.

Hatch et al. (2018) assessed the effect of sugar-sweetened beverages consumption in female and male together with fecundity within 3,828 women in preparation for pregnancy with other males in a cohort study in North America which was 1,045 in number. The confidence intervals and fecundity ratios of the sugar-sweetened beverages were performed with the aid of proportional probabilities regression. The result obtained revealed that ingestion of sugar-sweetened beverages could be associated with decreased fecundity, while the level of fecundity was further decreased among the people that took ≥7 servings/week of sugar-sweetened sodas. Their study showed that sugar-sweetened beverages most especially from sodas and energy drinks could be related to lower fecundity when compared to fruit juice and diet soda that show a little relationship.

Machtinger et al. (2017) determined the consequence of maternal beverage consumption and the result *of* in vitro fertilization outcomes. This was carried out among 340 women performing in vitro fertilization from 2014 to 2016 for infertility alongside pregenetic analysis for autosomal recessive disorders were registered in the process of ovarian stimulation and a questionnaire relating their typical tea intake. The in vitro fertilization outcomes were sourced from clinical records, while the sum of caffeine ingestion was assessed via evaluating the total caffeine value available for each major beverage when multiplied by the rate of consumption. The relationship between in vitro fertilization outcomes and the exact types of beverages was examined via logistic and Poisson regression models which allow the adjustment for probable confounders. The result obtained show that the consumption of sugared soda showed a relationship between developed and fertilized oocytes with top-quality embryo following ovarian stimulation. Also, the following observation was documented from the female who consumes sugared soda, 0.6 fewer top-quality fetus associated with women without ingesting sugared soda, the matured oocytes recuperated, and some oocytes were reclaimed when compared to women who ingest soda sugar. Additionally, likened to women deprived of sugared soda ingestion, the attuned percent modification in cycles resultant in a live birth for women ingesting 0.1 to a cup daily and >1 cup daily was −12% and −16%. However, there was no observed relationship discovered among the people that consumed caffeine, diet sodas, and coffee with in vitro fertilization outcomes. Their study shows that ingestion of sugar-containing tea independent of the amount of caffeine content might have an adverse effect on the level of caffeinated beverages and caffeine in the absence of sugar.

Hatch and Bracken (1993) assessed the relationship existing in caffeine ingestion and delayed conception resulting from intake of caffeine-containing tea in a community. It was carried out among married women in New Haven, Connecticut, during the period of 1980–1982. The result obtained using the logistic regression evaluation shows that consuming caffeine from tea, soft drinks, and coffee was related to an enhanced potential to extend the conception of 1 year or more. Moreover, it was observed that the women that consume the caffeine during pregnancy at the rate of 1–150 mg/day led to 1.39 delay in conception, ingesting of 151–300 mg daily of caffeine had ratio of 1.88, and 300 mg daily had a ratio of 2.24 after monitoring for the last technique of controlling conception, parity, and amount of cigarettes $day^{-1}$ when compared to no caffeine use. In each cycle studied using disconnected analog of proportional Cox hazards model, pregnant mothers who recounted drinking above 300 mg daily of caffeine had a 27% lesser probability of pregnancy for individual cycles, and women who consume lower than 300 mg daily had about 10% decrease in individual cycle pregnancy rates in comparison with non-caffeine ingested women. The danger of tea, colas, and coffee was evaluated concurrently in logistic models, and it was discovered that it had no effect in enhancing the fit of the model which enclosed a variable for overall caffeine consumption from diverse sources in the investigation.

Collier et al. (2009) evaluated the outcome of caffeine consumption through sodas, tea, chocolate, coffee, and medication containing caffeine during pregnancy. The evaluation was based on the National Birth Defects Prevention assessment carried out among a community with critical birth defect like chromosomal abnormalities and single gene disorder associated with the utilization of caffeine. Other disorders include progeny possessing or not having cleft palate but with the exception of infants having secondary to holoprosencephaly or amniotic band sequence. The evaluation was carried out for the period of 1997–2004. The result obtained shows that among 1,531 infants with cleft lip with or without cleft palate, 5,711 infants showed no major birth defects which serve as a control, 813 infants with cleft palate alone. The evaluation carried out among control mothers showed that 11 % ingest less than 10 mg/day caffeine, while intake of 8 % coffee was related as a high consumption (>or=3 servings per day), 4 % tea, and 15 % sodas, and 1 % among the despondent was reported to consume at least 100 mg dose containing caffeine. It was also observed that cleft palate was induced by ingesting minimum of 100 mg dose containing caffeine, while moderate caffeine consumption from all the tested beverages could be linked to the null in high caffeine amount. Their study shows that no relationship existed between orofacial clefts and intake of caffeine by mothers.

D'ius (1997) documented a detailed report on the caffeine content of common beverages like chocolates, tea, soft drinks, and coffee consumed by most people. Also, 3 mg/kg dose does not have any adverse effect on physiological and behavioral functions. It was also noted that new progeny also has the capability to metabolize caffeine gradually most especially in children whose age fell below 1 year, while adolescence has the capability to metabolize caffeine about twice as fast as non-smoking adults. Their study shows that ingesting of caffeine does not have any adverse effect on human well-being most especially when consumed from food and beverages.

Eskenazi et al. (1999) carried out a population-based research involving 7,855 live births in California's San Joaquin Valley. They evaluated maternal caffeinated and decaffeinated coffee consumption during the gestational period and during pregnancy development. This was carried out by taking a questionnaire from a mother in the hospital whenever they are filling a birth certificate. The result obtained was compared with maternal caffeinated and decaffeinated coffee, while some who ingest only non-caffeine-containing coffee exhibited no improved odds of small-for-gestational-age, birth, preterm delivery or decreased birth weight, reduced average birth weight, and reduced average gestational age. Also, it was detected that women who ingested only caffeinated coffee had 1.3 odds ratio, while women who consume decaffeinated and caffeinated coffee had odds of 2.3.

Karlson et al. (2003) evaluated the hazards related to the ingestion of coffee or decaffeinated coffee to the discovery of rheumatoid arthritis. The authors assessed decaffeinated coffee, total coffee, tea, and coffee. The data was collected with questionnaire based on the level of beverages consumption from 1980 to 1998. Similarly, 83,124 women filled the food frequency questionnaire at baseline, the analysis of the occurrence of rheumatoid arthritis during a period of between 1980 and 2000. This was completed by 480 women by collating medical records and connective tissue disease screening questionnaire following the American College of Rheumatology standards.

The relationship that existed between the level of beverages consumption and the hazards associated with the rheumatoid arthritis were evaluated using multivariate Cox proportional hazards models and age-adjusted models. Consequently, multivariate evaluations were performed using only information contained on beverages on assessments of previous reports. The result obtained revealed no significant correlation between the ingested caffeinated coffee containing >/=4 cups/day and incident of rheumatoid arthritis risk when tested using multivariate model or adjusted the multivariate model. Also, there is a correlation between rheumatoid arthritis risk and cumulative caffeinated coffee consumption. Furthermore, the total caffeine and total coffee ingestion do not have any relationship with any risk that could be linked to rheumatoid arthritis.

McCusker et al. (2006) estimated the level of caffeine content available in 19 carbonated sodas, 10 energy drinks, and 19 carbonated sodas. Moreover, they assessed the level of inconsistency in the amount of Coca-Cola fountain soda caffeine content. The level of caffeine present in all assessed samples using gas chromatography coupled with nitrogen-phosphorus detection and liquid-liquid extraction showed the level of caffeine present in the listed samples: caffeinated drinks (none detected −141.1 mg/serving), carbonated sodas (none detected −48.2 mg/serving), and beverages (< 2.7–105.7 mg/serving). They also discovered that the % coefficient of variation, intra-assay mean, and standard deviation available in the Coca-Cola were 6.64, 44.5, and 2.95 mg/serving, respectively.

## 6.5 ANTIMICROBIAL ACTIVITIES OF COCOA AND OTHER CAFFEINE PRODUCTS

Nurul et al. (2014) evaluated the antibacterial efficacy of instant coffee (NESCAFÉ) beverages against *Streptococcus species* causing pharyngitis. The in vitro experiment was performed by means of well diffusion, disc diffusion techniques, as well as Coffee Agar plate sensitivity testing. The level of antibacterial was detected using zone of inhibition after incubating the plates at 24 and 48 hours. The data obtained showed that 8 g/100 mL exhibited the greatest antimicrobial activity against all the four *Streptococcus* species with the exception of non-decaffeinated coffee exhibiting the greatest activity. The authors suggested that the highest concentration as well as the highest activity does not have any compatibility with human health. Furthermore, it was suggested that there is a need to critically determine the molecular processes of the antibacterial action through which coffee beverages exhibit their antimicrobial activity against pharyngitis-causing *Streptococcus* species used during this study.

Melsayed et al. (2018) assessed the consequence of the inhibitory influence of Arabica bean green coffee extract against *P. aeruginosa* strain ATCC 27853. The data obtained showed that Arabic bean green coffee extract exhibited the highest antimicrobial efficacy against the targeted strain with minimum bactericidal concentration and 0.25 and 0.2 mg/mL minimum inhibitory concentration (MIC), respectively. The data revealed that aqueous extract obtained from Arabic bean green coffee could be used in the management of drug-resistant pathogens, *P. aeruginosa* strain ATCC 27853, which has been highlighted as one of the major microorganisms observed to develop resistance against synthetic drugs and constituting a major health concern to the general public.

Nonthakaew et al. (2015) documented a detailed review on the antimicrobial influence of caffeine in foods. The authors discovered that concentration that ranged from 62.5 to >2,000 μg per ml could exhibit an inhibitory activity on bacteria, while concentration containing >5,000 μg per ml could suppress mold growth. The review highlighted the main sources of caffeine as coffee and tea which has been observed to be safe and could be utilized as an antimicrobial agent against pathogen causing microorganisms.

Lele et al. (2016) assessed the effect of caffeine against some strains of Methicillin-resistant *Staphylococcus aureus* (MRSA 1–4) which was compared with a standard strain MRSA. The in vitro assay was performed using a broth dilution method. The results observed showed that caffeine exhibited an inhibitory effect at 5 mg per ml concentration. Moreover, the sub-inhibitory concentrations utilized during this study were in different dilutions (1:2, 1:5 and 1:10) of the MIC of 2.5, 1, and 0.5 mg/mL, respectively. Also, the results obtained showed that 2.5 mg/mL exhibited the highest reduction in the biofilm formation (14 %–21 %), proteolytic activity (29 %–40 %), and hemolytic activity (9 %–19 %).

## 6.6 CONCLUSION AND FUTURE PERSPECTIVES

This chapter has provided insights into the composition, consumption, health benefits, in vivo and in vitro assay of caffeine, the antimicrobial action of caffeine, and the analytical techniques used for the analysis of caffeine-containing products. There is a need to sensitize the general public about the biological activity and health benefits of cocoa and caffeine in order to correct the erroneous belief against all caffeine-containing products. The application of metabolomics will also give a better insight into the array of biologically active metabolites that have numerous nutraceutical and health properties. There is a need to validate the safety and toxicological facts on the consumption of caffeine-containing products using in vivo and in vitro techniques. On the whole, this chapter has justified the significance of caffeine as an antimicrobial agent in foods and its numerous health benefits, while advocating to priority attention to its safe use.

## REFERENCES

Adetunji, C.O. and Anani O.A. (2021a). Recent advances in the application of genetically engineered microorganisms for microbial rejuvenation of contaminated environment. In: Adetunji, C.O., Panpatte, D.G., Jhala, Y.K. (eds) *Microbial Rejuvenation of Polluted Environment. Microorganisms for Sustainability*, vol 27. Springer, Singapore. https://doi.org/10.1007/978-981-15-7459-7_14

Adetunji, C.O., Anani, O.A., and Panpatte, D. (2021). Mechanism of actions involved in sustainable ecorestoration of petroleum hydrocarbons polluted soil by the beneficial microorganism. In: Panpatte, D.G., Jhala, Y.K. (eds) *Microbial Rejuvenation of Polluted Environment. Microorganisms for Sustainability*, vol 26. Springer, Singapore. https://doi.org/10.1007/978-981-15-7455-9_8

Adetunji, C.O. and Anani O.A. (2021b). Utilization of microbial biofilm for the biotransformation and bioremediation of heavily polluted environment. In: Panpatte, D.G., Jhala, Y.K. (eds) *Microbial Rejuvenation of Polluted Environment. Microorganisms for Sustainability*, vol 25. Springer, Singapore. https://doi.org/10.1007/978-981-15-7447-4_9

Ananya, M. (2019). Caffeine Occurrence. News-Medical. Retrieved on December 02, 2020 from https://www.news-medical.net/health/Caffeine-Occurrence.aspx.

Anani, O.A., Mishra, R.R., Mishra, P., Enuneku, A.A., Anani, G.A. and Adetunji, C.O. (2020). Effects of toxicant from pesticides on food security: Current developments. In: Mishra, P., Mishra, R.R., Adetunji, C.O. (eds) *Innovations in Food Technology.* Springer, Singapore. https://doi.org/10.1007/978-981-15-6121-4_22

Anani, O.A. and Adetunji, C.O. (2021c). Bioremediation of polythene and plastics using beneficial microorganisms. In: Adetunji, C.O., Panpatte, D.G., Jhala, Y.K. (eds) *Microbial Rejuvenation of Polluted Environment. Microorganisms for Sustainability,* vol 27. Springer, Singapore. https://doi.org/10.1007/978-981-15-7459-7_13

Bakker, R., Steegers, E. A., Obradov, A., Raat, H., Hofman, A., and Jaddoe, V. W. (2010). Maternal caffeine intake from coffee and tea, fetal growth, and the risks of adverse birth outcomes: The Generation R Study. *American Journal of Clinical Nutrition* 91(6):1691–1698.

Baggott, M. J., Childs, E., Hart, A. B., De Bruin, E., Palmer, A. A., Wilkinson, J. E., and de Wit, H. (2013). Psychopharmacology of theobromine in healthy volunteers. *Psychopharmacology* 228(1):109–118.

Brent, R. L., Christian, M. S. and Diener, R. M. (2011). Evaluation of the reproductive and developmental risks of caffeine. *Birth Defects Research. Part B, Developmental and Reproductive Toxicology* 92:152–187.

Collier, S. A., Browne, M. L., Rasmussen, S. A. and Honein, M. A. (2009). Maternal caffeine intake during pregnancy and orofacial clefts. *Birth Defects Research Part A Clinical and Molecular Teratology* 85(10):842–849. doi:10.1002/bdra.20600.

D'ius, P. B. (1997). Caffeine and children. *VoprPitan* 1:39–41.

Doré, A. S., Robertson, N., Errey, J. C., Ng, I., Hollenstein, K., Tehan, B., Hurrell, E., Bennett, K., Congreve, M., Magnani, F., Tate, C. G., Weir, M. and Marshall, F. H. (2011). Structure of the adenosine A(2A) receptor in complex with ZM241385 and the xanthines XAC and caffeine. *Structure* 19(9):1283–1293. doi:10.1016/j.str.2011.06.014.

Dwivedi, N., Dwivedi, S. and Adetunji, C.O. (2021). Efficacy of microorganisms in the removal of toxic materials from industrial effluents. In: Adetunji, C.O., Panpatte, D.G., Jhala, Y.K. (eds) *Microbial Rejuvenation of Polluted Environment. Microorganisms for Sustainability,* vol 27. Springer, Singapore. https://doi.org/10.1007/978-981-15-7459-7_15

Eskenazi, B., Stapleton, A. L., Kharrazi, M. and Chee, W. Y. (1999). Associations between maternal decaffeinated and caffeinated coffee consumption and fetal growth and gestational duration. *Epidemiology* 10(3):242–249.

Floegel, A., Pischon, T., Bergmann, M. M., Teucher, B., Kaaks, R. and Boeing, H. (2012). Coffee consumption and risk of chronic disease in the European prospective investigation into cancer and nutrition (EPIC)-Germany study. *American Journal of Clinical Nutrition* 95:901–908.

Franco, R., Oñatibia-Astibia, A. and Martínez-Pinilla, E. (2013). Health benefits of methylxanthines in cacao and chocolate. *Nutrients* 5(10):4159–4173. doi:10.3390/nu5104159.

Hatch, E.E., Wesselink, A.K., Hahn, K.A., Michiel, J.J., Mikkelsen, E.M., Sorensen, H.T., Rothman, K.J., and Wise, L.A. (2018). Intake of sugar-sweetened beverages and fecundability in a North American preconception cohort. *Epidemiology (Cambridge, Mass.)* 29(3):369–378. doi:10.1097/EDE.0000000000000812.

Heckman, M. A., Weil, J. and Gonzalez de Mejia, E. (2010). Caffeine (1, 3, 7-trimethylxanthine) in foods: A comprehensive review on consumption, functionality, safety, and regulatory matters. *Journal of Food Science* 75(3):77–87. doi:10.1111/j.1750-3841.2010.01561.x.

Hatch, E. E. and Bracken, M. B. (1993). Association of delayed conception with caffeine consumption. *American Journal of Epidemiology* 138(12):1082–1092.

Islam, S., Thangadurai, D., Adetunji, C.O., Nwankwo, W., Kadiri, O., Makinde, S., Michael, O.S., Anani, O.A., Adetunji, J.B. (2021). Nanomaterials and nanocoatings for alternative antimicrobial therapy. In: Kharissova, O.V., Martínez, L.M.T., Kharisov, B.I. (eds) *Handbook of Nanomaterials and Nanocomposites for Energy and Environmental Applications*. Springer, Cham. https://doi.org/10.1007/978-3-030-11155-7_3-1

Jensen, T. K., Henriksen, T. B., Hjollund, N. H. I., Scheike, T., Kolstad, H., Giwercman, A., Ernst, E., Bonde, J. P., Skakkebæk, N. E. and Olsen, J. (1998). Caffeine intake and fecundability: A follow-up study among 430 danish couples planning their first pregnancy. *Reproductive Toxicology* 12(3):289–295.

Karlson, E. W., Mandl, L. A., Aweh, G. N. and Grodstein, F. (2003). Coffee consumption and risk of rheumatoid arthritis. *Arthritis and Rheumatology* 48(11):3055–60.

Kuczkowski, K. M. (2009). Caffeine in pregnancy. *Archives of Gynecology and Obstetrics* 280:695–698.

Lele, O. H., Maniar, J. A., Chakravorty, R. L., Vaidya, S. P. and Chowdhary A. S. (2016). Assessment of biological activities of caffeine. *International Journal of Current Microbiology Applied Science* 5(5):45–53.

Machtinger, R., Gaskins, A. J., Mansur, A., Adir, M., Racowsky, C., Baccarelli, A. A., Hauser, R. and Chavarro, J. E. (2017). Association between preconception maternal beverage intake and *in vitro* fertilization outcomes. *Fertility and Sterility* 108(6):1026–1033. doi:10.1016/j.fertnstert.2017.09.007.

Martínez-Pinilla, E., Oñatibia-Astibia, A. and Franco, R. (2015). The relevance of theobromine for the beneficial effects of cocoa consumption. *Frontiers in Pharmacology* 6:30. doi:10.3389/fphar.2015.00030.

Maughan, R. J. and Griffin, J. (2003). Caffeine ingestion and fluid balance: A review. *Journal of Human Nutrition and Dietetics* 16(6):411–420.

McCusker, R. R., Goldberger, B. A. and Cone, E. J. (2006). Caffeine content of energy drinks, carbonated sodas, and other beverages. *Journal of Analytical Toxicology* 30(2):112–114.

Mesas, A. E., Leon-Munoz, L. M., Rodriguez-Artalejo, F. and Lopez-Garcia, E. (2011). The effect of coffee on blood pressure and cardiovascular disease in hypertensive individuals: A systematic review and meta-analysis. *American Journal of Clinical Nutrition* 94:1113–1126.

Melsayed, N., Saharthi, M., Mallehyani, N., and Nalotaibi, A. (2018). Determination of antibacterial activity of Green Coffee Arabica bean extract on Multidrug resistance *Pseudomonas aeruginosa* (ATCC 27853). *IOSR. Journal of Pharmacy* 8(12):33–38.

Mielgo-Ayuso, J., Marques-Jiménez, D., Refoyo, I., Del Coso, J., León-Guereño, P. and Calleja-González, J. (2019). Effect of caffeine supplementation on sports performance based on differences between sexes: A Systematic Review. *Nutrients* 11(10:2313. doi:10.3390/nu11102313.

Mishra, P., Mishra, R.R., and Adetunji, C.O. (eds.). (2020). *Innovations in Food Technology: Current Perspectives and Future Goals*. Springer, Singapore. https://doi.org/10.1007/978-981-15-6121-4.

Müller, C., Vetter, F., Richter, E. and Bracher, F. (2014). Determination of caffeine, myosmine, and nicotine in chocolate by headspace solid-phase microextraction coupled with gas chromatography-tandem mass spectrometry. *Journal of Food Science* 79(2):T251–T255. doi:10.1111/1750-3841.12339.

Nawrot, P., Jordan, S., Eastwood, J., Rotstein, J., Hugenholtz, A. and Feeley, M. (2003). Effects of caffeine on human health. *Food Additives and Contaminants* 20(1):1–30.

Nonthakaew, A., Matan, N., Aewsiri, T. and Matan, N. (2015). Caffeine in foods and its antimicrobial activity. *International Food Research Journal* 22(1):9–14.

Nurul, A. A. R., Siti, H. M. and Oduola, A. (2014). Antibacterial activity of NESCAFÉ instant coffee beverages and pharyngitis-causing Streptococcus species. *Brunei Darussalam Journal of Health* 5:70–79.

Nehlig, A. and Debry, G. (1994a). Potential teratogenic and neurodevelopmental consequences of coffee and caffeine exposure: A review on human and animal data. *Neurotoxicology and Teratology* 16(6):531–543.

Nehlig, A. and Debry, G. (1994b). Effects of coffee and caffeine on fertility, reproduction, lactation, and development. Review of human and animal data. *Journal of Gynecology Obstetrics and Human Reproduction* 23(3):241–256.

Olaniyan, O.T. and Adetunji, C.O. (2021). Biochemical role of beneficial microorganisms: An overview on recent development in environmental and agro science. In: Adetunji, C.O., Panpatte, D.G., Jhala, Y.K. (eds) *Microbial Rejuvenation of Polluted Environment. Microorganisms for Sustainability*, vol 27. Springer, Singapore. https://doi.org/10.1007/978-981-15-7459-7_2

Olmos, V., Bardoni, N., Ridolfi, A. S. and VillaamilLepori E. C. (2009). Caffeine levels in beverages from Argentina's market: Application to caffeine dietary intake assessment. *Food Additives and Contaminants. Part A, Chemistry, Analysis, Control, Exposure & Risk Assessment* 26(3):275–281. doi:10.1080/02652030802430649.

Oñatibia-Astibia, A., Franco, R. and Martínez-Pinilla, E. (2017). Health benefits of methylxanthines in neurodegenerative diseases. *Molecular Nutrition and Food Research* 61(6). doi:10.1002/mnfr.201600670.

Rudolph, E., Färbinger, A. and König, J. (2012). Determination of the caffeine contents of various food items within the Austrian market and validation of a caffeine assessment tool (CAT). *Food Additives and Contaminants. Part A, Chemistry, Analysis, Control, Exposure & Risk Assessment* 29(12):1849–1860. doi:10.1080/19440049.2012.719642.

Salinardi, T. C., Rubin, K. H., Black, R. M. and St-Onge, M. P. (2010). Coffee mannooligosaccharides, consumed as part of a free-living, weight-maintaining diet, increase the proportional reduction in body volume in overweight men. *The Journal of Nutrition* 140(11):1943–1948. doi:10.3945/jn.110.128207.

Sanchez, J. M. (2017). Methylxanthine content in commonly consumed foods in spain and determination of its intake during consumption. *Foods (Basel, Switzerland)* 6(12):109. doi:10.3390/foods6120109.

Souza, A. C., Dussán-Sarria, J., Souza, A. P., Caumo, W. and Torres, I. (2016). Caffeine teratogenicity in rats: Morphological characterization and hypothesized mechanisms. *Clinical and Biomedical Research* 36:179–186.

Thangadurai, D., Dabire, S.S., Sangeetha, J., Said Al-Tawaha, A.R.M., Adetunji, C.O., Islam, S., Shettar, A.K., David, M., Hospet, R., Adetunji, J.B. (2021). Greener composites from plant fibers: Preparation, structure, and properties. In: Kharissova, O.V., Martínez, L.M.T., Kharisov, B.I. (eds) *Handbook of Nanomaterials and Nanocomposites for Energy and Environmental Applications*. Springer, Cham. https://doi.org/10.1007/978-3-030-11155-7_21-1

Wierzejska, R. (2012). Caffeine--common ingredient in a diet and its influence on human health. *RoczPanstwZaklHig*. 63(2):141–7.

Wright, G. A., Baker, D. D., Palmer, M. J., Stabler, D., Mustard, J. A., Power, E. F., Borland, A. M. and Stevenson, P. C. (2013). Caffeine in floral nectar enhances a pollinator's memory of reward. *Science* 339 (6124):1202–1204. doi:10.1126/science.1228806.

# 7 Strategies for Yeast Strain Improvement through Metabolic Engineering

*Toochukwu Ekwutosi Ogbulie,*
*Augusta Anuli Nwachukwu, and*
*Priscilla Amaka Ogbodo*
Federal University of Technology Owerri

*Christiana N. Opara*
Federal University of Technology Otueke

## CONTENTS



## 7.1 INTRODUCTION

*Saccharomyces cerevisiae* has always been known over the years as a model organism widely employed in industrial fermentation with added advantage of promoting enrichment in nutritional values and health benefits in man. Baking, brewing and wine making using this yeast have been in practice for decades now because of its productive fermentation ability. Indeed the use of *S. cerevisiae* in traditional

DOI: 10.1201/9781003178378-7

fermentation of alcohol has obviously led to essential knowledge about the physiology, genetics and biochemistry of the yeast as well as the genetic engineering and fermentation technologies. Within the field of biotechnology, the applicability of *S. cerevisiae* is evident in its vulnerability to metabolic engineering (ME), facilitated by the availability of its complete genome sequence. In addition to the previously mentioned attributes, *S. cerevisiae* thrive on medium with increased concentration of ethanol and sugar and low pH values making it amenable to modification of its genome through genetic engineering. These traits of interest, together with yield improvement, elimination of by-products and process performance improvement, control and cellular properties, form the major target of ME; hence, biochemicals are produced which serve as food ingredients/additives in the industries.

Conventionally, improvement of strains of yeasts and other microorganisms of industrial importance depends on optimization of environmental and nutritional conditions: mutagenesis (methods not involving foreign DNA) and recombination (methods involving foreign DNA). The former actually involves modification of physical (temperature, agitation), chemical (pH, $O_2$ concentration) and biological (enzymes) parameters as well as carbon, nitrogen, mineral sources and precursors. Methods that do not involve foreign DNA (mutagenesis) basically involve treating the organisms to enhance improvement in their genotypic and phenotypic performances whereby spontaneous, direct, induced and site-directed mutations are employed (Steensels et al., 2014). Consequently, the improved strains are selected after random or rational screening. Furthermore, for methods involving foreign DNA, improvement is usually achieved through transduction (through bacteriophages), protoplast fusion (hybridization), transformation (taking up of DNA by competent cells), conjugation (natural transfer of DNA through plasmid) and genetic engineering (DNA fragment insertion and transfer using varying vectors constructs as plasmids, phasmids, cosmids and phages) (Adrio and Demain, 2006).

ME actually encompasses the modification of existing pathways or introduction of entirely new pathways through gene manipulation targeting yield, improved microbial products, elimination or reduction of undesirable side products or production of entirely new products. Strain improvement therefore ensures rapid growth of cells, genetic stability, non-toxicity to human, easy harvest from culture fluid, use of cheap substrates, increased productivity, low cultivation costs, improved use of carbon and nitrogen sources and production of compounds that inhibit contaminant microorganisms. All these attributes/traits of interest made *S. cerevisiae* a unique model and attractive microorganism for research and applications in industries. ME, therefore, is usually distinguished from classical applied molecular biology using direct approach, where the cell system is analyzed carefully and recombinant strain with traits of interest constructed (Figure 7.1). This elucidates the significant reproducibility of the optimal nutritional and environmental requirements Ostergaard et al. (2000). In other words, specific bioreactor systems, the influence of a single medium component (with constant factor) as well as the varying feed rate (continuous bioreactor system) on cellular function can be studied. This continuous bioreactor system, however, can be achieved through changing the rate of dilution of the medium which corresponds to the specific rate of growth at steady-state conditions.

Indeed, metabolic engineered new genes from foreign organisms can be introduced in yeast and homologous gene can be overproduced as well. In industrial fermentation,

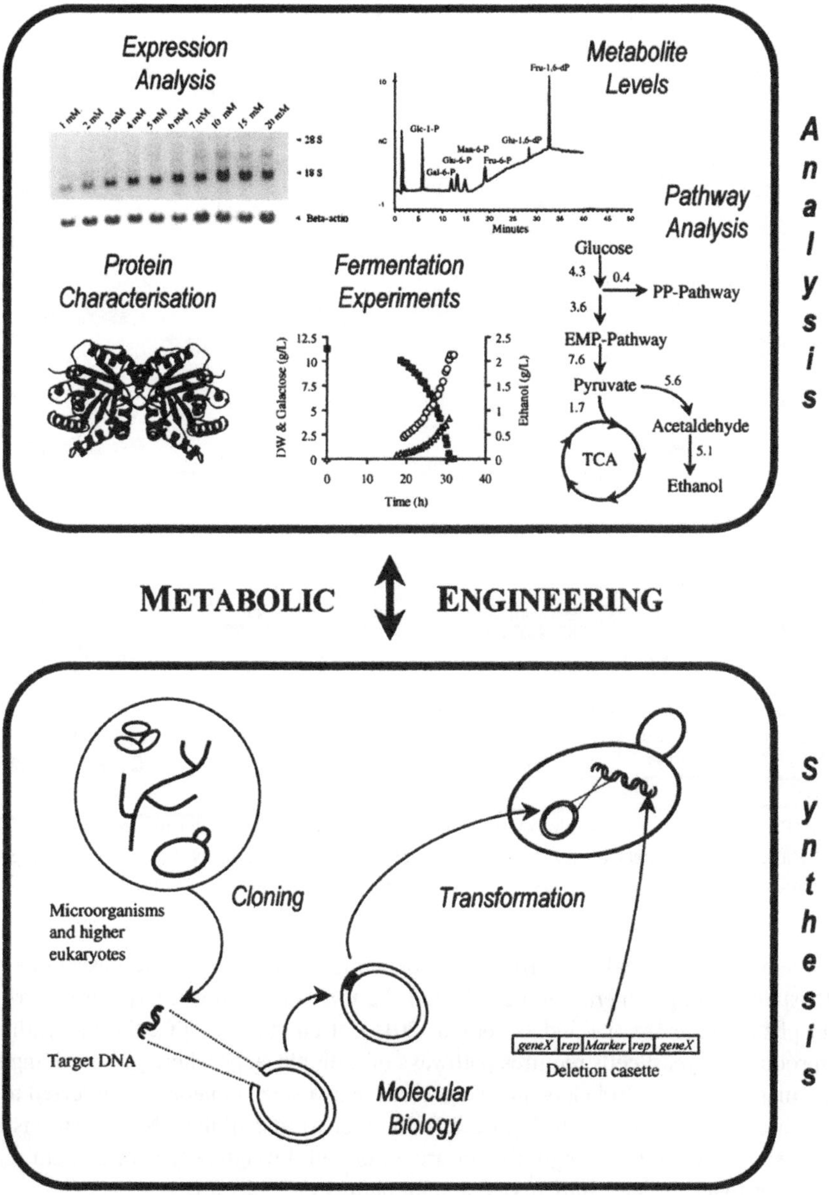


**FIGURE 7.1** The analytical and synthetic parts of ME (Ostergaard et al., 2000).

the yeast as earlier mentioned can relatively withstand low pH as well as high concentration of sugar and ethanol, the distinguishing features that helped in reduction of contamination risk during fermentation. The yeast also grows anaerobically and it is susceptible to genetic modification because of its complete genome sequence. Hence, the study of ME will surely involve the geneticists and molecular biologists,

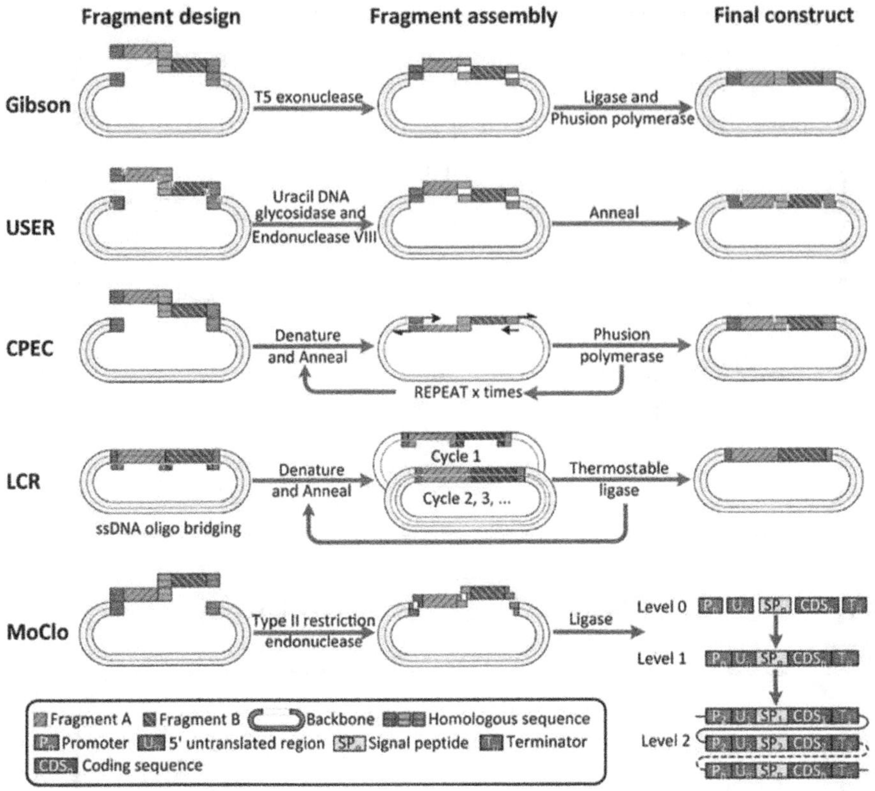


**FIGURE 7.2** ME studies elucidating the three-dimensional structure emphasizing the biosynthesis of products of yeast cells in pathways of multiple steps with genes coding for enzymes and the control elements (promoters, sensors, terminators). (Li and Borodina, 2015.)

biochemists, analytical chemists and biochemical engineers. A representative of ME studies using *S. cerevisiae* elucidated the three-dimensional structure of triose phosphate isomerase (Li and Borodina, 2015). It emphasized that the biosynthesis of products of yeast cells requires pathways of multiple steps where genes coding for enzymes and the control elements (promoters, sensors, terminators) are referred to as "BioBricks" of about ≤10 kb (Figure 7.2). In other words, all the DNA constructs are made synthetically from beginning or are amplified through polymerase chain reaction prior to assemblage into an operational pathway (Figure 7.2).

## 7.2 METABOLIC ENGINEERING: A POTENTIAL PROSPECT FOR ADVANCED RECOMBINANT DNA TECHNOLOGY

ME is an advanced tool aimed at creating new biological pathways and systems resulting in improvement of genotypic and phenotypic performances in microorganisms through the use of recombinant DNA technology. Here the cell factory is designed and constructed and specific and targeted genetic modification (gene deletions, overexpression or modulation) is identified followed by the use of molecular biology tools to implement this modification (Kern et al., 2007; Tyo et al., 2007; Patnaik, 2008). This basically leads to redirection of cellular pathways to ensure that a given product or organism is produced. Consequently, production of cost-effective valuable substances (novel and/or modified) on an industrial scale will be evident. Hence, biosynthetic pathways are modified to express some novel biochemical properties in the microbial cell.

Historically, ME of yeast was introduced in research for more than 20 years now during which the engineered yeast have been employed for the production of fuels, bulk organic chemicals, pharmaceuticals and nutraceutical ingredients (Figure 7.3). Some of the approaches to ME processes and their features are as shown in the table 7.1 below:

## 7.3 DIFFERENT TARGETS OF METABOLIC ENGINEERING OF *S. CEREVISIAE*

The different targets for ME of *S. cerevisiae* have been reported to include substrate utilization and its extension, efficiency and improvement of yield, by-products exclusion and profitable process performance, control and cellular properties. Utilization of certain range of substrates by *S. cerevisiae* for enhanced production of biotechnological products is illustrated using starch, lactose, raffinose, melibiose and xylose consumption.

### 7.3.1 Some Biotechnological Products Production through Substrate Utilization

Starch is usually stored as carbohydrate in plants. In biotechnological processes, starch is an important energy raw material for fuel bio-ethanol production. It is a polysaccharide, made up of long network of repeating monomeric units (glucose) chemically bonded by α-1,4-linkages at α-1,6-linkages at the branch point (Ostergaard et al., 2000). Many microorganisms, including wild-type *S. cerevisiae*, cannot directly degrade starch for their vegetative growth and fermentation due to lack of

**TABLE 7.1**
**Approaches to Metabolic Engineering Processes**

| Approaches (Advanced Recombinant DNA Technology) | Features |
|---|---|
| 1. Recombinant protein | Proteins encoded by the transgene are products of interest |
| 2. Metabolic engineering | Metabolites generated by the transgene-encoded enzymes are the products of interest |
| i. Product modification | The new enzymes modifies the product of existing biosynthetic pathway |
| ii. New substrate utilization | Inaccessible substrates converted into accessible form |
| iii. Completely new metabolite | All the genes of a new pathway transferred |
| iv. Enhanced metabolic production | Amplification of the gene encoding the enzyme whose activity is rate limiting |
| v. Enhanced growth | Enhanced substrate utilization |

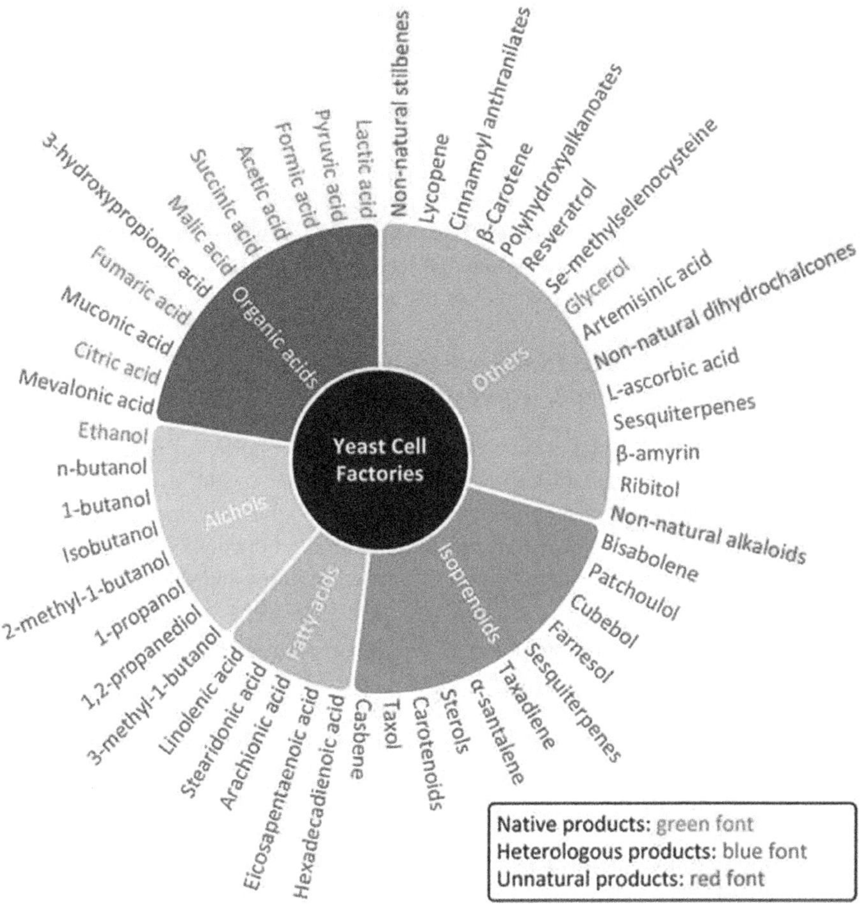


**FIGURE 7.3** Various native heterologous and unnatural synthesized products produced using yeast cells. (Li and Borodina, 2015.)

enzymes that degrade starch such as α-amylase, β-amylase and glucoamylase as well as pullulanase or isoamylase enzymes responsible for respective cleavage of α-1,4-glycosidic bonds, maltose and glucose units from the nonreducing ends of starch as well as hydrolysis of α-1,6-glycosidic bonds. The need to construct a recombinant strain from wild type through ME becomes necessary, to enable co-expression of these amylolytic enzymes for enhanced bio-ethanol production. In 1990, Hollenberg and Strasser showed that both AMYI (α-amylase gene) and GAMI (glucoamylase gene) originate from the yeast *Schwanniomyces occidentalis*. Furthermore, Janse and Pretorius, in 1995, carried out enzymatic decomposition of starch in a one-step reaction process using *S. cerevisiae* strain construct harboring genes coding for glucoamylase enzyme (*AMY1*, *STA2*) and pullulanase enzyme (*PUL1*) obtained, respectively, from *Bacillus amyloliquefaciens*, *S. cerevisiae* var. *diastaticus* and

*Klebsiella pneumoniae*. Their findings showed that the addition of the latter (*PUL1*) into the engineered yeast construct facilitated utilization of 99% starch substrate.

During production of cheese and yogurt in diary industries, lactose is usually trapped in the by-product of this production known as whey. This by-product has been reported to be produced in large volumes and represents a huge environmental problem during effluent discharge. This paved way to the quest for alternative and profitable use of this by-product. Domingues et al., in 2010, reported that lactose is not metabolized by many microorganisms as their sole carbon and energy sources, thus the high interest in development of strains of microorganisms that can efficiently utilize lactose and whey. Such strain development and improvement will definitely be of great importance in whey bioremediation and simultaneous transformation into useful products (e.g., bio-ethanol) leading to reduction in pollutant load of whey (Siso, 1996; Domingues et al., 2010; Zou et al., 2013). According to Prieto et al. (2006), *Escherichia coli*, *Lactobacillus* sp., *Leuconostoc* sp., *Streptococcus* sp. and *Pediococcus* sp. are examples of lactose-utilizing bacterial genome in nature. Filamentous fungi species like *Aspergillus niger*, *A. nidulans*, *Rhizopus oryzae* and *Neurospora crassa* often ferment lactose through different ways. While *A. niger* ferment by secreting β-galactosidase to extracellularly hydrolyze lactose in a medium, the other mentioned fungal genera takes up this disaccharide intracellularly first prior to hydrolysis (Domingues et al., 2010; Turner et al., 2016). Further, there are yeasts that can carry out lactose fermentation aerobically such as *Kluyveromyces lactis* (*K. fragilis*), *K. marxianus* and *Candida pseudotropicalis*, which are rare in nature. This is one of the major reasons for fermentative transformation of common distiller's yeast since they have special properties that represent higher benefits but cannot utilize lactose directly due to lack of β-galactosidase enzyme activity and lactose permease system for hydrolysis of lactose to glucose and galactose.

In 1986, Farahnaks et al. used protoplast fusion approach to obtain hybrids of yeast cells (*K. fragilis* and *S. cerevisiae*) for lactose hydrolysis, but the genetic stability of the fusants was their challenge. They however, bypassed this challenge through direct cloning of genes for utilization of lactose from three lactose consumers (*E. coli*, *K. fragilis* and *A. niger*) into *S. cerevisiae* through ME, which have been achieved with better yield in a research carried out by Domingues et al. in 2010. They achieved this by cloning the genes coding for lactose permease and β-galactosidase enzymes into *S. cerevisiae* to enable combined production of a specific lactose-transport system with intracellular β-galactosidase enzyme activity and by directing the secretion of β-galactosidase into the medium. Series of other similar breakthrough have been reported by Zou et al. (2013) and Turner et al. (2016).

Raffinose is a trisaccharide obtained from substrates such as honey, beet molasses, bean pulp, mallow and cotton seeds (Vincent et al., 1999; Prieto et al., 2006; Zhou et al., 2017). They are composed of one melibiose and one fructose in high quantity enough for economic extraction (Song et al., 2001; Zhou et al., 2017). They usually undergo hydrolysis by the invertase (β-fructosidases) enzyme leading to the production of melibiose and fructose. The melibiose is a disaccharide that contains glucose and galactose residues just like lactose with α-1,6-linkage bond.

Melibiose has been reported to have extensive applications as an additive in food, beverage and cosmetic industries (Heljo et al., 2013) and as a main ingredient in animal feed supplements, functional foods of man and pharmaceutical formulations as well due to their prebiotic features (Xiao et al., 2000; Tomita et al., 2007; Tanaka et al., 2014 and Zhou et al., 2017). Consequently, to obtain pure melibiose with limited by-products like sucrose, galactose, glucose and fructose, shift from the use of whole-cell bio-catalysis in the production reaction system should be considered. This, however, can be achieved through ME application whereby unwanted reactions such as β-galactosidase induction is inhibited resulting in the synthesis of the target product (melibiose) and fructose. Fructose synthesis can also be hindered through promotion of trans-membrane transportation through ME. To achieve this, strains can be improved by first screening for the expression of the heterologous fructose-specific co-transporter, encoded by *FSY1* gene in fructophilic yeast species such as *Zygosaccharomyces rouxii*, *Candida magnolia*, *Saccharomyces bayanus* and *S. carlsbergensis* (Zhou et al., 2017). This is followed by insertion of the encoded gene into yeast strain of interest to induce efficient and high affinity fructose-specific co-transformation operating system.

Furthermore, melibiose cannot be utilized by *S. cerevisiae* except after its hydrolysis for bio-ethanol production. This was evidently observed during industrial production using high melibiose-containing substrates as reported by Ostergaard et al. (2000) and Prieto et al. (2006). The inability of the yeast to directly use this disaccharide as carbon source causes reduction in the productivity of the cell multiplication process leading to increase in the organic content of industrial effluent, hence high biochemical oxygen demand (BOD) when discharged. Yeast strain of interest should, therefore, have β-galactosidase enzyme that will cleave the α-1–6-linkage bond of melibiose encoded by *MEL* gene which can be found in some species of yeast such as *S. bayanus*, *S. oleaginosus* and *S. carlsbergensis*. Ostergaard et al., in 2000, however, reported on a yeast construct through ME capable of melibiose consumption by expressing the *MEL1*-encoded gene. Model of yeast strains with complete and vigorous raffinose consumption ability had also been produced following combination of traits from strains of *S. bayanus* with resultant higher yields and significant BOD reduction from industrial effluents (Prieto et al., 2006).

After glucose, xylose is the second most abundant and prevalent monosaccharide (pentose sugar) found in lignocelluloses. When present in substrates, *S. cerevisiae* has been reported to be capable of utilizing the pentose sugar more efficiently than prokaryotes to produce fuel bio-ethanol (Hahn-Hägerdal et al., 2001). Their efficiency in carrying out this metabolic activity is credited to their ability to tolerate the inhibitory by-products of hydrolysis of lignocellulose although they are constrained by their inability utilize xylan, a xylose-based hemicellulose (especially in hardwood and some other agricultural residues) unless in its isomeric form xylulose. In other words, in yeast and filamentous fungi, reduction of xylose to xylitol by the enzyme xylose reductase (XR) is the initial step of xylose metabolism, followed by the oxidation of xylitol to xylulose by the enzyme xylitol dehydrogenase (XDH) (Ostergaard et al., 2000). However, for large-scale production of bio-ethanol, strains of the yeast of interest have to be improved through ME to express the heterologous genes that will efficiently convert xylose to xylulose via xylitol oxidation thereby increasing the yeast xylulose consumption and specific growth with resultant high bio-ethanol production. Xylulose consumption by this strain usually follows the pentose phosphate (PP) pathway of metabolism initiated by the intracellular enzyme xylulose

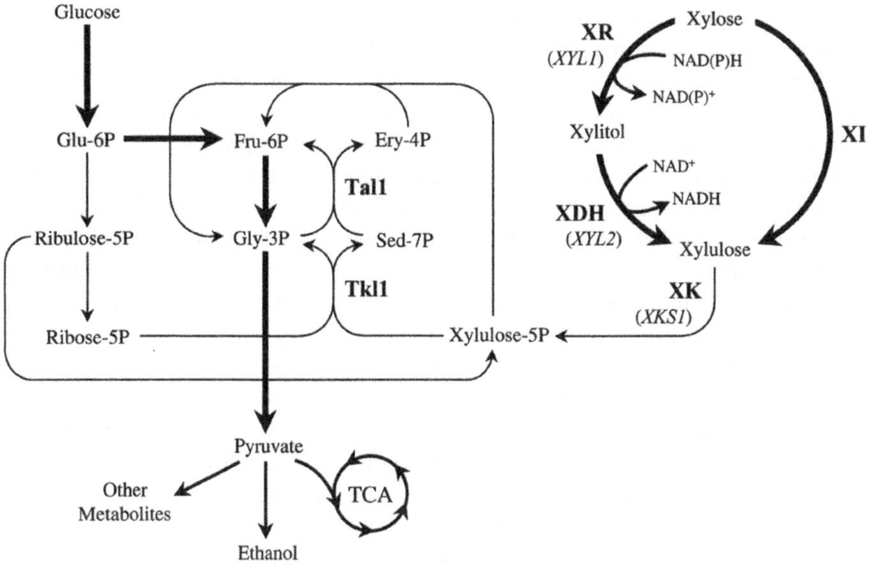


**FIGURE 7.4** The metabolic pathways for xylose metabolism (thick arrows) via PP pathway (thin arrows) after phosphorylation to bio-ethanol production through glycolysis. (Legend: TCA – tricarboxylic acid cycle; Tkl1 – transketolase; Tal1 – transaldolase; Glu-6P – glucose-6-phosphate; Fru-6P – fructose-6-phosphate; Gly-3P – glyceraldehyde-3-phosphate; Ery-4P – erythrose-4-phosphate; Sed-7P – sedoheptulose-7-phosphate. The encoded genes are represented in parentheses (Ostergaard et al., 2000).)

kinase (XK) of the yeast encoded by *XKS1* gene via phosphorylation as shown in Figure 7.4. Subsequent conversion to bio-ethanol from pyruvate will thereafter take place through glycolysis. Xylose utilization in yeast is targeted toward construction of improved strain of *S. cerevisiae* that expresses xylose reductase (*XYL1*), xylitol dehydrogenase (*XYL2*) and xylulose kinase (*XKS1*) enzyme activities from native xylose-metabolizing yeasts such as *Pichia stipitis*, *Candida shehatae* and *Pachysolen tannophilus* (Hahn-Hägerdal et al., 2001), which is the major strategy to be used in an attempt to construct new strain of *S. cerevisiae* for xylose metabolism.

### 7.3.2 Process Performance and Cellular Characteristics Improvement

Improvement of the cellular features of strains of interest as well as the process optimization is one of the key biotechnological processes for enhanced performance required during scaling up in industries. To achieve this, considerations should be made toward the design of bioreactor, the fermentation medium, sterilization techniques, fermentation process as well as process monitoring and control measures to be adopted (Raj and Karanth, 2006). Actually, the success of strain improvement in order to achieve the process objectives irrespective of the protocol employed is dependent on the reliable operations of the fermentation system which is functional based on the design of fermenter and process of fermentation.

In terms of bioreactor design, the objective to be considered is to enhance microbial enzyme catalytic anaerobic reaction system for frugal production of products which ascertain performance criteria. This means that the pattern/specification and the upscale consideration are achieved through provision of the test organism with trait of interest, the optimal condition for desired product production. This should uniformly be achieved in the reactor. Furthermore, the medium to be used, whether defined complex or technical medium, should meet up to certain standards as being able to give maximum yield, exclude undesirable product of metabolism, uphold compatible quality of products and reduce sterilization problems. Mode of process operation should, however, be dependent on choice and relies on the relative substrate metabolism and conversion into useful products.

Indeed, ME is a complex process and its involvement in industrial production through microbial enzyme catalytic process under anaerobic condition requires that control parameters be put in place. These parameters are very crucial in carrying out regular observations to detect defective signals and exercise influence over any prospective outcome for enhanced corrective measures during production process. This will ultimately enhance favorable conditions in accomplishing the process. The diverse control measures for processes could be biological ($CO_2$ production and $O_2$ uptake rates, respiration quotient, biomass, etc.), physical (speed of impeller; temperature, aeration rate and pressure) and chemical (acid-base indicator, dissolved $O_2$, redox potential, etc.). Further, the development and improvement of the appropriate methods and use of high-throughput methods to improve certain operational unit during biotechnological processes should not supersede the fact that the strain of interest and their capabilities for process performance improvement are crucial.

A case study was reported by Ostergaard et al. in 2000. They developed brewer's yeast capable of clumping during beer production (beer clarification). Of the two distinctive system designed for clumping that exist (i.e., the NewFlo phenotype, found in many brewer's yeast strains, and the Flo1 phenotype, found in laboratory strains), remarkable differences have been observed during the onset of the clumping process. Their team, however, successfully inserted the *FLO1* gene into the genome of a non-flocculent brewer's yeast strain with resultant yeast construct establishing clumping/flocculation activities. In other words, construction of yeast strains with such genetic system that have dominant gene encoding flocculation ability should form a competent focus in ME. Cantwell et al., in 1986, successfully incorporated and observed expression of the β-glucanase gene of *Bacillus subtilis*, *Trichoderma reesei* and barley in *S. cerevisiae*, with obvious secretion of active enzymes and no effect on beer quality but efficient degradation of β-glucans for improved filterability of beer.

## 7.4 METABOLIC ENGINEERING IN PRODUCTION OF FOOD INGREDIENTS

*S. cerevisiae* is indeed a model microorganism that holds a lot of promise for future applications in industrial production processes. It has the potential in production of useful chemical compounds from renewable sources, which are very crucial in production of food ingredients. These ingredients are actually substances that are added to food to obtain a desired freshness, taste or effect, improved safety and/or specific technical/functional purposes during food processing, packaging and storage. Some other expected effects or purposes for their use in food production include provision of flavor, appearance, color, physical stability or texture, nutritional values etc. Some of these ingredients are chemically/artificially produced; some exist naturally in plants or animal raw materials whereas some are obtained from microbiological or enzymatic processes. Due to the potential harmful effects/health risks of artificial ingredients on human, it is of great importance to focus on the use of biologically synthesized/natural ingredients in food. In the context of this chapter, the effect is geared toward yield optimization and productivity using ME to enhance economically viable up-scaling processes during production. In other words, aside health implication, there is also the need for evaluation of production cost and profits of use of microbial processes over chemical and enzymatic synthesis as part of process economic viability assessment. However, biological syntheses have been reported by Azelee et al. (2019) to have significant economic advantage over chemical/enzymatic syntheses due to the following reasons:

i. Most of the feedstock/substrates required during fermentation process are cheap and readily available even though the yield titer and productivity varies.
ii. Enzymatic process are costly due to high cost of enzyme immobilization for easy recovery and reuse.

The need for engineering yeast metabolic system becomes necessary to enhance production of products with low toxic by-product. An overview of some of the important chemical compounds used as food ingredients synthesized by engineered yeast strains is elucidated indicating, undoubtedly, the potentials these model organisms hold for the future.

### 7.4.1 Glycerol Production

Glycerol is a polyol compound with its backbone found in glycerides. It is an interesting chemical compound widely approved for use by the FDA not just for its antimicrobial activities and pharmaceutical applications, but also due to its use as sweeteners, solvent, humectants, preservatives and thickener in the food industry. Its synthetic preparation by enzymatic trans-esterification and/or recovery as by-product during biodiesel production have been reported to be non-economical and equally contains toxic by-products (Azelee et al., 2019). In addition to the above limitation, removal or separation of impurities from crude glycerol has been reported to be a challenging factor for microbial fermentation, hence the need for systematic engineering of yeast to improve strains for enhanced production of optically pure glycerol with low toxic by-products from cheap raw materials. A typical example is the use of the yeast *C. magnolia* which converts glycerol into mannitol, a sugar alcohol (Khan et al., 2009).

Steps in glycerol synthesis involve reduction of dihydroxy acetone phosphate (DAP), a product of intermediate glycolysis, into glycerol-3-phosphate (G3P). Thereafter, G3P is diphosphorylated to glycerol. These reactions involve enzymatic activities of NAD-dependent G3P-dehydrogenase enzyme encoded by glycerol phosphate dehydrogenase gene *GPD* (Figure 7.5), the expression of which is required for growth at high osmotic stress. According to Aslankoohi et al. (2015), *S. cerevisiae* plays significant role in glycerol production. However, previous studies have shown that constructing or developing mutant strains of *S. cerevisiae* with deletion or overexpression of *GPD* gene produced more glycerol than the wild type (Albertyn et al., 1994; Michnick et al., 1997; Cambon et al., 2006; Aslankoohi et al., 2015). Manipulating the genetic properties of wild yeast cells as well as controlling the culture condition through ME will surely enhance maximum glycerol production for industrial application. Cordier et al., in 2007, deduced an approach to improve glycerol yield in mutant strain by combining the deletion of triose phosphate isomerase gene, gene encoding the NAD+-dependent alcohol dehydrogenase enzyme, gene

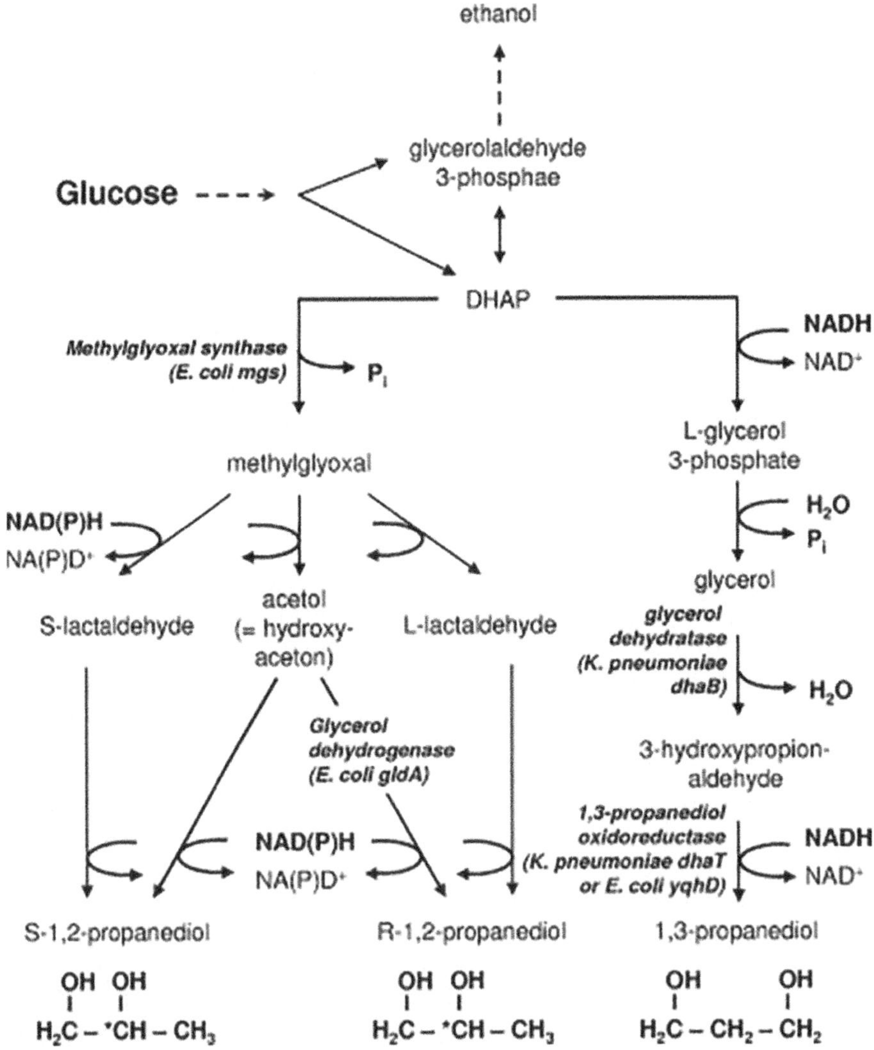


**FIGURE 7.5** Pathways for PD synthesis from sugars (Lee and DaSilva, 2006).

encoding an isomer of GDP and cytosolic $NAD^+$-dependent aldehyde dehydrogenase enzyme. Nevoigt, in 2008, also reported other approaches to significantly improve glycerol yield. However, in 2015, Aslankoohi et al. carried out a study on glycerol production in baker's yeast using mutant strain generated by deletion of *gpd1* and overexpression of *GPD1*. Their result showed that the strain with *gpd1* had impaired fermentation of dough and low glycerol production whereas the strain with *GPD1* increased rate of dough fermentation with improved gas retention.

### 7.4.2 Propanediol

Propanediol (PD) is a pure natural ingredient and carrier of flavor and extracts derived for food and beverage products. Its production is one of the major strategies that hold promise for utilization of glycerol through selective hydrogenolysis. Although the increase in biodiesel production made the price of pure and crude glycerol to drop (Tabah et al., 2016), the high cost of removal of unwanted glycerol from production plant is equally a huge constraint that influenced the development of eco-friendly solution to the fate of this valuable by-product that increasingly became a waste with an increased disposal cost (Lee et al., 2015). This in turn holds promise for sustainable biofuel market as well as economically valuable means of production of PD, a commercially valuable by-product from glycerol.

Furthermore, conversion of glycerol to PD through chemical process is faced with a lot of challenge such as high pressure and temperature requirements, use of chemical catalysts that are costly, production of toxic intermediate and unwanted products, use of toxic organic solvents during production, use of non-renewable materials as substrates and decrease in yield. Comparatively, biological conversion process is more eco-friendly, economically viable (i.e., less energy consumption and higher product yield) and generates nontoxic by-product (Jun et al., 2010; Dobson et al., 2012; Kaur et al., 2012). Glycerol metabolism to produce 1,3-propanediol has been studied using bacteria of the genera *Klebsiella*, *Citrobacter*, *Escherichia* and *Lactobacillus* (Papanikolaou et al., 2008; Rymowicz et al., 2009; Saxena et al., 2009; Szymanowska-Powalowska et al., 2013; Metsoviti et al., 2012). These microorganisms were reported to carry out the processes through respiratory metabolism where external electron acceptor is required but faced with drawback fermentatively (Schuller, 2003; Yazdani and Gonzalez, 2007). Extensive study made in 2014 by Rieckenberg et al. has shown that the use of these organisms during industrial applications was also faced with a lot of limitations resulting from their pathogenicity, culture conditions (need for strict anaerobic condition and rich nutrient supplements) as well as unavailability of the physiology and genetic tools necessary for manipulating them effectively. This necessitates the use of model microorganisms, *E. coli* and *S. cerevisiae*, that are amenable to industrial applications. Both microorganisms have been reported to utilize glycerol and produce a structural isomer of PD (1,2-propanediol) but can be genetically engineered to produce 1,3-propanediol. *S. cerevisiae* has a greater advantage as they can naturally ferment sugars to produce glycerol, a precursor of 1,3-propanediol and other useful products like citric acid, biosurfactants and carotenoids (Nevoigt, 2008). The production of 1,3-propanediol in mutant strain of *S. cerevisiae* with *mgs*, *dhaB* and dhaT/*yqhD* encoding genes for methylglyoxal synthase, glycerol dehydrogenase and

1,3-propanediol oxidoreductase from *Klebsiella pneumoniae* and *E. coli*, respectively, has been reported by Lee and DaSilva (2006) and Ma et al. (2007). Figure 7.5 shows the promising route of PD synthesis from sugars.

### 7.4.3 Organic Acids

Among the fermentative products in the global market, organic acids represent the third largest category. They are intermediates or products of many metabolic pathways like Krebs cycle and lactate and acetate fermentation processes. They have a long contributory history of conferring taste and flavor to fruits, vegetables and fermented foods and are equally used in beverage, food and animal feed production.. Some of the widely used additives are citric acid, acetic acid, pyruvic acid, succinic acid, lactic acid, tartaric acid, malic acid, benzoic acid, gluconic acid, fumaric acid and propionic acid. The low-molecular-weight organic acid which are synthesized microbiologically as part of primary metabolism of fermentation process are acetic acid, succinic acid, pyruvic acid, lactic acid and citric acid, whereas most of the others mentioned above are synthesized chemically on an industrial scale (Quitmann et al., 2014).

Biosynthesis of organic acid has recently been concentrated on the use of yeast due to the fact that yeast has high acidic tolerance than bacterial species which enhances reduction in microbial contamination of products. In other words, during biotechnological processing of organic acid, special nutrient supplements (complex nitrogen sources and vitamins) and neutralizing agents and pH control are unnecessary, which alleviates the problem of formation of amount of by-products that can affect the downstream processes (Nevoigt, 2008). Further, the ability of *S. cerevisiae* to synthesize organic acids like lactic acid requires the pyruvic acid pathway where pyruvate are reduced to lactic acid during glycolysis catalyzed by lactate dehydrogenase enzyme (LDH); the expression of heterologous LDH by this yeast strain is of utmost importance. Branduardi et al., in 2006, reported that the expression is also dependent on the physiology of the strain as well as the LDH gene.

Generally, organic acids are produced or generated through varying biosynthetic pathways where they serve as key intermediates or precursor to the formation of other substances (Figure 7.6). Through ME, the enzyme-dependent encoded genes required for their expression can be inserted into yeast for enhanced synthesis of desired product and yield.

### 7.4.4 Sugar Alcohols

Xylitol, lactitol, sorbitol, mannitol, erythritol, maltitol, starch hydrolysates and isomalt are sugar alcohols commonly found useful in the food industry (Mahian and Hakimzadeh, 2016). They are class of polyols with reduced carbonyl group corresponding to primary or secondary hydroxyl group. They are used as sweeteners and thickeners. Among the above listed types of sugar alcohol, xylitol and sorbitol are popularly used in processing of commercial (Schiweck et al., 2012). Further, Park et al. (2016) from their study reported that xylitol, mannitol, sorbitol and erythritol have gained much attention as emerging food ingredients with similar sensory properties of sucrose though of less caloric value. Their addition into food as ingredients

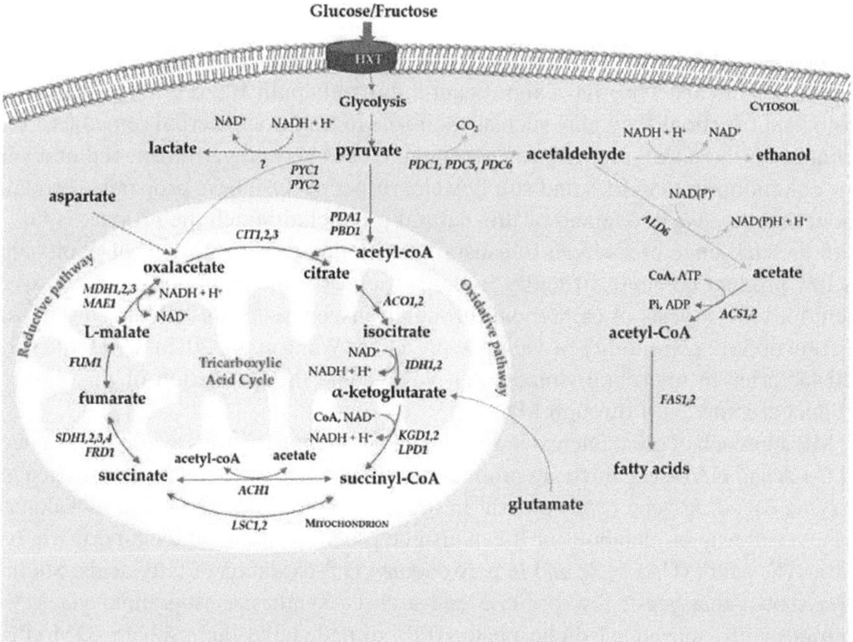


**FIGURE 7.6** Enzyme-catalyzed biosynthetic pathway of organic acids from glycolysis through reductive pathway of tricarboxylic acid cycle (TCA) indicating the encoding genes in each reaction steps. (Legend: PDC1,5,6—pyruvate decarboxylase; ADH—alcohol dehydrogenase; ALD6—aldehyde dehydrogenase; ACS1, ACS2—acetyl-CoA synthetase; FAS1, FAS2—fatty acid synthetase; PYC1,2-cytosol—pyruvate carboxylase; PDA1, PBD1—subunits of the pyruvate dehydrogenase complex; CIT 1,2,3—citrate synthase; ACO1,2—aconitase; IDH1,2—NAD(+)-dependent isocitrate dehydrogenase; KGD1,2, LPD1-α-ketoglutarate dehydrogenase complex; ACH1—protein for OH-CoA transfer from succinyl-CoA; LSC1,2—subunits of succinyl-CoA ligase; SDH1,2,3—subunits of succinate dehydrogenase; FRD1—fumarate reductase; FUM1—fumarase; MDH1—mitochondrial malate dehydrogenase; MAE1—mitochondrial malic enzyme (Ferreira and Mendes-Faia, 2020).)

improved nutritional profile of food products due to the health-promoting properties they possess. These properties, according to Mahian and Hakimzadeh (2016), include low glycemic index, insulin response (insulin-independent metabolism in human), noncarcinogenic nature and low caloric value. These features have greatly increased research focus on sugar acid production by equipping *S. cerevisiae* with necessary heterologous enzyme encoding genes to facilitate the biotechnological process industrially especially in biosynthesis of xylitol through ME. Typical example of such research was made in 2007 by Toivari et al. during their study on ME of *S. cerevisiae* to convert D-glucose to xylitol and sugar alcohols. Carneiro et al., in 2019, also carried out research on identification and comparison of new producing yeasts for xylitol production.

### 7.4.5 Isoprenoids

Isoprenoids, also called terpenoids, comprise a vast family of naturally occurring organic chemicals. They have significant industrial applications due to the diverse biological functions they play such as their role in traditional herbal remedies, contribution to the eucalyptus scent, cinnamon, cloves and ginger flavor, red and yellow coloration in tomatoes and sun flowers, respectively. These properties brought about the increase in demand of this natural product although the product is faced with the challenge of low yield/unsustainability due to slow growth of plants and its low product content, difficulty in cultivation of some plants or their improvement for biosynthesis of isoprenoid through transgenesis. This influenced the production of bulk commodity or valuable products (Wang et al., 2018). Consequently, cell factories of microbial isolates that will enable the generation of this natural product are now built through ME.

ME approach of great interest in isoprenoid production requires availability of acetyl Co-A and NADPH, which are produced from pyruvate by pyruvate dehydrogenase enzyme encoding gene (*pdh*) present in the cytosol of *S. cerevisiae* via mevalonate (MVA) pathway of metabolism. It can also be produced in the mitochondria via tricarboxylic acid (TCA) cycle and in peroxisome via β-oxidation of fatty acids. Studies have shown that yeast, few bacteria and archaea synthesize isoprenoid via MVA pathway with isopentenyl diphosphate (IPP), dimethylallyl diphosphate (DMAPP), geranyl diphosphate (GPP), long-chain isoprenyl diphosphate (LoPP), geranyl geranyl diphosphate (GGPP) and farnesyl diphosphate (FPP) as precursors (Liao et al., 2016; Frank and Groll, 2017; Cao et al., 2020). Further, increased availability of acetyl Co-A for sustainable production of isoprenoid can thus be achieved through insertion of acetyl Co-A synthetase encoding gene from *Salmonella enterica* (Shiba et al., 2007) into *S. cerevisiae* via engineering the pyruvate dehydrogenase bypass, leading to overproduction of acetaldehyde dehydrogenase. The acetyl Co-A synthetase then synthesizes enough acetyl Co-A, reduction of which enters the MVA pathway resulting in terpenoid synthesis. However, the different precursors according to Nevoigt (2008) and Wang et al. (2018) determine the class of terpenoids to be synthesized (Figure 7.7). Evidence of comprehensive approach to successful ME of yeast for enhanced and sustainable terpenoids synthesis has been reported (Maury et al., 2005; Chang et al., 2006; Zhang et al., 2017; Wang et al., 2018; Lyu et al., 2019; Cao et al., 2020).

### 7.4.6 Resveratrol

Resveratrol, a stilbenoid, is a product of plant secondary metabolism. It has attracted increasing interest following its proven nutritional, agricultural and health-promoting effects (Trantas et al., 2009); hence, its usefulness is of great importance in food processing, pharmaceutical and cosmetics industries. Chemical synthesis of this valuable product has been reported to be difficult caused by increase in cost, insufficiency in abundance and inaccessibility of plants. Its extraction from biological sources has also been inefficient. Due to its low extraction yield, microbial fermentation of resveratrol becomes a choice between many possibilities that holds promise to bypass the above-mentioned limitations encountered during product extraction from plants. They are, however, targeted for biosynthesis

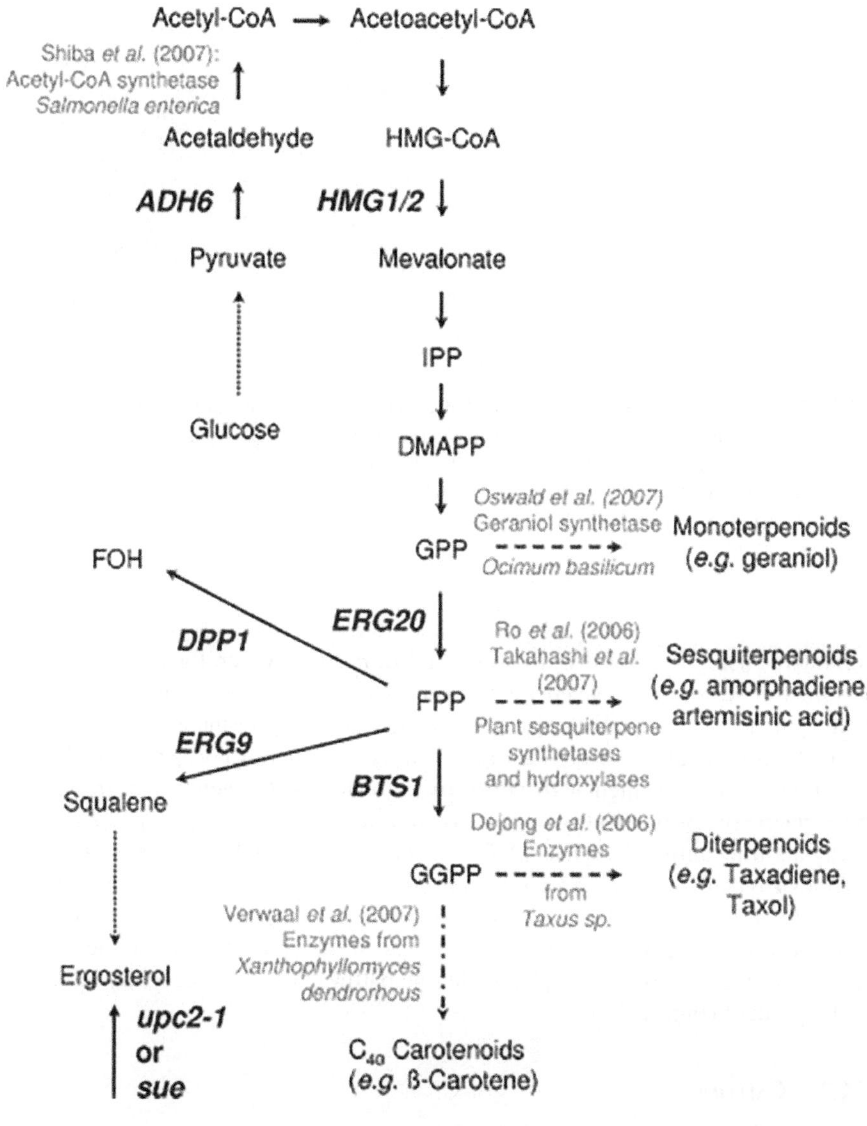


**FIGURE 7.7** Biosynthetic pathway for isoprenoid using engineered *S. cerevisiae*. (Legend: IPP—isopentenyl diphosphate; DMAPP—dimethylallyl diphosphate; FOH—farnesol, ADH6—alcohol dehydrogenase (NADP+); BTS1, GGPP—geranyl geranyl diphosphate synthase; DPP1, DGPP—geranyl phosphatase; ERG9—squalene synthase; ERG20—farnesyl diphosphate synthase; HMG1/2,3-hydroxy-3-methylglutaryl (HMG)-CoA reductase (Crowley et al., 1998; Takahashi et al., 2007).)

using engineered microorganisms to maximize yield and enhance sustainability of the product. With the use of biotechnology through genetic engineering, strategies employed in insertion of genes useful in resveratrol synthesis in yeasts were designed with the prospect of producing large quantities of resveratrol by engineering *S. cerevisiae*. Several research groups have produced resveratrol and its

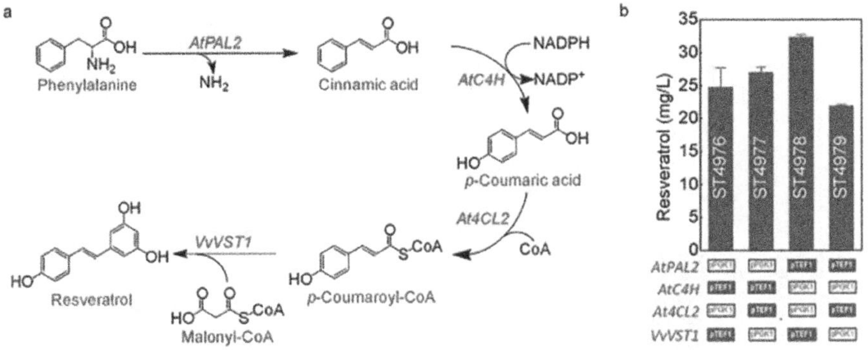


**FIGURE 7.8** (a) The resveratrol biosynthetic via phenylalanine pathway. (b) Expressing the four encoding genes during resveratrol synthesis by improved yeast strain and the titer values of the product concentration. (Legend: phenylalanine ammonia lyase (*AtPAL2*); cinnamic acid hydroxylase (*AtC4H*); p-coumaroyl-CoA; ligase (*At4CL2*); resveratrol synthase (*VvVST1*) (Li et al., 2016).)

derivatives using improved yeast strain and *E. coli*. However, those studies faced certain limitation as the need for the use of expensive precursors (p-coumaric acid, tyrosine or phenylalanine) to feed the strain during fermentation. In a study in 2012, Philip et al. reported that high quantity of resveratrol was produced using *S. cerevisiae* transformed with genes from tobacco (*4CL* gene) and grape vine (*STS* gene). In 2016, Li et al. reconstructed yeast by inserting genes encoding phenylalanine ammonia lyase (*AtPALz*), cinnamic acid hydroxylase (*AtC4H*), p-coumaroyl Co-A ligase (*At4CL2*) from *Arabidopsis thaliana* together with resveratrol synthase gene (*VvVST1*) from *Vitis vinifera* into the cell. They applied a "pull-push-block" engineering strategy on the strain which resulted to the overexpression of resveratrol via the phenylalanine pathway and obtained very high titer of the product (Figure 7.8).

### 7.4.7 Caffeine

Caffeine is a purine alkaloid which plays obvious role in pharmacology and food chemistry. It is most highly and widely consumed in beverages like chocolate, coffee, cola-like drinks and tea. In studies by Koshiishi et al. (2001) and Jin et al. (2014), caffeine (1,3,7-trimethylxanthine) was reported to exhibit a four-step pathway during biosynthesis involving methylation and cleavage reactions (Figure 7.9). Success of its high titer synthesis was also recorded in 2016 by McKeaguea et al., where microbial platform was metabolically engineered for de novo biosynthesis of intermediate of caffeine synthesis. In their study, they constructed a xanthine-to-xanthosine conversion pathway in native yeast such as microbial host. This induced central metabolism to increase flux in endogenous purine production of 7-methylxanthine, an intermediate precursor to caffeine biosynthesis.

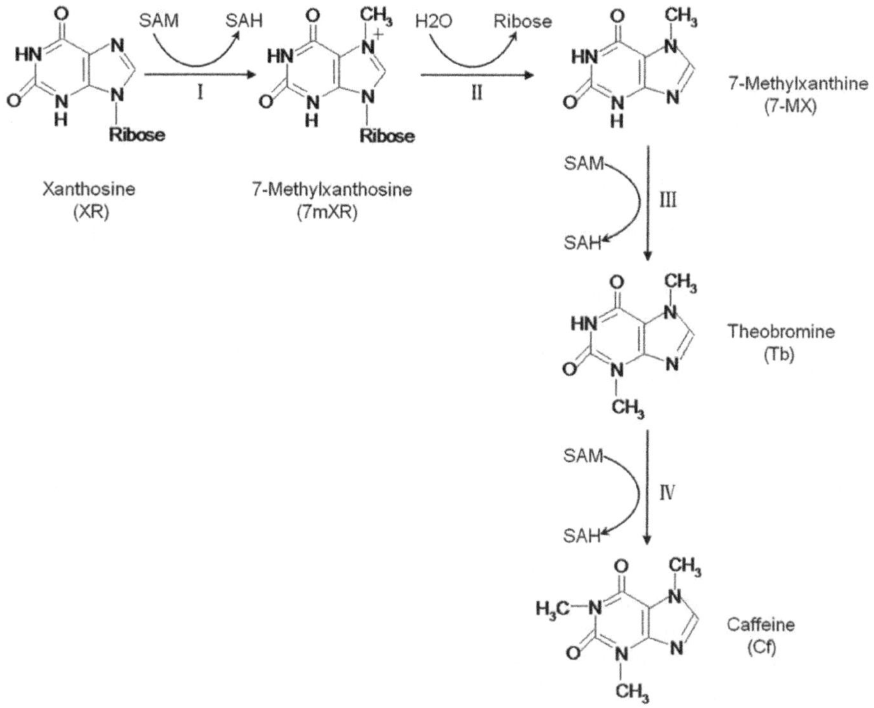


**FIGURE 7.9** Caffeine biosynthetic pathway depicting the three methylation steps (I, III, IV) and cleavage/hydrolysis step (II) (Koshiishi et al., 2001).

## REFERENCES

Adrio, J. L., & Demain, A. L. (2006). Genetic improvement of processes yielding microbial products. *FEMS Microbiology Review*, 30, 187–214.

Albertyn, J., Hohmann, S., Thevelein, J. M., & Prior, B. (1994). GPD1, which encodes glycerol-3-phosphate dehydrogenase, is essential for growth under osmotic stress in *Saccharomyces cerevisiae*, and its expression is regulated by the high-osmolarity glycerol response pathway. *Molecular Cell Biology*, 14, 4135–4144.

Aslankoohi, E., Rezael, M. N., Vervoort, Y., Courtin, C. M., & Verstrepen, K. J. (2015). Glycerol production by fermenting yeast cells essential for optimal bread dough fermentation. *PLoS One*, 10(3), e0119364. doi:10.1371/journal.pone.0119364.

Azelee, N. I. W., Ramli, A. N. M., Manas, N. H. A., Salamun, N., Man, R. C., & El-Enshasy, H. (2019). Glycerol in food, cosmetics and pharmaceutical industries: basics and New applications. *International Journal of Scientific and Technology Research*, 8(12), 553–558.

Cambon, B., Monteil, V., Remize, F., Camarasa, C., & Dequin, S. (2006). Effects of GPD1 over expression in *Saccharomyces cerevisiae* commercial wine yeast strains lacking ALD6 genes. *Applied Environmental Microbiology*, 72, 4688–4694.

Cantwell, B. A., Brazil, G., Murphy, N., McConnell, D. J. (1986). Comparison of expression of the endo-beta-1,3-1,4-glucanase gene from *Bacillus subtilis* in *Saccharomyces cerevisiae* from the *CYC1* and *ADH1* promoters. *Journal of Genetics*, 11, 65–70.

Cao, X., Yanga, S., Cao, C., & Zhou, Y. J. (2020). Harnessing sub-organelle metabolism for biosynthesis of isoprenoids in yeast. *Synthetic and Systems Biotechnology*, 5, 179–186. doi:10.1016/j.synbio.2020.06.005.

Carneiro, C. G. C., Silva, F. C., & Almeida, J. R. M. (2019). Xylitol production: identification and comparison of new producing yeasts. *Microorganisms*, 7, 484–498. doi:10.3390/microorganisms7110484.

Chang, M. C., & Keasling, J. D. (2006). Production of isoprenoid pharmaceuticals by engineered microbes. *Nature Chemical Biology*, 2, 674–681.

Cordier, H., Mendes, F., Vasconcelos, I., & Francois, J. M. (2007). A metabolic and genomic study of engineered *Saccharomyces cerevisiae* strains for high glycerol production. *Journal of Metabolic Engineering*, 9, 364–378.

Crowley, J. H., Leak, F. W., Shianna, K. V., Tove, S., & Parks, L. W. (1998). A mutation in a purported regulatory gene affects control of sterol uptake in *Saccharomyces cerevisiae*. *Journal of Bacteriology*, 180, 4177–4183.

Dobson, R., Gray, V., & Rumbold, K. (2012). Microbial utilization of crude glycerol for the production of value-added products. *Journal of Industrial Microbiology and Biotechnology*, 39(2), 217–226 doi:10.1007/s10295-011-1038-0.

Domingues, L., Guimarães, P. M. R., & Oliveira, C. (2010). Metabolic engineering of *Saccharomyces cerevisiae* for lactose/whey fermentation. *Bioengineered Bugs*, 1(3), 164–171. doi:10.4161/bbug.1.3.10619.

Farahnak, F., Seki, T., Ryu, D. D., & Ogrydziak, D. (1986). Construction of lactose-assimilating and high ethanol producing yeasts by protoplast fusion. *Applied Environmental Microbiology*, 51, 362–367.

Ferreira, A. M., & Mendes-Faia, A. (2020). The role of yeasts and lactic acid bacteria on the metabolism of organic acids during winemaking. *Foods*, 9, 1231. doi:10.3390/foods9091231.

Frank, A., & Groll, M. (2017). The methylerythritol phosphate pathway to isoprenoids. *Chemical Review*, 117, 5675–5703. doi:10.1021/acs.chemrev.6b00537.

Hahn-Hägerdal, B., Wahlbom, C. F., Gárdonyi, M., van Zyl, W. H., Otero, R. R. C., & Jönsson, L. J. (2001). Metabolic engineering of *Saccharomyces cerevisiae* for Xylose utilization. *Advances in Biochemical Engineering and Biotechnology*, 73, 56–84.

Heljo, V. P., Sainio, J., Shevchenko, A., Kivikero, N., Lakio, S., & Juppo, A. M. (2013). The effect of relative humidity on the physical properties of two melibiose monohydrate batches with differing particle size distributions and surface properties. *Journal of Pharmaceutical Science*, 102(1), 195–203. doi:10.1002/jps.23366.

Hollenberg, C. P., & Strasser, A. W. M. (1990). Improvement of baker's and brewer's yeast by gene technology. *Journal of Food Biotechnology*, 4, 527–534.

Janse, B. J. H., & Pretorius, I. S. (1995). One-step enzymatic hydrolysis of starch using a recombinant strain of *Saccharomyces cerevisiae* producing α-amylase, glucoamylase and pullulanase. *Applied Microbiology and Biotechnology*, 42, 878–883.

Jeandet, P., Delaunois, B., Aziz, A., Donnez, D., Vasserot, Y., Cordelier, S., & Courot, E. (2012). Metabolic engineering of yeast and plant for the production of biological active hydroxystilbene, resveraltrol. *Journal of Biomedicine and Biotechnology*, 3, 1–14.

Jeong, Y. S., & Vieth, W. R. (1991). Fermentation of lactose to ethanol with recombinant yeast in an immobilized yeast membrane bioreactor. *Journals of Biotechnology Bioengineering*, 37, 587–590.

Jin-Hwan, P., Kwang, H., Tae-Yong, K., & Sang, Y. L. (2007). Metabolic engineering of *E. coli* for the production of L-valine based on transcriptome analysis and in silicon gene knockout stimulation. *Proceedings of National Academy of Science*, 104(19), 7797–7802.

Jin, L., Bhuiya, W. M., Mengmeng, L., Xiang, Q. L., Jixiang, H., Weiwei, D., Min, W., Oliver, Y., & Zhengzhu, Z. (2014). Metabolic engineering of saccharomyces cerevisiea for caffeine and thrombine production. *Public Library of Science Journal*, 9(8), 11–20.

Jun, S., Moon, C., Kang, C., Kong, S., Snag, B., & Um, Y. (2010). Microbial fed-batch production of 1,3-propanediol using raw glycerol with suspended and immobilized Klebsiella pneumonia. *Applied Biochemistry and Biotechnology*, 161(1–8), 491–501. doi:10.1007/s12010-009-8839-x.

Kaur, G., Srivastava, A. K., & Chand, S. (2012). Advances in biotechnological production of 1,3-propanediol. *Biochemical Engineering Journal*, 64, 106–118. doi:10.1016/j.bej.2012.03.002.

Kern, A., Tilley, E., Hunter, I. S., Legisa, M., & Glieder, A. (2007). Engineering primary metabolic pathways of industrial micro-organisms. *Journal of Biotechnology*, 129 (1), 6–29.

Khan, A., Blide, A., & Gadre, R. (2009). Mannitol production from glycerol by resting cells of *Candida magnolia. Bioresource Technology*, 100(20), 4911–4913.

Koshiishi, C., Kato, A., Yama, S., Crozier, A., & Ashihara, H. (2001). A new caffeine biosynthetic pathway in tea leaves: utilisation of adenosine released from the Sadenosyl-L-methionine cycle. *FEBS Letter*, 499, 5.

Kumar, V., Ramakrishnan, S., Teeri, T. T., Knowles, J. K. C., & Hartley, B. S. (1992). *Saccharomyces cerevisiae* cells secreting an *Aspergillus niger* p-galactosidase grow on whey permeate. *Biotechnology*, 10, 82–85.

Lee, W., & DaSilva, N. A. (2006). Application of sequential integration for metabolic engineering of 1,2-propanediol production in yeast. *Journal of Metabolic Engineering*, 8, 58–65.

Lee, C. S., Aroua, M. K., Daud, W. M., Cognet, P., Pérès-Lucchese, Y., Fabre, P. L., Reynes, O., & Latapie, L. (2015). A review: conversion of bioglycerol intol,3-propanediol via biological and chemical method. *Renewable Sustainable Energy Reviews*, 42, 963–972. doi:10.1016/j.rser.2014.10.033.

Li, M., & Borodina, I. (2015). Application of synthetic biology for production of chemicals in yeast *Saccharomyces cerevisiae. FEMS Yeast Research*, 15, 1–12.

Li, M., Schneider, K., Kristensen, M., Borodina, I., & Nielsen, J. (2016). Engineering yeast for high-level production of stilbenoid antioxidants. *Scientific Reports*, 6, 1–8. doi:10.1038/srep36827.

Lyu, X., Lee, J., & Chen, W. N. (2019). Potential natural food preservatives and their sustainable production in yeast: terpenoids and polyphenols. *Journal of Agriculture and Food Chemistry*, 67, 4397–4417. doi:10.1021/acs.jafc.8b07141.

Ma, Z., Rao, Z. M., Shen, W., Fang, H. Y., & Zhuge, J. (2007). Construction of recombinant *Saccharomyces cerevisiae* producing 1,3-propanediol by one step method. *Wei Sheng Wu Xue Bao*, 47, 598–603.

Mahian, R. A., & Hakimzadeh, V. (2016). Sugar alcohols: a review. *International Journal of PharmTech Research*, 9(7), 407–413.

Maury, J., Asadollahi, M. A., Moller, K., Clark, A., & Nielsen, J. (2005). Microbial isoprenoid production: an example of green chemistry through metabolic engineering. *Advances Biochemical Engineering and Biotechnology*, 100, 19–51.

McKeaguea, M., Wanga, Y.-H., Cravensa, A., Win, M. N., & Smolk, C. D. (2016). Engineering a microbial platform for de novo biosynthesis of diverse methylxanthines. *Metabolic Engineering*, 38, 191–203. doi:10.1016/j.ymben.2016.08.003.

Metsoviti, M., Paraskevaidi, K., Koutinas, A. A., & Zeng, A. P. (2012). Papanikolaou: production of 1,3-propanediol, 2,3-butanediol and ethanol by a newly isolated *Klebsiella oxytoca* strain growing on biodiesel-derived glycerol based media. *Process Biochemistry*, 47, 1872–1882. doi:10.1016/j.procbio.2012.06.011.

Michnick, S., Roustan, J. L., Remize, F., Barre, P., & Dequin, S. (1997). Modulation of glycerol and ethanol yields during alcoholic fermentation in Saccharomyces cerevisiae strains overexpressed or disrupted for GPD1 encoding glycerol 3-phosphate dehydrogenase. *Yeast*, 13, 783–793.

Nevoigt, E. (2008). Progress in metabolic engineering of *Saccharomyces cerevisiae. Microbiology and Molecular Biology Reviews*, 72(3), 379–412. doi:10.1128/MMBR.00025-07.

Ostergaard, S., Olsson, L., & Nielsen, J. (2000). Metabolic engineering of *Saccharomyces cerevisiae. Microbiology and Molecular Biology Reviews*, 64(1), 34–50.

Papanikolaou, S., Fakas, S., Fick, M., Chevalot, I., Galiotou-Panayotou, M., Komaitis, M., Marc, I., & Aggelis, G. (2008). Biotechnological valorisation of raw glycerol discharged after bio-diesel (fatty acid methyl esters) manufacturing process: production of 1,3-propanediol, citric acid and single cell oil. *Biomass & Bioenergy*, 32 (1), 60–71.

Park, Y., Oh, E. J., Jo, J., Jin, Y., & Seo, J. (2016). Recent advances in biological production of sugar alcohols. *Current Opinion in Biotechnology*, 37, 105–113. doi:10.1016/j.copbio.2015.11.006.

Patnaik, R. (2008). Engineering complex phenotypes in industrial strains. *Biotechnology Progress*, 24(1), 38–47.

Penttila, M. E., Suihko, U., Lehtinen, M., Nikkola, M., & Knowles, J. K. C. (1987). Construction of brewer's yeasts secreting fungal endo-β-glucanase. *Journal of Genetics*, 12, 413–420.

Prieto, J. A., Aguilera, J., & Randez-Gil, F. (2006). Genetic engineering of baker's yeast: challenges and outlook. In K. Shetty, G. Paliyath, A. Pometto, & R. E. Levin (Eds.), *Food Biotechnology* (pp 266–300), 2nd Edition. Boca Raton, Taylor & Francis Group, LLC.

Priya, S. S., Kumar, V. P., Kantam, M. L., Bhargava, S. K., & Chary, K. V. (2014). Catalytic performance of Pt/$AIPO_4$ catalysts for selective hydrogenolysis of glycerol to 1, 3, -propanediol in the vapour phase. *RSC Advances*, 4, 51893–51903.

Quitmann, H., Fan, R., & Czermak, P. (2014). Acidic organic compounds in beverage, food, and feed production. *Advanced Biochemistry and Engineering Biotechnology*, 143, 91–141. doi:10.1007/10_2013_262.

Raj, A. E., & Karanth, N. G. (2006). Fermentation technology and bioreactor design. In K. Shetty, G. Paliyath, A. Pometto, & R. E. Levin (Eds.), *Food Biotechnology* (pp 55–108), 2nd Edition. Boca Raton, Taylor & Francis Group, LLC.

Rieckenberg, F., Ardao, I., Rujananon, R., & Zeng, A. P. (2014). Cell-free synthesis of 1,3-propanediol from 1199 glycerol with a high yield. *Engineering Life Science*, 14, 380–386.

Rymowicz, W., Rywijska, A., Zlarowska, B., & Wojtatowicz, M. (2009). Biosynthesis of citric acid from glycerol by acetate mutants of yarrowia lipolyticain fed-batch fermentation. *Food Technology and Biotechnology*, 47, 1–6.

Saxena, R. K., Anand, P., Saran, S., & Isar, J. (2009). Microbial production of 1,3-propanediol: recent developments and emerging opportunities. *Biotechnology Advances*, 27(6), 895–913. doi:10.1016/j.biotechadv.2009.07.003.

Schiweck, H., Bär, A., Vogel, R., Schwarz, E., Kunz, M., Dusautois, C., ........., Peters, S. (2012). Sugar alcohols. *Ullmann's Encyclopedia of Industrial Chemistry*. Weinheim, Wiley-VCH. doi:10.1002/14356007.a25_413.pub3.

Schuller, H. J. (2003). Transcriptional control of non-fermentative metabolism in the yeast Saccharomyces cerevisiae. *Current Genetics*, 43, 139–160.

Shiba, Y., Paradise, E. M., Kirby, J., Ro, D. K., & Keasling, J. D. (2007). Engineering of the pyruvate dehydrogenase bypass in *Saccharomyces cerevisiae* for high-level production of isoprenoids. *Journal of Metabolic Engineering*, 9, 160–168.

Siso, M. G. I. (1996). The biotechnological utilization of cheese whey: a review. *Bioresource Technology*, 57, 1–11.

Song, C., Zhao, L., Ono, S., Shimasaki, C., & Inoue, M. (2001). Production of poly (3-hydroxybutyrate-co-3-hydroxyvalerate) from cottonseed oil and valeric acid in batch culture of *Ralstonia* sp. strain JC-64. *Applied Microbiology and Biotechnology*, 94(2), 169–178. doi:10.1385/ABAB:94:2:169.

Steensels, J., Snoek, T., Meersman, E., Nicolino, M. P., Voordeckers, K., & Verstrepen, K. J. (2014). Improving industrial yeast strains: exploiting natural and artificial diversity. *FEMS Microbiology Reviews*, 38, 947–995.

Szymanowska-Powalowska, D., Drozdzynska, A., & Remszel, N. (2013). Isolation of new strains of bacteria able to synthesize 1,3-propanediol from glycerol. *Advances in Microbiology*, 3, 171–180. doi:10.4236/aim.2013.32027.

Tabah, B., Varvak, A., Pulidindi, I. N., Foran, E., Banin, E., & Gedanken, A. (2016). Production of 1, 3-propanediol from glycerol via fermentation by *Saccharomyces cerevisiae. Green Chemistry*, 18, 4657–4666.

Takahashi, S., Yeo, Y., Greenhagen, B. T., McMullin, T., Song, L., Maurina-Brunker, J., Rosson, R., Noel, J. P., & Chappell, J. (2007). Metabolic engineering of sesquiterpene metabolism in yeast. *Biotechnology and Bioengineering Journal*, 97, 170–181.

Tanaka, S., Shinoki, A., & Hara, H. (2014). Melibiose, a non-digestible saccharide, promotes absorption of quercetin glycosides in rat small intestine with a novel mechanism. *FASEB*, 28(1), 1044–1049.

Toivari, M. H., Ruohonen, L., Miasnikov, A.N., Richard, P., Penttila, M. (2007). Metabolic engineering of *Saccharomyces cerevisiae* for conversion of d-glucose to xylitol and other five-carbon sugars and sugar alcohols. *Journal of Applied and Environmental Microbiology*, 73, 5471–5476.

Tomita, K., Nagura, T., Okuhara, Y., Nakajima-Adachi, H., Shigematsu, N., & Aritsuka, T. (2007). Dietary melibiose regulates the cell response and enhances the induction of oral tolerance. *Bioscience Biotechnology and Biochemistry*, 71(11), 2774–2780. doi:10.1271/bbb.70372.

Trantas, E., Panopoulos, N., & Ververidis, F. (2009). Metabolic engineering of the complete pathway leading to heterologous biosynthesis of various flavonoids and stilbenoids in *Saccharomyces cerevisiae. Metabolic Engineering*, 11(6), 355–366.

Turner, T. L., Kim, E., Hwang, C., Zhang, G., Liu, J., & Jin, Y. (2016). Conversion of lactose and whey into lactic acid by engineered yeast. *Journal of Dairy Science*, 100, 124–128. doi:10.3168/jds.2016-11784.

Tyo, K. E., Alper, H. S., & Stephanopoulos, G. N. (2007). Expanding the metabolic engineering toolbox: more options to engineer cells. *Trends in Biotechnology*, 25(3), 132–137.

Vincent, S. F., Bell, P. J. L., Bissinger, P., & Nevalainen, K. M. H. (1999). Comparison of melibiose utilizing baker's yeast strains produced by genetic engineering and classical breeding. *Journal of Applied Microbiology*, 2, 148–152.

Wang, C., Liwei, M., Park, J.-B., Jeong, S.-H., Wei, G., Wang, Y., & Kim, S.-W. (2018). Microbial platform for terpenoid production: Escherichia coli and yeast. *Frontier Microbiology*, 9, 2460–2467. doi:10.3389/fmicb.2018.02460.

Xiao, M., Tanaka, K., Qian, X. M., Yamamoto, K., & Kumagai, H. (2000). High-yield production and characterization of α-galactosidasen form *Bifdobacterium breve* grown on raffnose. *Biotechnology Letters*, 22(9), 747–751. doi:10.1023/A:1005626228056.

Yazdani, S. S., & Gonzalez, R. (2007). Anaerobic fermentation of glycerol: a path to economic viability for the biofuels industry. *Current Opinion in Biotechnology*, 18, 213–219.

Zhang, Y., Nielsen, J., & Liu, Z. (2017). Engineering yeast metabolism for production of terpenoids fro use as perfume ingredients, pharmaceuticals and biofuels. *FEMS Yeast Research*, 17(8), 1–11.doi:10.1093/femsyr/fox080.

Zhou, Y., Zhu, Y., Dong, Y. C., Sun, Y., & Zhang, J. (2017). Construction of engineered Saccharomyces cerevisiae strain to improve whole-cell biocatalytic production of melibiose from raffinose, *Journal of Industrial Microbioliogy and Biotechnology*, 44, 489–501. doi:10.1007/s10295-017-1901-8.

Zou, J., Guo, X., Shen, T., Dong, J., Zhang, C., & Xiao, D. (2013). Construction of lactose-consuming *Saccharomyces cerevisiae* for lactose fermentation into ethanol fuel. *Journal of Industrial Microbiology and Biotechnology*, 40, 353–363. doi:10.1007/s10295-012-1227-5.

# 8 The Role of an Intelligent Feedback Control System in the Standardization of Bio-fermented Food Products

*Charles Oluwaseun Adetunji, Wilson Nwankwo, Samuel Makinde, Kingsley Eghonghon Ukhurebor, Olugbemi Tope Olaniyan, and Florence U. Masajuwa*
Edo State University Uzairue

*Benjamin Ewa Ubi*
Ebonyi State University
Tottori University

## CONTENTS



DOI: 10.1201/9781003178378-8

## 8.1 INTRODUCTION

Fermentation of food products which normally involves the incorporation of edible microbes is an age-long practice throughout the globe. It has been identified as a sustainable methodology for the extension of shelf life of these food products (Steinkraus, 1986, 1994, 1997, 2009). Fermented foods normally improve the nutritional status in human diet and provide adequate nourishing texture, flavour, aroma, etc. It has been highlighted that beverages and fermented foods sum up to about one-third of the diet humans consume daily (Campbell-Platt, 1994).

Typical examples of fermented foods include sour-dough breads, sausages, alcoholic beverages, vegetable protein amino acid/peptide sauces and pastes, yoghurts, leavened bread and pickled vegetables (Steinkraus, 1997). The useful mechanisms that could be associated with microbial fermentation entail the biochemical transformation of sugars into carbon dioxide, alcohols, simple acids, for carbon metabolism as well as the biotransformation reactions such as the elimination of glycol-side residues which introduces health-beneficial activities, elimination of anti-nutrients from food substrates and effective conveyance of probiotics (van Hylckama Vlieg et al., 2011). Moreover, the biological action of these beneficial microorganisms could generate an array of metabolites that could enhance the preservation of these biological products as well as prevent several detrimental effect on human health (Adetunji and Anani, 2021a,b,c; Adetunji et al., 2021; Dwivedi et al., 2021; Olaniyan and Adetunji, 2021, Mishra et al., 2020; Thangadurai et al., 2021; Anani et al., 2020; Islam et al., 2021).

Moreover, the consumption of fermented foods incorporating live beneficial microorganisms has been identified as a significant dietary approach that could improve the normal human health (Marco et al., 2017). Also, lactic acid bacteria derived from numerous genera which include *Leuconostoc*, *Streptococcus*, and *Lactobacillus* are the major microorganisms mostly present in fermented foods, but there might be presence of fungi and yeast which might also contribute to food fermentation. Some commercially generated fermented food products could serve as vehicle for probiotic bacteria. Furthermore, despite the high awareness of these fermented foods and the potential public health benefits present in these foods, there is still a need to establish the safety of these fermented food products (Slashinski et al., 2012; Marco et al., 2017).

In view of the aforementioned, there is need to establish a robust framework that would ease production, distribution, safe consumption of fermented food products as well as policy regulation of all these highlighted activities. Hence, the application of an intelligent monitoring system could play an active role towards the collation of significant information on standards, procedures and available products vis-à-vis the health challenges posed by such products. This will go a long way towards the provision of technological frameworks that could provide intelligent feedbacks for adequate identification of fake products as well as to prevent it circulating along the distribution chain. Therefore, this study provides information on the coordinated intelligent feedback control systems that could boost the production of quality bio-fermented products that have been in existence in Nigeria.

### 8.1.1 Standardization of Bio-fermented Food Products

It has been highlighted that different food biotechnology or local fermented food industries utilized various forms of integrating mechanism during their production processes majorly from product to process development transfer (Ettlie, 1995). Moreover, considerable effort has been expended on the evaluation of necessary organizational techniques which entail employee rotation, cross-functional teams and personnel integration. These are all necessary for incorporation of necessary developmental units which could enhance adequate technology transfer (Liker et al., 1999). Cross-functional integration in new product development could be referred to as level of interaction and communication, the level of coordination, the rate of disseminating information and the rate of collaborative involvement among various functions that are involved in targeting new developmental projects tasks (Clark and Fujimoto, 1990; Wheelwright and Clark, 1992).

The major reasons behind this could be linked to wider functional diversity, the level of information and data available to members of the company or industry producing bio-fermented products which will increase drastically. Therefore, a more understanding on numerous challenges that revolves around the product development can present necessary solutions to mitigate challenges from small scale to mass production of fermented products (Milliken and Martins, 1996). It must be noted that standardization is a complement to cross-functional integration. Standardization might be referred to as the level of compatibility of new products in the contest of innovation (Besen and Farrell, 1994). Standardization enhances the synchronization of terminologies, the harmonization of measurement and evaluating techniques, as well as impeccable data exchange at boundaries (Blind and Gauch, 2009). Scientifically, there are five different types of standard which are applicable to fermented food production in industries. This includes quality, terminology, interface, measurement and testing, and compatibility standards (Blind and Gauch, 2009). These highlighted standards could help in overcoming several challenges that could be possible during the production of fermented food products, most especially in overcoming uncertainties such as organizational, technological and operational determinants, which could influence the transfer of product to process development.

## 8.2 EXAMINING THE BENEFITS AND CHALLENGES OF BIO-FERMENTED FOODS

### 8.2.1 Health Benefits

The benefits of bio-fermented foods have gained recognition globally in recent times. Alluding to the foregoing is the increasing rate at which individuals and food processing companies are redefining, re-emphasizing and utilizing fermentation in food preparation, processing, preservation, boosting of shelf life and advancement of flavour (Frías et al., 2016; Linares et al., 2017; Pessione and Cirrincione, 2017; Sanlier et al., 2017; Milini et al., 2019). Consequent upon these health benefits, bio-fermented foods currently play an integral part in the diet of several households in developing nations where they are an essential component of customary/traditional food preparations (Frías et al., 2016; Linares et al., 2017; Pessione and Cirrincione, 2017; Sanlier et al., 2017; Milini et al., 2019). Over time, research studies have shown that fermentation has a nexus with healthy living (Sanlier et al., 2017; Sivamaruthi et al., 2018a).

Presently, some microbial agents of fermentation have been linked to some health benefits, and as such, these microorganisms have also become elemental points of emphasis (Sanlier et al., 2017; Sivamaruthi et al., 2018b; Rocchetti et al., 2019). As reported by Sanlier et al. (2017), lactic acid bacteria are one of the commonly studied microbes in this regard. According to them, through fermentation, these microbes manufacture vitamins, minerals, peptides and enzymes like peptidase and proteinase while eliminating some non-nutrients. It is also reported that biologically energetic peptides produced by those bacteria offer significant health benefits (Beltrán-Barrientos et al., 2018a; Çabuk et al., 2018; Sivamaruthi et al., 2018a). Table 8.1 shows the various sources and several microorganisms involved in the production of bio-fermented foods in Nigeria.

Like the peptides, conjugated linoleic acids have also been reported to have blood pressure lowering characteristics (Frías et al., 2016; Pessione and Cirrincione, 2017; Sanlier et al., 2017; Beltrán-Barrientos et al., 2018b; Milini et al., 2019); exopolysaccharides display prebiotic features; bacteriocins demonstrate antibiotic effects; sphingolipids have anti-carcinogenic and anti-microbial features; and bioactive peptides display anti-oxidant, anti-microbial, opioid adversary, anti-allergenic and blood pressure reducing effects (Frías et al., 2016; Pessione and Cirrincione, 2017; Sanlier et al., 2017; Beltrán-Barrientos et al., 2018a; Milini et al., 2019).

Bio-fermented foods provide several health benefits which include "anti-inflammatory, anti-oxidant, anti-microbial, anti-diabetic, anti-fungal, and anti-atherosclerotic activity". According to Milini et al. (2019) and Zhang et al. (2018), some benefits exhibited by bio-fermented foods include but not limited to the following:

a. Lowering of blood cholesterol levels.
b. Immune boosting.
c. Protection against pathogens.
d. Anti-carcinogenic, anti-diabetic, anti-obese, anti-allergic and anti-atherosclerosis and reduce osteoporosis.
e. Alleviate symptoms of lactose intolerance.

**TABLE 8.1**
**Some Bio-fermented Foods in Nigeria**

| Major Groups of Fermented Foods | Specific Examples of Fermented Food Products | Microorganisms Involved in the Fermentation Processes | References |
|---|---|---|---|
| Fermented cereals | Ogi derived from Maize, sorghum, millet | *Penicillium* sp., *Lb. plantarum*, *Aspergillus* sp., *Lb. pantheris*, *Fusarium* sp., *Lb. vaccinostercus*, *Cephalosporium* sp., *Corynebacterium* sp., *Aerobacter* sp., *Rhodotorula* sp., *Sacch. cerevisiae*, *Aerobacter* sp., *Lactobacillus plantarum*, *Streptococcus lactis* | Omemu (2011) |
| Fermented legumes | Dawadawa, Iru, derived from Locust bean | *B. pumilus*, *Lysininbacillus sphaericus*, *B. licheniformis*, *B. mojavensis*, *B. firmus*, *B. mojavensis*, *B. amyloliquefaciens* | Odunfa (1981); Adewumi (2014); Owusu-Kwarteng (2020) |
| Fermented roots/ tubers | Fufu, Garri derived from Cassava | *Candida* sp., *Bacillus* sp., *Corynebacterium* sp., *Lb. plantarum*, *Leuconostoc* sp., *Leuconostoc Mesenteroides*, *Klebsiella* sp., *Lactobacillus brevis*, *Lactobacillus cellobiosus*, *Lactobacillus coprophilus*, *Lc. lactis* | Adetunji (2017) |
| Fermented milk products | Wara (locally produced cheese) | *Leuconostoc lactis* subsp. *cremoris*, *Leuconostoc lactis* subsp. *lactis*, *Lactobacillus delbrueckii* subsp. *delbrueckii*, *Lactobacillus delbrueckii* subsp. *lactis*, *Lactobacillus helveticus*, *Lactobacillus casei*, *Lactobacillus plantarum*, *Lactobacillus salivarius*, *Leuconostoc* spp., *Strep. thermophilus*, *Ent. durans*, *Ent. faecium*, and *Staphylococcus* spp., *Brevibacterium linens*, *Propionibacterium freudenreichii*, *Debaryomyces hansenii*, *Geotrichum candidum*, *Penicillium camemberti*, *P. roqueforti* | Sangoyomi (2010) |
| Fermented meat product | Suya | *W. paramesenteroides*, *Lactobacillus pentosus*, *Weissella cibaria*, *Lactobacillus plantarum*, *Ped. pentosaceus*, *Ped. stilesii*, *Lactobacillus paracasei*, *Lactobacillus fermentum*, *Ped. acidilactici*, *Ped. pentosaceus*, *Lactobacillus. namurensis*, *Lactobacillus lactis*, *Leuconostoc. citreum*, *Leuconostoc fallax* | Orpin (2018) |
| Fermented, dried and smokedfish products | | *Saccharomycopsis* sp., *Lact. lactis*, *Lb. plantarum*, *Lb. pobuzihii*, *Lb. fructosus*, *Clostridium irregular*, *Azorhizobium caulinodans*, *Candida* sp., *Lb. amylophilus*, *Lb. coryniformis*, *Ent. faecium*, *B. subtilis*, *B. pumilus*, *B indicus*, *Micrococcus* sp., *Staphy. cohnii* subsp. *cohnii*, *Staphy. carnosus*, *Tetragenococcus halophilus* subsp. *flandriensis* | Adeyeye (2015); Ayeloja et al. (2018); Babalola (2018) |
| Alcoholic beverages | Burukutu, Kunu | *Kloeckera apiculata*, *Kl. thermotolerans*, *Hanseniaspora uvarum*, *bacteria* (*Oenococcus oeni*, *Lactobacillus plantatum*), *Saccharomyces* spp., *Candida colliculosa*, *C. stellata*, *Torulaspora delbrueckii*, *Metschnikowia pulcherrima*, *Pichia fermentans*, *Schizosaccharomyces pombe* | Yusuf (2020) |

The health benefits of bio-fermented foods may be ascribed to the bioactive peptides produced following microbial breakdown of proteins, vitamins and other composites during fermentation (Frías et al., 2016; Pessione and Cirrincione, 2017; Sanlier et al., 2017; Beltrán-Barrientos et al., 2018b; Çabuk et al., 2018; Zhang et al., 2018; Milini et al., 2019). One of these bioactive peptides is "angiotensin-1-converting enzyme inhibitor peptides" synthesized during the fermentation of milk. Such peptides especially valyl-prolyl exhibit some antihypertensive effects (Sanlier et al., 2017; Milini et al., 2019). It is submitted that notwithstanding the seemingly beneficial qualities of these bio-fermented products, more research should be conducted as to unravel the quantity and quality of these products that would confer optimum health benefits to the end consumer (Frías et al., 2016; Pessione and Cirrincione, 2017; Sanlier et al., 2017; Milini et al., 2019).

## 8.2.2 Health Problems Posed by Bio-fermented Foods

Anticipated health benefits, taste and customary beliefs are major drivers behind the consumption of bio-fermented foods. In most climes, consumers are not mindful of the attendant problems and health challenges such products may pose (Gobbetti et al., 2014; Bell et al., 2017; Aslam et al., 2018; Beltrán-Barrientos et al., 2018a; Galli et al., 2018; Giau et al., 2018; Sivamaruthi et al., 2018b; Zhang et al., 2018; Mohammed, 2019; Mustafa et al., 2019). Accordingly, some health risks/hazards attributed to bio-fermented foods are many, some of which are presented as follows.

### 8.2.2.1 Bloating

This is a momentary upsurge in gas and distending of the gastrointestinal tract following the consumption of bio-fermented foods (Galli et al., 2018; Zhang et al., 2018; Mohammed, 2019). This occurs as a result of superfluous gas being generated after probiotics destroy injurious gut bacteria and fungi. Probiotics secrete anti-microbial peptides that destroy injurious pathogenic microorganisms such as *Salmonella* spp. and *E. coli.* Recently it has been reported that anti-microbial effects of probiotic Lactobacilli strains are found in commercial yoghurt (Mohammed, 2019). Though to bloat after intake of some bio-fermented foods probiotics appears to be a good indication that the injurious bacteria are being eliminated from the gut, some persons could witness essence distending, which could sometimes be so painful. Intake of excess kombucha tea could also lead to superfluous sugar and calorie consumption, which might also cause bloating.

### 8.2.2.2 Headaches, Hemicrania and Migraine

Bio-fermented foods such as kimchi, sauerkraut and yoghurt contain a lot of probiotics, such as biogenic amines formed during fermentation (Gobbetti et al., 2014; Galli et al., 2018; Mohammed, 2019). Amines are the products of microbial breakdown of amino acids present in bio-fermented foods. Histamine and tyramine are abundant in bio-fermented foods (Mohammed, 2019). Studies have shown that some individuals are allergic to histamine and other amines, and such persons could have headaches following the ingestion of bio-fermented foods owing to excessive stimulation of the central nervous system. Such neurotransmissions covertly affect blood flow, blood pressure and intracranial pressure with resultant headaches/migraines (Gobbetti et al., 2014; Galli et al., 2018; Mohammed, 2019).

#### 8.2.2.3 Histamine Allergy

Bio-fermented foods contain histamine in significant quantities. In some cases, these substances are catalytically broken down by specific enzymes within the body. However, such enzymes are not present in every individual (Mohammed, 2019). It follows that in such individuals histamine would circulate freely in the bloodstream. This may predispose the affected person to a chain of histamine-induced allergies leading to urticaria, headaches/migraines, rhinitis, hyperaemia, tiredness, hives and some digestive anomalies (diarrhoea, biliousness, nausea, etc.). Histamine-induced allergy could also cause asthma, hypotension, dysrhythmia, circulatory insufficiency, insomnia, nervousness, ferociousness, dizziness and concentration deficiency (Beltrán-Barrientos et al., 2018b; Mohammed, 2019).

#### 8.2.2.4 Food-Borne Infection

Even though most bio-fermented foods are innocuous, there is every tendency to be contaminated with infectious microbes, as have been reported in the USA and South Korea (Mohammed, 2019).

#### 8.2.2.5 Compromise of Probiotics

In some very rare cases, probiotics could predispose their hosts to infection especially in immunocompromised persons (Aslam et al., 2018; Giau et al., 2018; Mohammed, 2019). A prominent case has been documented in London where liver abscess developed in a diabetic patient following the consumption of probiotic (Mohammed, 2019).

#### 8.2.2.6 Antibiotic Resistance

Probiotics can transmit genes that offer resistance to some antibiotics (Sivamaruthi et al., 2018a; Mohammed, 2019). Such genes could be transferred to other microbes present in the food and/or in the gastrointestinal tract through horizontal gene transmission (Mohammed, 2019). Resistance to erythromycin and tetracycline through bio-fermented food has been noted (Mohammed, 2019). Researchers have found resistant probiotic strains in commercially available dietary supplements, which could mean resistance to several common antibiotics used to treat serious bacterial infections (Aslam et al., 2018; Sivamaruthi et al., 2018b; Mohammed, 2019).

A study conducted in Malaysia revealed that probiotic *Lactobacilli* in *kefir* exert resistance to numerous antibiotics, including ampicillin, tetracycline, and penicillin. These drugs had long been used for the treatment of serious infections including pneumonia, gonorrhoea, meningitis, etc. (Sivamaruthi et al., 2018a; Mohammed, 2019). Despite that there are an eclectic variety of health benefits derivable from consuming bio-fermented foods, it is worthy to note that some research studies have also shown that bio-fermented foods also pose serious health risks and hazards (Mohammed, 2019).

## 8.3 REGULATORY FRAMEWORK AND SUBSTANTIVE LEGISLATIONS GOVERNING THE PRODUCTION AND CONSUMPTION OF BIO-FERMENTED FOOD PRODUCTS

In various jurisdictions, public authorities are legitimately created often by an act (of parliament) or laws, edicts, decrees, bye-laws, etc. to regulate and control the production and sale of food products. To complement the efforts of such public authorities, some countries extend the rights of control and regulation to the private sector. In the USA and Canada, there are more than one public food control and inspection agency. The same is true in other jurisdictions like the Netherlands, Germany, etc. In such jurisdictions, the rational underlying the existence of dual agencies is not far-fetched. The authorities are meant to complement each other and where there exist private agents, they provide a somewhat independent quality assessment and this makes quality assurance an all-encompassing venture unlike in scenarios where it is left in the hands of public bureaucracy alone which would often exhibit greater tendencies towards compromising standards.

Regulatory frameworks are critical towards ensuring that standards and quality practices are duly adhered to by the vendors or producers of such food and allied products. They come in various formats such as policies, guidelines, legislations, executive and/or judicial orders. Irrespective of the form such documents take, one thing is aimed at, that is the control and regulation of products.

Policies are legitimate instruments that are operated or executed by agencies duly mandated by an applicable legislation to carry out some defined activities within the scope of the said policy. Such policies may not originate from the agency but from the supervising ministry and duly approved by the executive government. Guidelines on the other hand are subordinate legislations or rules emanating from the agency that issues it by virtue of the powers vested on the agency by virtue of the law establishing it.

For instance, in Nigeria, agencies such as National Agency for Food and Drug Administration (NAFDAC) operate policies on food and drug control communicated to it by its supervisor, the Federal Ministry of Health (FMoH) whose Minister is the Chair of the National Council on Health (NCH). The NCH is responsible for the making and approval of all health-related policies in Nigeria; however, the NCH is not the final authority as whatever policy it makes or validates must be presented to the Federal Executive Council (FEC) whose chairman is the President and Commander-in-chief of the armed forces. Once the policy is authenticated and endorsed by FEC, it is set for implementation.

### 8.3.1 Constitutional Basis for Food Health

The right to health may be inferred from Chapter 2 of the Constitution of the Federal Republic of Nigeria (1999) which dwells on the State Policy. Section 17(3)(d) of this chapter is to the effect that the government shall ensure adequate medical and health facilities in the country for all of its citizenry. Though this section is not justiciable (Okogie, 1981; Okeke and Okeke, 2013; Awolowo, 2017), government at the three tiers (federal, state and local) has consistently made efforts to establish facilities and agencies aimed at driving constitutional healthcare objectives at the various levels of government.

### 8.3.2 National Health Act (NHA)

The NHA bill was endorsed in 2014 and became effective January 2015. The NHA has been praised in various quarters as the most comprehensive health legislation Nigeria has ever had. Some of the remarkable provisions of the said legislation are:

a. Robust framework that includes development, regulation and management of the nation's Health System (NHS) that integrates public, private and traditional healthcare institutions.
b. As there is no tangible constitutional provision on healthcare, the NHA became the principal health legislation providing for standardization and provision of care throughout the country.
c. A structure for governance and policy in health delivery.
d. Substantial provisions on healthcare provision fund.
e. Framework for the establishment of robust health information system.
f. Quality assurance system for healthcare services across private and public health subsectors that provide for standards for certification, registration and monitoring.
g. A comprehensive legal framework that would drive affordable quality healthcare delivery in Nigeria.

### 8.3.3 Food, Drugs and Related Products (Registration, etc.) Act

This legislation provides the necessary governance, regulatory and enforcement framework for ensuring that only processed foods duly registered by NAFDAC in line with the provisions of this Act are to be produced, imported and sold in the common markets and consumed by the citizenry (Section 1(1)). Section 11(1) of the Act established the Food and Drug Registration Committee saddled with the task of evaluating the formation, preparation method, packaging, labelling, efficacy, safety, etc. of drugs, medical devices, food products, cosmetics and drug products, for which vendors, distributors, importers and producers have filed an application for (Section 11(2(a)). This legislation criminalizes actions made by producers, importers and distributors of processed food or drug in contravention to the provisions of the Act. NAFDAC has the mandate to prosecute individual or organization engaged in any contravening activity against the said provisions (Awolowo, 2017).

### 8.3.4 Food and Drug Act (FDA)

Like its counterpart discussed in Section 8.2.3 above, the FDA provides substantial framework to regulate the production, distribution, sale of food products, etc. The legislation prohibits the production, manufacturing, importation, distribution and sale of any unwholesome or substandard food and food-related products. Section 5(a) is of the effect that all manner of misleading practices and misrepresentations as to food/drug/cosmetic labelling, packaging, processing, distribution, advertising, etc. in a misleading manner in respect of composition, quality, value, safety, etc. is prohibited (Food and Drug Act Cap F33 Laws of the Federal Republic of Nigeria, 2004).

### 8.3.5 Counterfeit and Fake Drugs and Unwholesome Processed Foods (Miscellaneous Provisions) Act

This federal legislation criminalizes and prohibits the production, importation, manufacturing, distribution, sale, possession, displaying for sale, of any substandard, fake, adulterated, banned, unwholesome processed food regardless of the nature or form (Section 1(a)). In the same vein, anyone who aids another person or organization to do the same is culpable and may be convicted in accordance with the provisions of the law (Food and Drug Act Cap F32 Laws of the Federal Republic of Nigeria, 2004).

To aid enforcement of the said provisions this law confers jurisdiction on the Federal High Court for the trial of all offences defined therein three enforcement authorities are established:

a. Federal Task Force (FTF) (Section 5).
b. State Task Force (STF) (Section 7).
c. Nigeria Police Force (NPF) Squad (Section 9).

### 8.3.6 National Health Policy (NHP) 2016

The NHP is another milestone in the delivery of healthcare services including the control and regulation of food products. The NHP 2016 replaces the NHP 2004. The NHA is an offshoot of the NHA and its objectives are directed towards ensuring universal health coverage in the country.

### 8.3.7 National Policy of Food Safety (NFSP)

The NFSP has the following goals:

a. Establishment of an institutional framework that consolidates existing food safety and/or control systems towards a unified food safety standard.
b. Provides and maintains minimum benchmarks to be adhered to by food producers, vendors, distributors, etc. to ensure consumer protection against hazards and health problems arising from food products.
c. Promote consumption of safe food products.
d. Support public health programmes towards reducing incidences of food-borne diseases across the nation.
e. Ensure consumer protection through programmes that promote reduction of exposure to food hazards.

### 8.3.8 NAFDAC Act

The NAFDAC Act LFN 2004 established the NAFDAC, the sole regulatory and control agency mandated by the Act to drive programmes and health policies designed to ensure quality of drugs (medicaments), food, etc. produced locally, imported, exported, and/or advertised for, sold, or distributed in Nigeria. According to the NHP, notwithstanding NAFDAC's efforts in keeping substandard and fake food and drug products at bay, there are still enormous challenges.

### 8.3.9 Standards Organization of Nigeria (SON) Act 2015

The SON Act 2015 replaced the previous legislation SON Act 2004. This legislation provided extensive legal and regulatory framework and established SON, Standards Council of Nigeria, and its other machineries including task forces, etc. Section 5 of the Act empowers the agency to (Counterfeit and Fake Drugs and Unwholesome Processed Foods (Miscellaneous Provisions) Act Cap C34 LFN 2004):

a. Establish, define and approve standards relating to materials, commodities, processes, measurements, etc. in Nigeria.
b. Enforce the said standards as to product quality, among others.
c. Certify products that attain set standards.
d. Inspect the standard and/or quality of products/processes and facilities deployed for production purposes.
e. Seize, destroy and prohibit the sale of products that are inconsistent with approved standards in Nigeria.

### 8.3.10 National Health Promotion Policy 2006

This framework provides an integrated platform wherein every stakeholder in the sector is recognized as a key player in furtherance ensures that contributions from every quarter count. It provides grassroots community structures (committees like ward development, village development, health facility, etc.) that participate in demand-creation, health services monitoring, community mobilization, health awareness campaigns, etc.

The essence of this intensive integration is to afford every member of the public to report or partake in health promotion activities. For instance, where food products in circulation within a community, village, etc. are suspected or proven to have caused some harm, it is the duty of the stakeholders within that locality to report such matters for further actions by the authorities.

## 8.4 EXISTING FEEDBACK MECHANISMS FOR BIO-FERMENTED FOOD PRODUCTS

### 8.4.1 Mechanisms Designed by NAFDAC

The identification of fake bio-fermented food products like other fake products is a heinous task. In Nigeria, NAFDAC is the only agency saddled with this responsibility. Note that NAFDAC is not a standards issuance authority. NAFDAC's mission is to protect the public against food and consumable products that could jeopardize the health of the public. The mechanisms put in place to check such fake products include:

a. Licensing and registration of products.
b. Periodic inspection of vendor/Producer's production site for recertification.
c. Ad hoc visits to markets, stalls, shops, etc. for on-the-spot inspection and assessment of products.

d. Use of the Task forces and the NPF Squad to inspect, seize, arrest and prosecute offenders.
e. Seizure of products deemed not to conform to regulatory requirements.
f. Use of regional laboratories for analysis of product samples.
g. Arrest and prosecution of offenders.
h. Public awareness programmes and campaigns.

### 8.4.2 Mechanisms of Operation of SON

The issue of what constitutes standard of locally produced or imported products is within the jurisdiction of the SON. This implies that SON has a critical role in matters of standardization and could prosecute any defaulter. As regards food products, it is common knowledge that raw materials, processes and vehicles could be involved. If the quality or standard of such materials are inconsistent with set standard, it follows that the product realized from such would be substandard ab initio. SON is mandated by the law establishing it to work with relevant agencies locally and internationally to achieve its objects. This explains the synergy that exists between NAFDAC and SON. Overtly, SON applies similar strategies as NAFDAC in ensuring that its mandate of standards maintenance is not brought into question.

### 8.4.3 Consumer Protection Council (CPC)

Like SON and NAFDAC, the CPC is a public agency with the mandate of protecting consumers against sharp practices by vendors of consumable products not limited to food. The CPC helps aggrieved persons and victims in obtaining justice either through mediation, conciliation, arbitration or adjudication at the courts. The agency provides platforms that relates with the industry food producers, distributors, marketers, professional associations, etc. on such matters as directed towards quality standards development and enforcement especially as it affects the interests of the end consumer.

### 8.4.4 Adequacy of Existing Mechanisms

Various legislations and authorities have been discussed in the preceding sections. Whereas the legislations and authorities are crucial requirements to ensure safety of food produced, imported, distributed and consumed in a country like Nigeria, there is reasonable doubt on the capacity of the existing mechanisms to provide adequate campaign and control against fake foods including bio-fermented foods which are on the increase owing to their relatively low cost of production.

Another point is the strength and resources of the agencies and their existing modus operandi. It may be submitted that two national agencies with combined staff strength of less than 200,000 cannot police food and drug production, importation, distribution and regulation in a country of 200 million people. Such a situation calls for an emergency in the sector especially in the development of complementary media/channels/mechanisms that could help save the country from fake food products not just bio-fermented food products.

## 8.5 PROPOSING AN INTELLIGENT FEEDBACK CONTROL SYSTEM (IFCS)

Owing to the numerous health concerns and deaths that may ensue following bio-fermented food consumption, and the attendant inadequacy of existing mechanisms of control, an intelligent stakeholder-centred feedback system is proposed. This proposition is hinged on the growing acceptance and use of modern information technologies (smart phones, tablets, notebooks, etc.) as part of the day-to-day socio-economic activities in all parts of the country regardless of the degree of urbanization.

Feedback control systems have an age-long history in the health delivery cycle. Surveillance and digital decision-support systems have been used for the data gathering, disease detection, prognosis, forecasting, etc. (Thacker and Berkelman, 1988; White and McDonnell, 2000; Nsubuga et al., 2006). The NHA and NHP both harp on the need for a health management information system (HMIS). The HMIS system is intended to be an all-encompassing integrated platform that would drive all health delivery efforts. However, at present the realization of the HMIS could at best be described as on-going. There is no certainty as to when it would be commissioned. Moreover, if it is accomplished in meeting the demands of strategic agencies like NAFDAC in food regulation and standardization, it may pose another issue, thus the need to have a system dedicated to standardization which could also be incorporated into a larger system.

Figure 8.1 shows some vital technologies, authorities and communication channels in the establishment of a digital feedback control system to check the menace posed by fake bio-fermented foods in Nigeria. In the diagram, NAFDAC is the coordinator of the system whereas every concerned stakeholder has access to either submit complaints or make inquiries on certain issues while the system responds in real time. All stakeholders in the healthcare delivery chain are connected to the system.

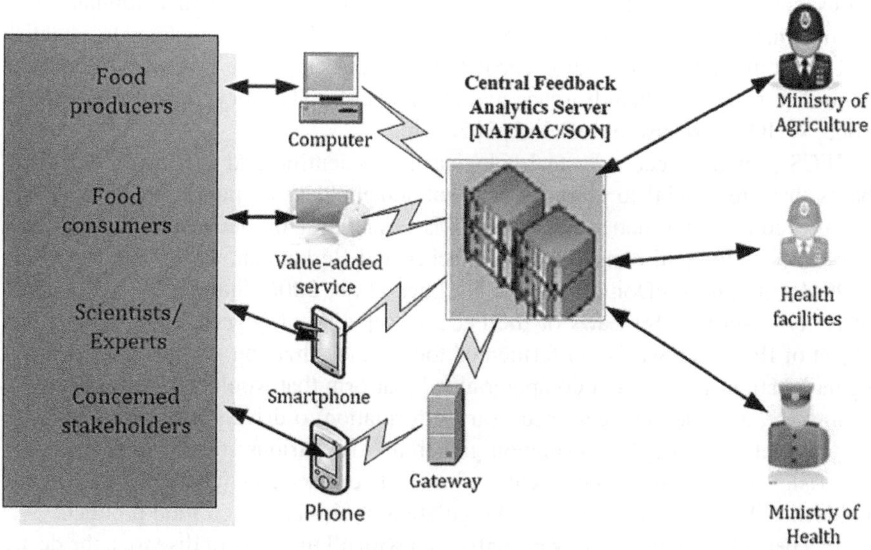


**FIGURE 8.1** Technologies, and authorities in a food product feedback control system.

### 8.5.1 Rationale for IFCS in Bio-fermented Food Standardization

IFCS is an essential component in every functional information-driven sector of an economy. In the health sector, effective public health information exchange is the bane of a functional health management. Health has a very strong nexus with food consumed. Therefore, it may be safe to submit that the quality of food consumed to a large extent defines the health indices of the population. To ensure safety and health of the citizenry, a system that complements the legislative and policy frameworks; and existing enforcement machineries are advocated.

The IFCS is a potential solution which supports public health intelligence through the provision of resources for data gathering, investigation, fuzzy analysis, inference, storage and exchange thereby enabling the prevention, and management of health issues (Thacker and Berkelman, 1988; Thacker et al., 1996; White and McDonnell, 2000; Nsubuga et al., 2006; Standards Organization of Nigeria Act Cap S9 Laws of the Federal Republic of Nigeria, 2015). Supposedly, the IFCS is a subset of the public health promotion system whose objectives and dealings are required in making appropriate interventions as to food and related products in circulation, which could exert significant impact on the health subsector.

According to Nsubuga et al. (2006), the main reason for IFCS is to strengthen the necessary stakeholders in their management procedures for effective efficiency provision, since these systems could assist in evaluation of the essential interventions as well as the measurement of such interventions. Gradually, relevant health stakeholders and agencies in developing nations are now recognizing the fact that data from IFCS could enable the appropriate channelling of efforts, resources and programmes (Thacker and Berkelman, 1988; Thacker et al., 1996; White and McDonnell, 2000; Nsubuga et al., 2006; Standards Organization of Nigeria Act Cap S9 Laws of the Federal Republic of Nigeria, 2015). The foregoing is consistent with the guidelines for executing the 2005 International Health Regulations (IHR) which stipulates that member nations of the World Health Organization (WHO) are mandated to establish and maintain facilities including capacity that could support responses and feedback control in respect of health issues and emergencies (Thacker et al., 1996; USAID, 2005; WHO, 2016; Nwankwo et al., 2020a, b).

IFCS provides necessary technological and scientific and high precision databases that are crucial to alert the relevant stakeholders on public health concerns and procedures, and such may include all issues relating bio-fermented food production, distribution and consumption (Thacker and Berkelman, 1988; Thacker et al., 1996; White and McDonnell, 2000; Nsubuga et al., 2006; Standards Organization of Nigeria Act Cap S9 Laws of the Federal Republic of Nigeria, 2015). The main object of IFCS as regards bio-fermented food standardization is to provide an intelligent location-independent computer-aided platform that would assist authorities to synthesize, analyse and relay necessary information to drive specification of quality parameters (through information gathered from various stakeholders including experts), control and management, refinement, checks and feedbacks that would guide distribution, consumption and regulation of bio-fermented food products.

In a period when the society is confronted with all manners of diseases, the desire to integrate international links is incontestable and research on what way these apprehensions are tackled is indispensable. Partnership amid all the relevant stakeholders both nationally and internationally is needed to evolve functional IFCS to ensure that every produced and distributed bio-fermented food product meets the stipulated standards in order to ensure such a drive and technological solutions.

### 8.5.2 Data Specifications for an IFCS

Standardization of any bio-fermented food product requires accurate and reliable data which must exhibit measurable characteristics like completeness, known sources and validation by an appropriate authority. In the same vein, feedback control regardless of its intelligent capacity is a function of data on the target phenomenon. Accordingly, to evolve a functional system appropriate data specifications and algorithm relevant to the context must be defined. In this section attempt is made at defining the relevant data items in the context of inputs and outputs. The business logic is expected to be implemented using fuzzy logic system owing to multi-valued inputs (uncertain, vague, imprecise, multifarious, etc.) that may be harvested from the actors. The system implementation details are not discussed in this chapter.

#### 8.5.2.1 Actors

The actors include primary stakeholders (vendors, producers, consumers and sellers), food experts, policy makers/implementers (NAFDAC/SON) and observers. The IFCS provides a common platform for all the actors to interact. All actors are also valid sources of data. However, data provided by the actors must be cleaned and validated by the logic and inference engine of the IFCS prior to use in updating any existing data or creating a new piece.

#### 8.5.2.2 Objectification of the Stakeholder Relationships

Here, the various objects are defined and their interrelationships defined. Figure 8.2 is a class diagram that reflects such relationships. The central hub in the diagram is the *infobasket* which is an object that provides interface-short messages (SMS), web, mobile, etc. through which data for the inference/logic engine are supplied.

## 8.6 CONCLUSION

In this chapter, the sources, usefulness and problems associated with bio-fermented food products were adequately reviewed. It was noted that several policies, legal frameworks and machineries already exist which the authors consider suitable in its own right. Notwithstanding the foregoing, experience shows that a gap exists among the various stakeholders. Of utmost worry is the absence of functional and comprehensive Information and Communication Technology (ICT)-driven platform (as clearly adumbrated by the NHA) such as an IFCS that provides intelligent feedbacks on food products for easy identification, tracking of fake bio-fermented products, and at the same time, support data harvesting from experts, consumers and other informed stakeholders thereby developing and entrenching necessary standards.

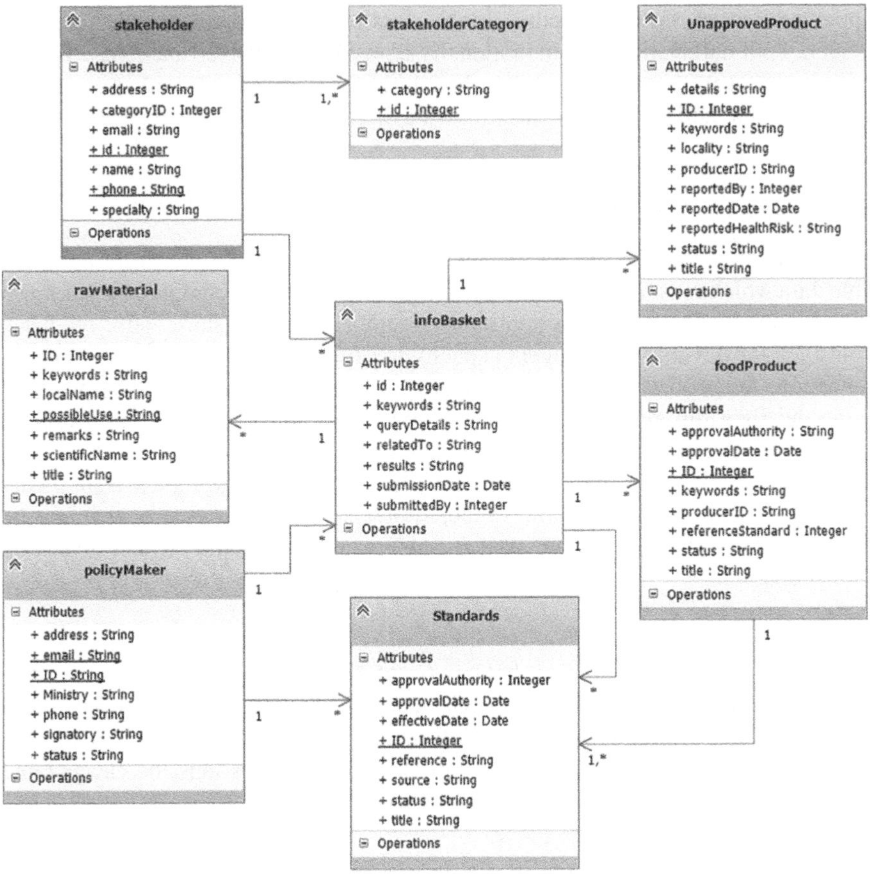


**FIGURE 8.2** Data specification for the IFCS.

The authors propose a coordinated IFCS and conclude that such is the only ideal platform that would boost the standardization, production and safe consumption of bio-fermented products based on real-time collective inputs.

## REFERENCES

Adetunji, C.O., Akande, S.A., Oladipo, A.K., Salawu, R.A., Onyegbula, A.F. (2017). Determination of the Microbiological Quality and Proximate Composition of Fermented Cassava Food Products Sold in Ilorin-West Local Government Area, Nigeria. *Ruhuna Journal of Science*, 8, 76–89. DOI: 10.4038/rjs.

Adetunji, C.O., Anani, O.A. (2021a). Recent advances in the application of genetically engineered microorganisms for microbial rejuvenation of contaminated environment. In: Adetunji, C.O., Panpatte, D.G., Jhala, Y.K. (eds). *Microbial Rejuvenation of Polluted Environment. Microorganisms for Sustainability*, vol 27. Springer, Singapore. https://doi.org/10.1007/978-981-15-7459-7_14

Adetunji, C.O., Anani, O.A., Panpatte, D. (2021). Mechanism of actions involved in sustainable ecorestoration of petroleum hydrocarbons polluted soil by the beneficial microorganism. In: Panpatte, D.G., Jhala, Y.K. (eds). *Microbial Rejuvenation of Polluted Environment. Microorganisms for Sustainability*, vol 26. Springer, Singapore. https://doi.org/10.1007/978-981-15-7455-9_8

Adetunji, C.O., Anani, O.A. (2021b). Utilization of microbial biofilm for the biotransformation and bioremediation of heavily polluted environment. In: Panpatte, D.G., Jhala, Y.K. (eds). *Microbial Rejuvenation of Polluted Environment. Microorganisms for Sustainability*, vol 25. Springer, Singapore. https://doi.org/10.1007/978-981-15-7447-4_9

Anani, O.A., Adetunji, C.O. (2021c). Bioremediation of polythene and plastics using beneficial microorganisms. In: Adetunji, C.O., Panpatte, D.G., Jhala, Y.K. (eds). *Microbial Rejuvenation of Polluted Environment. Microorganisms for Sustainability*, vol 27. Springer, Singapore. https://doi.org/10.1007/978-981-15-7459-7_13

Anani, O.A., Mishra, R.R., Mishra, P., Enuneku, A.A., Anani, G.A., Adetunji, C.O. (2020). Effects of toxicant from pesticides on food security: Current developments. In: Mishra, P., Mishra, R.R., Adetunji, C.O. (eds). *Innovations in Food Technology*. Springer, Singapore. https://doi.org/10.1007/978-981-15-6121-4_22

Adewumi, G.A., Oguntoyinbo, F.A, Romi, W., Singh, T.A., Jeyaram, K. (2014). Genome Subtyping of Autochthonous Bacillus Species Isolated from Iru, a Fermented *Parkia biglobosa* Seed. *Food Biotechnology*, 28(3), 250–268.

Adeyeye, S.A.O., Oyewole, O.B., Obadina, A.O., Omemu, A.M. (2015). Microbiological Assessment of Smoked Silver Catfish (Chrysichthys nigrodigitatus). *African Journal of Microbiology Research*, 5, 1821–1829.

Aslam, H., Green, J., Jacka, F.N., Collier, F., Berk, M., Pasco, J., Dawson, S.L. (2018). Fermented Foods, the Gut and Mental Health: A Mechanistic Overview with Implications for Depression and Anxiety. *Nutritional Neuroscience*, 11, 1–13.

Awolowo, O. (2017). Fundamental Objectives and Directive Principles of State Policy as Panacea for National Transformation and Sustainable Development in Nigeria. *Journal of Law, Policy and Globalization*, 65, 23–27.

Ayeloja, A.A., George, F.O.A., Jimoh, W.A., Shittu, M.O., Abdulsalami, S.A. (2018). Microbial Load on Smoked Fish Commonly Traded in Ibadan, Oyo State, Nigeria. *Journal of Applied Science and Environmental Management*, 22, 493–497. DOI: 10.4314/jasem.v22i4.9.

Babalola, B.J., Odebode, J.A., Ojomo, K.Y., Ogungbemile, O.A., Jonathan, S.G. (2018). Mycological Evaluation and Nutritional Composition of Smoked-Dried Fish from Igbokoda Market in Ondo State, Nigeria. *Archives of Basic and Applied Medicine*, 6, 51–53.

Bell, V., Ferrão, J., Fernandes, T. (2017). Nutritional Guidelines and Fermented Food Frameworks. *Foods*, 6, 65.

Beltrán-Barrientos, L.M., González-Córdova, A.F., Hernández-Mendoza, A., Torres-Inguanzo, E.H., Astiazarán-García, H., Esparza-Romero, J., Vallejo-Cordoba, B. (2018a). Randomized Double-Blind Controlled Clinical Trial of the Blood Pressure-Lowering Effect of Fermented Milkwith Lactococcus lactis: A Pilot Study. *Journal of Dairy Science*, 101, 2819–2825.

Beltrán-Barrientos, L.M., Hernández-Mendoza, A., González-Córdova, A.F., Astiazarán-García, H., Esparza-Romero, J., Vallejo-Córdoba, B. (2018b). Mechanistic Pathways Underlying the Antihypertensive Effect of Fermented Milk with Lactococcus lactis NRRL B-50571 in Spontaneously Hypertensive Rats. *Nutrients*, 10, 262.

Besen, S.M., Farrell, J. (1994). Choosing how to Compete: Strategies and Tactics in Standardization. *The Journal of Economic Perspectives*, 8(2), 117–131.

Blind, K., Gauch, S. (2009). Research and Standardisation in Nanotechnology: Evidence from Germany. *The Journal of Technology Transfer*, 34(3), 320–342.

Çabuk, B., Nosworthy, M.G., Stone, A.K., Korber, D.R., Tanaka, T., House, J.D., Nickerson, M.T. (2018). Effect of Fermentation on the Protein Digestibility and Levels of Non-Nutritive Compounds of Pea Protein Concentrate. *Food Technology and Biotechnology*, 56, 257–264.

Campbell-Platt, G. (1994). Fermented Foods - A World Perspective. *Food Research International*, 27(3), 253–257.

Clark, K.B., Fujimoto, T. (1990). The Power of Product Integrity. *Harvard Business Review*, 68(6), 107–118.

Constitution of the Federal Republic of Nigeria, 1999 (as Amended). (1999). Federal Government Press, Nigeria, pp. 1–169.

Counterfeit and Fake Drugs and Unwholesome Processed Foods (Miscellaneous Provisions) Act Cap C34 LFN. (2004), pp. 1–9.

Dwivedi, N., Dwivedi, S., Adetunji, C.O. (2021). Efficacy of microorganisms in the removal of toxic materials from industrial effluents. In: Adetunji, C.O., Panpatte, D.G., Jhala, Y.K. (eds). *Microbial Rejuvenation of Polluted Environment. Microorganisms for Sustainability*, vol 27. Springer, Singapore. https://doi.org/10.1007/978-981-15-7459-7_15

Ettlie, J.E. (1995). Product-Process Development Integration in Manufacturing. *Management Science*, 41(7), 1224–1237.

Food and Drug Act Cap F32 Laws of the Federal Republic of Nigeria. (2004). Federal Government Press, Nigeria, pp. 1–26.

Food, Drugs and Related Product (Registration etc.) Act Cap F33 Laws of the Federal Republic of Nigeria (LFN). (2004). Federal Government Press, Nigeria, pp. 1–8.

Frías, J., Martínez-Villaluenga, C., Peñas, E. (2016). *Fermented Foods in Health and Disease Prevention*. 1st Edition. Academic Press, Boston, MA.

Galli, V., Mazzoli, L., Luti, S., Venturi, M., Guerrini, S., Paoli, P., Vincenzini, M., Granchi, L., Pazzagli, L. (2018). Effect of Selected Strains of Lactobacilli on the Antioxidant and Anti-Inflammatory Properties of Sourdough. *International Journal of Food Microbiology*, 286, 55–65.

Giau, V.V., Wu, S.Y., Jamerlan, A., An, S.S.A., Kim, S.Y., Hulme, J. G. (2018). Microbiota and their neuroinflammatory implications in Alzheimer's disease. *Nutrients* 10(11), 1765. doi: 10.3390/nu10111765. PMID: 30441866; PMCID: PMC6266223.

Gobbetti, M., Rizzello, C.G., Di Cagno, R., De Angelis, M. (2014). How the Sourdough May Affect the Functional Features of Leavened Baked Goods. *Food Microbiology*, 37, 30–40.

Islam, S., Thangadurai, D., Adetunji, C.O., Nwankwo, W., Kadiri, O., Makinde, S., Michael, O.S., Anani, O.A., Adetunji, J.B. (2021). Nanomaterials and nanocoatings for alternative antimicrobial therapy. In: Kharissova, O.V., Martínez, L.M.T., Kharisov, B.I. (eds). *Handbook of Nanomaterials and Nanocomposites for Energy and Environmental Applications*. Springer, Cham. https://doi.org/10.1007/978-3-030-11155-7_3-1

Liker, J.K., Collins, P.D., Hull, F.M. (1999). Flexibility and Standardization: Test of a Contingency Model of Product Design-manufacturing Integration. *Journal of Product Innovation Management*, 16(3), 248–267.

Linares, D.M., Gómez, C., Renes, E., Fresno, J.M., Tornadijo, M.E., Ross, R.P., Stanton C. (2017). Lactic Acid Bacteria and Bifidobacteria with Potential to Design Natural Biofunctional Health-Promoting Dairy Foods. *Food Microbiology*, 8, 846.

Marco, M.L., Heeney, D., Binda, S., Cifelli, C.J., Cotter, P.D., Foligné, B., Gänzle, M., Kort, R., Pasin, G., Pihlanto, A., Smid, E.J., Hutkins, R. (2017). Health Benefits of Fermented Foods: Microbiota and Beyond. *Current Opinion in Biotechnology*, 2(44), 94–102. DOI: 10.1016/j.copbio.2016.11.010. Epub 2016 Dec 18. PMID: 27998788.

Milini, F., Milini, V., Luziatelli, F., Ficca, A.G., Ruzzi, M. (2019). Health-Promoting Components in Fermented Foods: An Up-to-Date Systematic Review. *Nutrients*, 11(5), 1189.

Milliken, F.J., Martins, L.L. (1996). Searching for Common Threads: Understanding the Multiple Effects of Diversity in Organizational Groups. *The Academy of Management Review*, 21(2), 402–433.

Mishra, P., Mishra, R.R., Adetunji, C. O. (eds.). (2020). *Innovations in Food Technology: Current Perspectives and Future Goals*. Springer, Singapore. pp. 143–162. 1st Edition. DOI: 10.1007/978-981-15-6121-4

Mohammed, M. (2019). Why Fermented Foods Could Cause Serious Harm to Your Health. *News*, Monday 18 November 2019, *Independent UK*.

Mustafa, S.M., Chua, L.S., El-Enshasy, H.A., Abd Majid, F.A., Hanapi, S.Z., Abdul Malik, R. (2019). Effect of Temperature and pH on the Probiotication of Punica granatum Juice using Lactobacillus Species. *Journal of Food Biochemistry*. DOI: 10.1111/jfbc.12805.

Nsubuga, P. White, M.E., Thacker, S.B., Anderson, M.A., Blount, S.B., Broome, C.V., Trostle M. (2006). Public health surveillance: A tool for targeting and monitoring interventions. In: Jamison, D.T., Breman, J.G., Measham, A.R., et al. (eds). *Disease Control Priorities in Developing Countries*. 2nd Edition. Washington (DC): The International Bank for Reconstruction and Development/The World Bank, Chapter 53, Oxford University Press, New York, pp. 1–22.

Nwankwo, W., Adetunji, C.O., Ukhurebor, K.E., Makinde, A.S., Ubi, B. (2020a). The Precursory Machinery of Internet of Things (IoT) in the Platform for Harmonizing Bio-Mined Data. *Nigerian Research Journal of Engineering and Environmental Sciences*, 5(2), 786–796.

Nwankwo, W., Olayinka, A.S., Ukhurebor, K.E. (2020b). Nanoinformatics: Why Design of Projects on Nanomedicine Development and Clinical Applications may Fail? *Proceeding of the 2020 International Conference in Mathematics, Computer Engineering and Computer Science (ICMCECS)*, Lagos, Nigeria, IEEE Xplore, 1–7.

Odunfa, S.A. (1981). Microorganisms Associated with Fermentation of African Locust Bean (Parkia filicoidea) during Iru Preparation. *Journal of Plant Foods*, 3(4), 245–250. DOI: 10.1080/0142968X.1981.11904236.

Okeke, G.N., Okeke, C. (2013). The Justiciability of the Non-Justiciable Constitutional Policy of Governance in Nigeria. *IOSR Journal of Humanities and Social Science*, 7(6), 9–14.

Okogie, A.V. (1981). Attorney General, Lagos State, 1981 2 NCLR337.

Omemu, A.M. (2011). Fermentation Dynamics during Production of Ogi, a Nigerian Fermented Cereal Porridge. *Report and Opinion*, 3(4), 8–17.

Orpin, J.B., Mzungu, I., Osuji, C.G. (2018). Isolation and Identification of Bacteria Associated with Suya (Roasted Meat Product) Sold in Dutsinma Local Government Area, Kastina State. *Journal of Advances in Biology & Biotechnology*, 20(2), 1–8.

Owusu-Kwarteng, J., Parkouda, C., Adewumi, G.A., Ouoba, L.I.I., Jespersen, L. (2020). Technologically relevant *Bacillus* species and microbial safety of West African traditional alkaline fermented seed condiments. *Critical Reviews in Food Science and Nutrition*. 1–18. doi: 10.1080/10408398.2020.1830026. Epub ahead of print. PMID: 33030021.

Olaniyan, O.T., Adetunji, C.O. (2021). Biochemical role of beneficial microorganisms: An overview on recent development in environmental and agro science. In: Adetunji, C.O., Panpatte, D.G., Jhala, Y.K. (eds). *Microbial Rejuvenation of Polluted Environment. Microorganisms for Sustainability*, vol 27. Springer, Singapore. https://doi.org/10.1007/978-981-15-7459-7_2

Pessione, E., Cirrincione, S. (2016). Bioactive Molecules Released in Food by Lactic Acid Bacteria: Encrypted Peptides and Biogenic Amines. *Frontiers in Microbiology*, 7, 876.

Rocchetti, G., Miragoli, F., Zacconi, C., Lucini, L., Rebecchi, A. (2019). Impact of Cooking and Fermentation by Lactic Acid Bacteria on Phenolic Profile of Quinoa and Buckwheat Seeds. *Food Research International*, 119, 886–894.

Sangoyomi, T.E., Owoseni, A.A., Okerokun, O. (2010). Prevalence of Enteropathogenic and Lactic Acid Bacteria Species in Wara: A Local Cheese from Nigeria. *African Journal of Microbiology Research*, 4(15), 1624–1630.

Sanlier, N., Gökcen, B.B., Sezgin, A. (2017). Health Benefits of Fermented Foods. *Critical Reviews in Food Science and Nutrition*, 59(1), 1–22. DOI: 10.1080/10408398.2017.1383355.

Sivamaruthi, B.S., Kesika P., Prasanth, M.I., Chaiyasut, C. (2018a). A Mini Review on Antidiabetic Properties of Fermented Foods. *Nutrients*, 10, 1973.

Sivamaruthi, B.S., Kesika, P., Chaiyasut, C. (2018b). Impact of Fermented Foods on Human Cognitive Function—A Review of Outcome of Clinical Trials. *Scientia Pharmaceutica*, 86, 22.

Slashinski, M.J., McCurdy, S.A., Achenbaum, L.S., Whitney, S.N., McGuire, A.L. (2012). "Snake-oil," "Quack Medicine," and "Industrially Cultured Organisms:" Biovalue and the Commercialization of Human Microbiome Research. *BMC Medical Ethics*, 13, 28.

Standards Organization of Nigeria Act Cap S9 Laws of the Federal Republic of Nigeria. (2015). Federal Government Press, Nigeria, pp. 1–14.

Steinkraus, K.H. (1986). Fermented Foods, Feeds, and Beverages. *Biotechnology Advances*, 4, 219–243.

Steinkraus, K.H. (1994). Nutritional Significance of Fermented Foods. *Food Research International*, 27(3), 259–267.

Steinkraus, K.H. (2009). Fermented Foods, Encyclopedia of Microbiology. Academic Press, London, pp. 45–53.

Steinkraus, K.H. (1997). Classification of Fermented Foods: Worldwide Review of Household Fermentation Techniques. *Food Control*, 8, 311–317.

Thangadurai, D., Dabire, S.S., Sangeetha, J., Said Al-Tawaha, A.R.M., Adetunji, C.O., Islam, S., Shettar, A.K., David, M., Hospet, R., Adetunji, J.B. (2021). Greener composites from plant fibers: Preparation, structure, and properties. In: Kharissova, O.V., Martínez, L.M.T., Kharisov, B.I. (eds). *Handbook of Nanomaterials and Nanocomposites for Energy and Environmental Applications*. Springer, Cham. https://doi.org/10.1007/978-3-030-11155-7_21-1

Thacker, S.B., Berkelman, R.L. (1988). Public Health Surveillance in the United States. *Epidemiologic Reviews*, 10, 164–190.

Thacker, S.B., Stroup, D.F., Parrish, R.G., Anderson, H.A. (1996). Surveillance in Environmental Public Health: Issues, Systems, and Sources. *American Journal of Public Health*, 86(5), 633–638.

United States Agency for International Development (USAID). (2005). Infectious Disease and Response Strategy 2005, USAID, Washington, DC.

van Hylckama Vlieg, J.E.T., Veiga, P., Zhang, C., Derrien, M., Zhao, L. (2011). Impact of Microbial Transformation of Food on Health - From Fermented Foods to Fermentation in the Gastro-intestinal Tract. *Current Opinion in Biotechnology*, 22(2), 211–219.

Wheelwright, S.C., Clark, K.B. (1992). Creating Project Plans to Focus Product Development. *Harvard Business Review*, 70(2), 70–82.

White, M.E., McDonnell, S.M. (2000). Public health surveillance in low- and middle-income countries. In: Teutsch, S.M., Churchill, R.E. (eds). *Principles and Practices of Public Health Surveillance*. Oxford University Press, New York, 287–315.

WHO. (2016). International Health Regulations (2005), 3rd Edition, Geneva.

Yusuf, A.B., Gulumbe, B.H., Kalgo, Z.M., Aliyu, B., Haruna, M. (2020). Microorganisms Associated with the Production of Burukutu (An Alcoholic Beverage) in Kebbi State, Nigeria. *Equity Journal of Science and Technology*, 7(1), 67–73.

Zhang, Z.P., Ma, J., He, Y.Y., Lu, J., Ren, D.F. (2018). Antioxidant and Hypoglycemic Effects of Diospyros Lotus Fruit Fermented with Microbacterium flavum and Lactobacillus plantarum. *Journal of Bioscience and Bioengineering*, 125, 682–687.

# 9 Algae Biotechnology for Novel Foods

*Mathias A. Chia*
Ahmadu Bello University

*Emeka G. Nwoba*
Murdoch University

*James Chukwuma Ogbonna*
University of Nigeria

## CONTENTS



## 9.1 INTRODUCTION

Food is any substance that humans and animals consume to sustain life, health, development, and reproduction. Human food (diet) is made of macro- and micronutrients obtained from plant and animal sources. Proteins, carbohydrates, and lipids are macronutrients, and vitamins, minerals, and organic acids represent micronutrients (Sedlmeier et al., 2018). Others are essential nutrients such as amino acids, vitamins, and long-chain polyunsaturated fatty acids that cannot be synthesized de novo by humans. Following the international guidelines for human nutrition, optimum proportions of macro- and micronutrients vary according to the age, physical activities, sex, health condition, and pregnancy state of individuals (Udokanma and Emeahara, 2017). These variations in proportions of nutrients sustain normal

DOI: 10.1201/9781003178378-9

cellular, tissue, organ, and system function, and body homeostasis of living organisms. The consumption of foods with the wrong nutrient composition constitutes a significant dietary risk because it can adversely affect human health. Poor diet causes cancers, cardiovascular and circulatory diseases, diabetes, and obesity (Knudsen et al., 2019). Recent studies reveal that dietary risk now surpasses tobacco smoking and alcohol consumption, constituting one of the most pressing public health challenges (Knudsen et al., 2019; Udokanma and Emeahara, 2017). This risk has led to the search for alternative and novel sources of food to provide the nutrients needed by humans, and algae and cyanobacteria are excellent alternatives.

According to the Food Standards Australia New Zealand (FSANZ), novel foods are unconventional (nontraditional) foods that need to be subjected to public health risk and safety assessment (in view of the potential for adverse effects in humans, composition/structure of the food, the process by which the food has been prepared, source from which the food is derived, and patterns and levels of consumption of the food) before inclusion in the food supply/chain (FSANZ, 2020). Traditional foods constitute those foods or substances derived from foods or source from which the substance is derived, which have a history of human consumption in the region or country. According to the European Commission, novel foods represent newly developed, innovative foods, food ingredients/nutrients (e.g., synthetic zeaxanthin), foods produced through new technologies/production processes (e.g., nanotechnology or fruit preparations from high-pressure pasteurization), extracts from existing foods (e.g., rapeseed proteins), or foods traditionally consumed outside of the European Union (e.g., chia seeds, Ngali nuts, Noni juice), which have not been widely consumed to a significant degree by humans in the EU prior to 15 May 1997, when the first novel foods legislation came into action (EC, 2020). The potential classes of novel foods are plants/animals and components derived from them, plants/animal extracts, herbs and their extracts, single chemical entities, dietary macro-components, microorganisms, fungi, algae, cell cultures, probiotics, minerals or engineered nanomaterials, foods with intentionally modified primary structures, and foods derived from new sources/ingredients or through processes not already used for food production. In general, novel foods are substances (with microorganisms inclusive) that lack a history of safe consumption as foods, foods that have been manufactured, prepared, preserved, or packaged through processes that either have not been previously used for the food or bring about major changes to the food, foods that are derived from plants, animals, or microorganisms that have been genetically modified in such ways that they exhibit qualities that were not formerly associated with them or no longer show associated qualities or certain qualities are no longer within the expected range for the organism. While food additives, flavorings, including extraction solvents utilized in food production processes are not included as novel foods, scientific risk assessment and authorization must be carried out before they are allowed into the market. Genetically modified (GM) foods are currently excluded and not regulated under the definition of novel foods but are specifically regulated under GM foods and feeds (Saarela, 2007). Functional foods are not necessarily novel foods but many functional products such as phytosterols or phytostanols-containing products have been authorized in line with novel foods regulation procedures.

Algae and cyanobacteria are cosmopolitan organisms that have a rich biochemical composition with several industrial applications (Chia et al., 2013). Characteristics such as rapid growth, ability to thrive in diverse environments and industrial wastewater, and the modification of biochemical composition in response to changes in their environment make algae ideal candidates for the production of specific compounds (Liang et al., 2009). In several parts of the world such as the Chad region, China, and Japan, algae have been consumed without much preparation since ancient times. The consumption of algae has been driven by the fact that their rich biochemical composition provides crucial health benefits to humans, including the prevention and treatments of several cardiovascular diseases and cancers (Udenigwe and Aluko, 2012). Thus, the quality and quantity of polysaccharides, vitamins, bioactive peptides, polyunsaturated fatty acids, pigments, and antioxidants determines their applications as functional foods, food supplements, and novel foods in general. For example, the quality of meat products such as sausages, steaks, and frankfurters is improved using algae as additives (Cottin et al., 2011, Nwoba et al., 2020a). Also, the enrichment of food by algae extends to fish and fish products, and cereal-based products (e.g., bread and flour). This chapter explores the characteristics of algae as an excellent source of novel foods. The chapter will highlight the growth characteristics of algae, biomolecule composition, secondary metabolites, and other high-value products contributing to the use of these organisms as a source of novel foods.

## 9.2 ALGAE AS A NOVEL FOOD SOURCE

Algae represent a poorly explored novel food source concerning their enormous prospect as a bioresource for sustainable production of food commodities such as fats, polyunsaturated fatty acids, edible oil, natural dyes, carbohydrates, pigments, proteins, antioxidants, high-value bioactive compounds, and other fine chemicals (Mata et al., 2010) (Figure 9.1). Under certain conditions, the proteins and carbohydrates content of algae can reach 60% dry weight of biomass (Bondioli et al., 2012). Algae have unique chemical compositions and thus, can be used to improve the nutritional components of traditional human diet with a positive impact on health. Table 9.1 presents the nutritional value of algae and their comparison with traditional protein sources. Microalgae have a high protein content, with an amino acid composition similar to conventional protein sources that meets the WHO/FAO requirements (Draaisma et al., 2013; Spolaore et al., 2006). While traditional protein sources such as soybean have low essential amino acids such as methionine (1.2 g amino acid per 100 g proteins), algae synthesize all amino acids with methionine level ranging between 1.4 and 4.5 g amino acid per 100 g protein (Fleurence et al., 2012). On the other hand, macroalgae especially seaweeds have low protein content (5%–25% of dry weight) (Fleurence et al., 2012). However, red seaweeds such as *Porphyra yezoensis* and *Palmaria palmata* have high protein contents of 47% and 35% of dry weight, respectively (Fleurence et al., 2012). Therefore, algae have the potential to be used as an unconventional protein source in the forms of single cell proteins for microalgae and vegetal proteins for macroalgae. The indigenous utilization of algae in food is well documented in Asia, with reports describing its use as a gift to the king in the 6th century in China (Fujiwara-Arasaki et al., 1984; http://naturalknowledge247.

**TABLE 9.1**
**Chemical Composition of Algae and Selected Conventional Foods (Chacón-Lee and González-Mariño, 2010; Suganya et al., 2016)**

| Algae | Habitat | Proteins (% Dry Weight) | Carbohydrates (% Dry Weight) | Lipids (% Dry Weight) |
|---|---|---|---|---|
| | | **Traditional foods** | | |
| Meat | - | 43 | 1 | 34 |
| Baker's yeast | - | 39 | 38 | 1 |
| Egg | - | 47 | 4 | 41 |
| Milk | - | 26 | 38 | 28 |
| Rice | - | 8 | 77 | 2 |
| Soybean | - | 37 | 30 | 20 |
| | | **Microalgae** | | |
| *Scenedesmus obliquus* | Freshwater | 50–56 | 10–17 | 12–14 |
| *Scenedesmus quadricauda* | Freshwater | 47 | - | 1.9 |
| *Scenedesmus dimorphus* | Freshwater | 8–18 | 21–52 | 16–40 |
| *Chlamydomonas rheinhardtii* | Freshwater | 48 | 17 | 21 |
| *Chlorella vulgaris* | Freshwater | 51–58 | 12–17 | 14–22 |
| *Chlorella pyrenoidosa* | Freshwater | 57 | 26 | 2 |
| *Dunaliella bioculata* | Marine | 49 | 4 | 8 |
| *Dunaliella salina* | Marine | 57 | 32 | 6 |
| *Euglena gracilis* | Freshwater | 39–61 | 14–18 | 14–20 |
| *Prymnesium parvum* | Marine | 28–45 | 25–33 | 22–39 |
| *Tetraselmis maculata* | Marine | 52 | 15 | 3 |
| *Porphyridium cruentum* | Marine | 28–39 | 40–57 | 9–14 |
| *Spirulina platensis* | Freshwater | 46–63 | 8–14 | 4–9 |
| *Spirulina maxima* | Freshwater | 60–71 | 13–16 | 6–7 |
| *Synechococcus* sp. | Marine | 63 | 15 | 11 |
| *Aphanizomenon flos-aquae* | Freshwater | 62 | 17 | 21 |
| *Nannochloropsis* spp. | Marine | 28.8 | 35.9 | 18.4 |
| *Haematococcus pluvialis* | Marine | 48 | 27 | 15 |
| *Isochrysis galbana* | Marine | 27.0 | 17.0 | 17.2 |
| *Anabaena cylindrica* | Marine | 43–56 | 25–30 | 4–7 |
| *Spirogyra* sp. | Freshwater | 6–20 | 33–64 | 11–21 |
| | | **Macroalgae/seaweeds** | | |
| *Hypnea valentiae* | Marine | 11.8–12.6 | 11.8–13.0 | 9.6–11.6 |
| *Acanthophora spicifera* | Marine | 12.0–13.2 | 11.6–13.2 | 10.0–12.0 |
| *Laurencia papillosa* | Marine | 11.8–12.9 | 12.0–13.3 | 8.9–10.8 |
| *Ulva lactuca* | Marine | 11.4–12.6 | 11.6–13.2 | 9.6–10.5 |
| *Caulerpa racemosa* | Marine | 11.8–12.5 | 16.0 | 9.0–10.5 |
| *Ulva reticulate* | Marine | 12.8 | 16.9 | 8.5 |
| *Enteromorpha compressa* | Marine | 7.3 | 24.8 | 11.5 |

(*Continued*)

**TABLE 9.1 (*Continued*)**
**Chemical Composition of Algae and Selected Conventional Foods (Chacón-Lee and González-Mariño, 2010; Suganya et al., 2016)**

| Algae | Habitat | Proteins (% Dry Weight) | Carbohydrates (% Dry Weight) | Lipids (% Dry Weight) |
|---|---|---|---|---|
| *Chaetomorpha aerea* | Marine | 10.1 | 31.5 | 8.5 |
| *Chaetomorpha antennina* | Marine | 10.1 | 27.0 | 11.5 |
| *Chaetomorpha linoides* | Marine | 9.5 | 27.0 | 12.0 |
| *Cladophora fascicularis* | Freshwater | 15.5 | 49.5 | 15.7 |
| *Microdictyon agardhianum* | Marine | 20.9 | 27.0 | 9.4 |
| *Boergesenia forbesii* | Marine | 7.4 | 21.4 | 11.4 |
| *Valoniopsis pachynema* | Marine | 8.8 | 31.5 | 9.1 |
| *Dictyosphaeria cavernosa* | Marine | 6.0 | 42.8 | 10.5 |
| *Caulerpa cupressoides* | Marine | 7.4 | 51.8 | 11.0 |
| *Caulerpa peltata* | Marine | 6.4 | 45.0 | 11.4 |
| *Caulerpa laetevirens* | Marine | 8.8 | 56.3 | 8.8 |
| *Caulerpa racemosa* | Marine | 8.8 | 33.8 | 10.6 |
| *Caulerpa fergusonii* | Marine | 7.8 | 23.6 | 7.2 |
| *Caulerpa sertularioides* | Marine | 9.1 | 49.5 | 7.0 |
| *Halimeda macroloba* | Marine | 5.4 | 32.6 | 9.9 |
| *Codium adhaerens* | Marine | 7.3 | 40.5 | 7.4 |
| *Codium decorticatum* | Marine | 6.1 | 50.6 | 9.0 |
| *Codium tomentosum* | Marine | 5.1 | 29.3 | 7.2 |
| *Palmaria palmata* | Marine | 8–32 | 11–60 | 5–13 |
| *Porphyra yezoensis* | Marine | 44–47 | 45 | 2 |
| *Chondrus crispus* | Marine | 11–18 | 50–58 | 2–4 |
| *Laminaria digitata* | Marine | 8–15 | 49 | 1–2 |
| *Himanthalia elongata* | Marine | 6–20 | 40–60 | 1–5 |
| *Saccharina latissima* | Marine | 8 | 60 | 6 |
| *Fucus vesiculosus* | Marine | 12–13 | 54 | 8 |
| *Undaria pinnatifida* | Marine | 12–14 | 48 | 1–3 |
| *Sarcodiotheca gaudichaudii* | Marine | 17 | 33 | 5–6 |
| *Ascophyllum nodosum* | Marine | 5–10 | 13–56 | 2–9 |

com/seaweed-a-brief-history/). The nutritional components of algae (carbohydrates, proteins, lipids, vitamins, minerals, antioxidants) are species-specific, seasonal, and depend on culture conditions (Lerat et al., 2018). Today, microalgae are sold as health foods in tablets, capsules, or liquid extracts (Figure 9.2). Both micro- and macroalgae can be added into pastas, snacks, gums, candies, drinks, beverages, or used as supplements or natural food colors (Nwoba et al., 2020b). Hence, conventional protein sources such as soybean, eggs, meat, and milk can be replaced with algal biomass and this presents a promising sustainable production route.

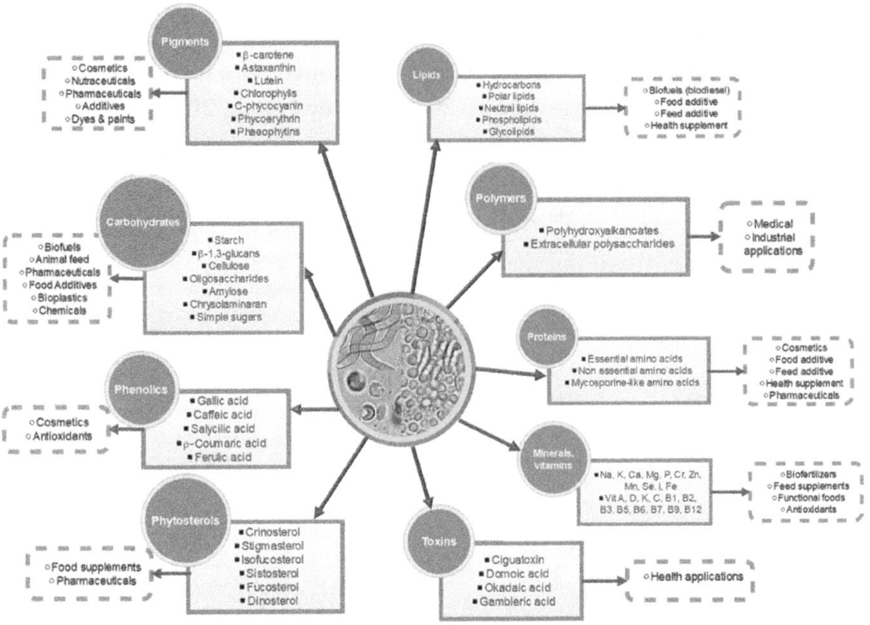


**FIGURE 9.1** Bioproducts from microalgae with applications (Nwoba et al., 2019).

## 9.3 HISTORY OF ALGAE AS FOOD

The use of algae as food is not new and has a very long history dated to the 6th century in China (Fujiwara-Arasaki et al., 1984). It is estimated that over 30,000 species of microalgae exists (Radmer, 1996), only four freshwater microalgal species, *Chlorella*, *Nostoc*, *Spirulina*, and *Aphanizomenon*, have been consumed as food for thousands of years (Jensen, 2001). *Chlorella vulgaris* was used as alternative medicine in the Far East of North America and formed a traditional food in the orient, since primordial times (Sousa et al., 2008). However, commercial large-scale production of microalgae began at the dawn of the 1960s, with Nihon Chlorella Co. Ltd, Japan, growing Chlorella cultures (Richmond, 2008). The proximate composition of dried Chlorella product were proteins (45%), fat (20%), carbohydrate (20%), fiber (5%), and minerals and vitamins (10%) (Spolaore et al., 2006). Hence, in the 1960s, human health food made from algae became marketed as tablets, drinks, granules, capsules, powders, and even as additives in Japan (Belasco, 1997). By the 1970s, a Spirulina plant was started in Mexico by Sosa Texcoco, S.A. (Spolaore et al., 2006). Before this time, Spirulina, a cyanobacterium also called Arthrospira, was consumed as a food in China and Africa (Abdulqader et al., 2000) (Figure 9.3). Spirulina has significant health benefits with a high content of protein and balanced nutritional profiles. Its cultivation for the nutraceuticals market is commercially a success story. Big companies such as Earthrise Nutritionals LLC and Cyanotech Corporation, Hawaii, USA, produce Spirulina at large scale and sell it as a superfood

**FIGURE 9.2** Food products from algae (Nwoba et al., 2020b).

to be consumed whole. Currently, Spirulina production happens extensively in China, Japan, and India. *Aphanizomenon flos-aquae* is another cyanobacterium with up to 60% protein content on a dry weight basis and has been harvested from natural freshwater lakes in Oregon (USA) for nutraceuticals (Carmichael et al., 2000).

**FIGURE 9.3** Kanembu women harvesting Spirulina in Lake Chad.

The number of algae consumed globally has risen to about 145 species, and a greater proportion of this consumption happens in Asian countries, where they are consumed as vegetables (Fleurence et al., 2012). The popular Hoshi-nori, Wakame, Kombu, Pioca, Dulse, Sea Spaghetti, Sushi, etc., were delicacies made from different seaweeds and formed daily foods of Japanese people since the beginning of civilization (Fleurence et al., 2012). Currently, seaweeds are an essential resource (no longer weeds) extensively being exploited by the industrial sector for the production of agar, alginate, phycolloids, carrageenans, and fucerellans (Fleurence et al., 2012).

Based on the EU definition of novel foods (see Section 9.1), several algae (microalgae and seaweeds) species have been established for human consumption, with *Odontella aurita* included in the list via legislation ANSES 2001-SA-0082, CE 258/97, using the novel food procedures. As at 2016, the following algal species were permitted for human consumption: *Ascophyllum nodosum, Fucus vesiculosus, Fucus serratus, Himanthalia elongata, Undaria pinnatifida, Laminaria digitata, Laminaria saccharina, Laminaria japonica (kombu), Alaria esculenta, Palmaria palmata, Pyropia umbilicalis, Pyropia tenera, Porphyra yezoensis, Pyropia dioica, Pyropia purpurea, Pyropia laciniata, Pyropia leucosticta, Chondrus crispus, Gracilaria verrucosa, Lithothamnium calcareum, Ulva* sp., *Enteromorpha* sp., *Spirulina* sp., *Odontella aurita*, and *Chlorella* sp. (Wells et al., 2017). These algae are guaranteed to be safe. The alga, *Schizochytrium* sp. has a high content of docosahexaenoic acid (DHA) (up to 32%), and oil from this species and that from *Ulkenia* sp. were approved by legislation 2003/427/CE – Saisine AFSSA 2008-SA-0316 and 2001-SA-0095 and CE 2009/777/CE, respectively (Lerat et al., 2018). These oils can be supplemented with *Haematococcus pluvialis* and sold as dry biomass or combined with the astaxanthin as an oil extract. Synthetic astaxanthin is not permitted for human consumption but can be used for feed application. *Dunaliella* sp. has a significant amount of beta carotene and can be marketed as dry biomass for food supplementation. The following microalgae have the GRAS (generally regarded as safe) status by the United States Food and Drug Administration

(USFDA): *Chlorella* sp., *Spirulina* sp., *Dunaliella* sp., *Haematococcus* sp. and *Schizochytrium* sp. (Chacón-Lee and González-Mariño, 2010).

## 9.4 ALGAL BIOTECHNOLOGY: CHARACTERISTICS OF ALGAE THAT MAKE THEM IMPORTANT SOURCES OF NOVEL FOODS

Microalgae are capable of producing many primary and secondary metabolites that are used as food and food supplements. In comparison with animals and higher plants, they have many distinct characteristics that make them important sources of novel foods. One of these characteristics is their rapid growth rates. The growth rates of microalgae vary with the species as well as the growth conditions. However, on the whole, their growth rates can be an order of magnitude higher than those of higher plants (Raja et al., 2014). Although the photosynthetic mechanism of green algae is similar to that of higher plants, several reports have shown that the photosynthetic efficiencies of most species of microalgae are higher than those of higher plants (Karthikeyan, 2012). The theoretical maximum photosynthetic efficiency of microalgae ranges from 9% to 10% (Melis, 2009) while those of C3 and C4 plants are 4.6% and 6%, respectively (Zhu et al., 2010, 2008). Photosynthetic efficiency of 9–10% corresponds to about 80 g of biomass/m$^2$/day (280 ton/ha/year) (Melis, 2009) but in conventional large-scale photobioreactors, the efficiency is around 2.5%. Even then, a yield of 140 tons/ha/year (Rodolfi et al., 2009) is more than an order of magnitude higher than that of any food crop. Furthermore, only a very small percentage of the photosynthetic product of higher plants are used for food while, depending on the species, 100% of algae biomass (e.g., Spirulina) can be used as food or feed. Aside from the photosynthetic growth, many species of microalgae can also grow in dark, using organic carbon as the source of energy and carbon (Ogbonna et al., 1998) or in light in the presence of organic carbon in which case they grow mixotrophically (photoheterotrophically) using both inorganic and organic carbon sources (Ogbonna et al., 2002). Thus, biomass productivities of microalgae are much higher than those of higher plants.

Another important characteristic of microalgae that make them useful as a source of novel foods is the diversity of their habitats. There are psychrophilic microalgae that grow in very cold regions (Cvetkovska et al., 2017), thermophiles that grow in very hot weather (Ghozzi et al., 2013), acidophiles that grow in acidic environments (Johnson, 2012), as well as alkaliphiles that grow at high pH values (Liu et al., 2019). There are many freshwater algae (Rashid et al., 2019) as well as marine species (Mourelle et al., 2017). Thus, microalgae can be cultivated under various environmental conditions across all the ecological zones of the world. These diverse habitats mean that microalgae can be cultivated on marginal lands, including arid regions (Ogbonna and Ogbonna, 2015), to avoid competition with food crops for arable lands. Another important attribute of microalgae is that they are non-seasonal and thus can be cultivated all year round.

The main nutrients required by microalgae include the macronutrients and micronutrients. The macronutrients, required in large quantities (> g/L), include P, N, C, H, and O that are usually supplied in the form of their salts. The micronutrients, on the other hand, are required in small quantities (~mg/L) and include Ca, Mg, Na, K, Fe, Mn, S, Zn, Cu, and Co. The cost of supplying all these nutrients represents a significant percentage of the total production cost of microalgae. Fortunately, all these nutrients are abundant in industrial wastes, especially in agro-industrial wastes, which are in abundant supply all-round the year. According to the Food and Agriculture Organization of the United Nations (FAO), about one-third of the edible foods in the world, representing about 1.3 billion tons, is wasted annually (Losses and Waste, 2011). These industrial wastes, if not treated, pose a lot of environmental problems all over the world. Various species of microalgae are used for simultaneous treatment of these wastes and production of many useful metabolites in a process often referred to as valorization of wastes. For example, cultivation of microalgae on digestate effluents serves dual purposes of waste treatment and valorization (Chuka-ogwude et al., 2020). Their cultivation in these agro-industrial wastewater leads to recycling of valuable nutrients, treatment of the wastes, reduction in the requirements for fresh water for the culture, and production of many valuable products (Markou et al., 2018). Other types of wastes that have been used for cultivation of various species of microalgae include dairy manure, digested manure, industrial wastewater, activated sludge, municipal wastewater, domestic wastewater, sewage sludge, piggery wastewater, piggery wastewater, anaerobically digested food wastewater, swine wastewater, hydrolysate of food waste, brewer fermentation waste, crude glycerol, soybean processing wastewater, hydrolysate of cyperus esculentus waste, sugarcane bagasse hydrolysate, enzymatic hydrolysates of sweet sorghum and rice straw, as well as palm oil mill effluent (Sibi, 2018).

## 9.5 ALGAE AS FUNCTIONAL FOODS

Functional food is a food prepared with "scientific intelligence" to provide the body with the required amount of vitamins, fats, proteins, carbohydrates, etc., needed for healthy living (El Sohaimy, 2012). Per appearance, functional foods are not different from conventional foods but have added health benefits (El Sohaimy, 2012). They expatiate the interactions between nutrition, sensory satisfaction, fortification, and modulation of physiological systems to facilitate defined fortified food products with special constituents (Bigliardi and Galati, 2013). Functional foods are prepared to improve general body conditions (e.g., prebiotics and probiotics), decrease the risk of some diseases (e.g., cholesterol-lowering products), and cure some illnesses (Bigliardi and Galati, 2013). Algae are excellent natural sources of unique bioactive compounds such as polysaccharides, peptides, polyunsaturated fatty acids (PUFAs), and antioxidants that are used as functional ingredients (Murray et al., 2018).

### 9.5.1 Polysaccharides

Algal polysaccharides are the most widely and regularly unsuspectingly consumed food of algal origin. Polysaccharides comprise of polymers made of simple sugars (monosaccharides) linked together by glycosidic bonds. They have diverse commercial functions in products such as food, feed, stabilizers, thickeners, emulsifiers, and beverages (Mišurcová et al., 2012). Specifically, small amounts are incorporated into beverages, meat and dairy products, and fillers (Cofrades et al., 2008; Griffin, 2015) at levels generally deemed to be beneficial and safe by regulatory agencies (Fleurence et al., 2012). The cell wall of algae is different from that of terrestrial plants due to the presence of special polyuronides and polysaccharides that may be methylated, acetylated, pyruvylated, or sulfated (Pal et al., 2014; Stiger-Pouvreau et al., 2016). Marine algae are rich in sulfated polysaccharides (SPs) which include carrageenans in red algae, fucoidans in brown algae, and ulvans in green algae. SPs have antioxidant, anti-allergic, anti-human immunodeficiency virus, anticancer, and anticoagulant activities (Ngo and Kim, 2013). Also, carrageenans are extensively used in the food industry for their excellent gelling, thickening, and stabilizing properties; fucoidans help to slow blood clotting; and ulvan serves as dietary fiber that promotes gastrointestinal health and reduces the chances of having chronic diseases (Raposo et al., 2013).

Edible macroalgae such as *Codium reediae* and *Gracilaria* spp. have high dietary fiber content (23.5%–64.0% dry weight, DW), which exceeds that of wheat bran (Benjama and Masniyom, 2012; Wells et al., 2017). Similarly, Wells et al. (2017) found total polysaccharide concentrations ranging from 4% to 76% DW in the seaweed species such as *Ascophyllum*, *Porphyra*, *Palmaria*, and *Ulva*. Dietary fibers obtained from seaweeds have several health benefits like reduced risk of hypertension, diabetes, and cardiac heart disease (Wells et al., 2017).

Microalgae are also important sources of polysaccharides when consumed whole or as extracted polysaccharides. These polysaccharides influence several remarkable biological processes (Chacón-Lee and González-Mariño, 2010; Pulz and Gross, 2004). Some of the microalgal genera commonly considered as beneficial dietary supplements include *Chlorella*, *Arthrospira* (Spirulina), *Dunaliella*, *Haematococcus*, *Scenedesmus*, *Aphanizomenon*, *Odontella*, and *Porphyridium*. *Chlorella* spp. are recognized to be very rich in polysaccharides compared to other species (Chacón-Lee and González-Mariño, 2010). The bioactivities of microalgal polysaccharides recognized are anticancer properties, macrophage activation, anti-inflammatory effects, cytokine modulation, and inhibition of protein tyrosine phosphatase (Hsu et al., 2010; Wells et al., 2017). Also, algal polysaccharide extracts possess strong immunomodulating effects as demonstrated by in vitro and in vivo studies (Suárez et al., 2010). Kwak et al. (2012) found that 5 g of *Chlorella* had immunostimulatory effects in 30 Korean volunteers compared with patients given a placebo in a double-blinded 8-week trial. Chlon A and RespondinTM supplements are acidic polysaccharide extracts from *Chlorella pyrenoidosa* with antitumor and immunostimulating properties (Kralovec, 2005; Kralovec et al., 2005).

### 9.5.2 Bioactive Peptides

Bioactive peptides are short-chain proteinogenic amino acid residues linked by peptide bonds. They are also produced by enzymatic, alkali and acidic cleavage of proteins during food digestion, microbial fermentation, food processing, and exogenous hydrolysis (Ejike et al., 2017; Udenigwe and Aluko, 2012). Innovation in food functional ingredients highlights the rapid development of protein-derived bioactive peptides. Peptides are crucial in the management of human health conditions such as hypertension, inflammation, oxidative stress, cancers, immune disorders, and diabetes (Udenigwe and Aluko, 2012). Precursors of bioactive peptides derived from primary human food proteins are not sustainable, considering the current and projected global food security challenges (Ejike et al., 2017).

Seaweeds have for a long time shown great potential as a sustainable protein source of bioactive peptides because some species have ca. 47% of their dry weight as proteins (Table 9.1), and this value changes per season and species. Lately, there has been a lot of interest in using marine algae as a bioactive peptide source due to their health promotion and disease therapy potentials (Udenigwe, 2014). For example, they are used in therapeutic drug and functional food formulations in the treatment and management of cardiovascular diseases and diabetes. Isolated seaweed peptides inhibit the activities of angiotensin-converting enzyme, thereby enhancing their antihypertensive properties. Also, they are excellent antioxidative and antidiabetics functional food additives (Admassu et al., 2018). Similar to macroalgae, peptides resulting from enzymatic hydrolysis of microalgal proteins possess antioxidative and antihypertensive properties that promote human health (da Silva Vaz et al., 2016; Ejike et al., 2017). Thus, peptides derived from microalgae demonstrate great potential for inclusion in the production of functional foods. The use of microalgae has several advantages including high biomass per unit area, rapid growth, ease of isolation and cultivation, high levels of proteins and other bioactive compounds, and less competition with other food sources as a raw material (Ejike et al., 2017).

### 9.5.3 Polyunsaturated Fatty Acids (PUFA) Enriched Food

Fatty acids (FAs) are a part of the building blocks of lipids that are key components of human diets. All living organisms have lipids with varying fatty acid compositions, but some long-chain PUFAs such as linoleic acid (LA) and α-linolenic acid (ALA) cannot be synthesized de novo by humans. Long-chain PUFAs are generally categorized as omega-3 and omega-6 with their first double bond located at the third and the sixth carbon atoms, respectively, as well as the presence of a cis configuration at the methyl end (Nagy and Tiuca, 2017; Wells et al., 2017). Both omega-3 and omega-6 fatty acids have similar biosynthetic pathways that are driven by the same enzymes (Nagy and Tiuca, 2017). They have essential functional, structural, and biological roles, and they represent an important source of energy in the form of adenosine triphosphate (ATP) for metabolic and physiological processes.

The main sources of omega-3 and omega-6 fatty acids are plants including algae. Long-chain PUFAs comprise a significant portion of marine algal lipids, with planktonic algae as the most predominant source of omega fatty acids to fish (Nwoba et al., 2020c). Eicosapentaenoic acid (EPA; 20:5 n-3) is the most important essential fatty acid along with DHA (22:6 n-3), and their respective precursors LA and ALA (Bruneel et al., 2013; Cottin et al., 2011). The primary product of ALA in the biosynthetic pathway to C20–22 PUFAs is stearidonic acid (SA, 18:4 n-3), which forms a significant portion of PUFAs in some edible macroalgae (sea vegetables). EPA is the most predominant PUFA in many macroalgae in conjunction with arachidonic acid (20:4 n-6), particularly the red algae (Ortiz et al., 2009). For example, EPA comprises up to 50% of the total fatty acid content of *Palmaria palmata* (van Ginneken et al., 2011).

Several controlled interventions and epidemiological trials of DHA and EPA derived from fish oils and algal sources revealed that they have several health benefits to humans (Cottin et al., 2011). Triacylglycerol containing DHA extract from the dinoflagellate *Crypthecodinium cohnii* has proven cardioprotective effects (Mendes et al., 2009). Also, *Schizochytrium* sp., a thraustochytrid stramenopile (Li et al., 2009), is used as a supplement containing DHA and marketed as nutritional infant formula, whole infant foods, and in foods (e.g., dairy, bakery, eggs, and nonalcoholic beverages) (Cottin et al., 2011). Diet supplementation of laying hens to improve the level of n-3 PUFA of egg yolk with several sources, such as flaxseed, fish oil, DHA Gold (heterotrophic microalgae), or autotrophic microalgae, has proven to be very effective (Fraeye et al., 2012; Lemahieu et al., 2015).

### 9.5.4 Antioxidants

A healthy cell function requires a balance between reactive oxygen species (ROS) and endogenous antioxidants (Miyashita, 2014). An increase in ROS content disturbs this balance and initiates subcellular changes with pathological consequences (Miyashita, 2014). Polyphenols and pigments from algae protect against chronic-degenerative diseases (Seifried et al., 2007). Polyphenolic compounds exhibit a range of health benefits such as antioxidant, anticancer, antimicrobial, anti-inflammatory, and antidiabetic activities, as well as other bioactivities associated with their antioxidant properties. Marine polyphenols and pigments such as phycocyanin, neoxanthin, phycoerythrin, zeaxanthin, fucoxanthin, and carotenoids protect cellular constituents against oxidative stress and reduce tissue injuries due to potentials antioxidant activities either by direct free radical scavenging or through enhancing the actions of endogenous reducing agents (Fernando et al., 2016; Murray et al., 2018). The mechanisms of action reveal that the antioxidants alter the expression of specific genes and proteins, leading to their beneficial biological properties (Miyashita, 2014). Polyphenols and pigments are used in nutraceutical as additives and food colorants (Pulz and Gross, 2004). In addition, pigments such as ß-carotene, phycobiliproteins, and astaxanthin are commonly used as supplements in the production of ice creams, noodles, biscuits, tea, drinks, bread, and beers (Liang et al., 2004).

## 9.6 SAFETY CONSIDERATIONS

The use of algae as foods requires that they are safe for human consumption and must meet certain regulatory standards or requirements. Algal foods are novel; hence, consideration of consumer safety of the composition is important. Hence, algal products must be subjected to physicochemical characterization to ensure safety of the individual ingredients. Products derived from algae must be characterized to establish the potential for toxicity, such as mutagenicity, multigeneration and systemic toxicity, presence of allergens, toxins, heavy metals, and harmful concentrations of pathogenic microbes. In line with regulation CE-258/97, CE-1881/2006, and updated with CE-629/2008, algae for human consumption must demonstrate meeting the quality specifications (maximum allowable limits) of potentially toxic metals (values stated as per kg dry weight of algal biomass): Arsenic $<3$ mg; Cadmium $<0.5$ mg ($<3$ mg for food supplements); Mercury $<0.1$ mg; Lead $<5$ mg; Tin $<5$ mg; and Iodine $<2000$ mg. Microbiologically, the biomass must conform to the following standards: $<10^5$ aerobic mesophilic bacteria $g^{-1}$, $<10$ fecal coliforms $g^{-1}$, $<100$ anaerobic sulfite-reducing bacteria $g^{-1}$, $<1$ *Clostridium perfringens* $g^{-1}$, and $<1$ *Salmonella* 25 $g^{-1}$ of dried algal biomass (Lerat et al., 2018).

While the quality of algae is highly dependent on the quality of the culture medium used for its production, the safety of the organisms can be used to evaluate their product toxicity. Therefore, the history of safe use becomes important to measure the safety of the product (Constable et al., 2007). Using this approach combined with available toxicological data, algae, and foods derived from them can be prepared to promote human consumption. Other concerns on the algae especially microalgae for a novel protein source is the presence of high concentrations of deoxyribonucleic acid.

## 9.7 CONCLUSION

Algae represent an underexplored and -utilized resource in Nigeria compared to the rest of the world—general low awareness and skilled personnel to promote the industry. Considering the rich diversity of algae and the already known biochemical composition of edible algae, they bridge the gap as a food source and provide excellent health benefits and disease therapies. Employing best safety practices, already characterized algae can be safely introduced into Nigerian and African markets as novel food sources.

## REFERENCES

Abdulqader, G., Barsanti, L., Tredici, M.R., 2000. Harvest of *Arthrospira platensis* from Lake Kossorom (Chad) and its household usage among the Kanembu. *Journal of Applied Phycology* 12, 493–498.

Admassu, H., Gasmalla, M.A.A., Yang, R., Zhao, W., 2018. Bioactive peptides derived from seaweed protein and their health benefits: Antihypertensive, antioxidant, and antidiabetic properties. *Journal of Food Science* 83, 6–16.

Belasco, W., 1997. Algae burgers for a hungry world? The rise and fall of *Chlorella* cuisine. *Technology and Culture* 38, 608–634.

Benjama, O., Masniyom, P., 2012. Biochemical composition and physicochemical properties of two red seaweeds (*Gracilaria fisheri* and *G. tenuistipitata*) from the Pattani Bay in Southern Thailand. *Sonklanakarin Journal of Science and Technology* 34, 223.

Bigliardi, B., Galati, F., 2013. Innovation trends in the food industry: The case of functional foods. *Trends in Food Science and Technology* 31, 118–129.

Bondioli, P., Della Bella, L., Rivolta, G., Zittelli, G.C., Bassi, N., Rodolfi, L., Casini, D., Prussi, M., Chiaramonti, D., Tredici, M.R., 2012. Oil production by the marine microalgae *Nannochloropsis* sp. F&M-M24 and *Tetraselmis suecica* F&M-M33. *Bioresource Technology* 114, 567–572.

Bruneel, C., Lemahieu, C., Fraeye, I., Ryckebosch, E., Muylaert, K., Buyse, J., Foubert, I., 2013. Impact of microalgal feed supplementation on omega-3 fatty acid enrichment of hen eggs. *Journal Functional Foods* 5, 897–904.

Carmichael, W.W., Drapeau, C., Anderson, D.M., 2000. Harvesting of *Aphanizomenon flos-aquae* Ralfs ex Born. & Flah. var. flos-aquae (Cyanobacteria) from Klamath Lake for human dietary use. *Journal Applied Phycology* 12, 585–595.

Chacón-Lee, T.L., González-Mariño, G.E., 2010. Microalgae for "healthy" foods—possibilities and challenges. *Comprehensive Review in Food Science and Food Safety* 9, 655–675.

Chia, M.A., Lombardi, A.T., Gama Melão, M.D.G., 2013. Calorific values of *Chlorella vulgaris* (Trebouxiophyceae) as a function of different phosphorus concentrations. *Phycological Research* 61. doi:10.1111/pre.12026.

Chuka-ogwude, D., Ogbonna, J., Moheimani, N.R., 2020. A review on microalgal culture to treat anaerobic digestate food waste effluent. *Algal Research* 47, 101841.

Cofrades, S., López-López, I., Solas, M.T., Bravo, L., Jiménez-Colmenero, F., 2008. Influence of different types and proportions of added edible seaweeds on characteristics of low-salt gel/emulsion meat systems. *Meat Science* 79, 767–776.

Constable, A., Jonas, D., Cockburn, A., Davi, A., Edwards, G., Hepburn, P., Herouet-Guicheney, C., Knowles, M., Moseley, B., Oberdörfer, R., 2007. History of safe use as applied to the safety assessment of novel foods and foods derived from genetically modified organisms. *Food Chemistry and Toxicology* 45, 2513–2525.

Cottin, S.C., Sanders, T.A., Hall, W.L., 2011. The differential effects of EPA and DHA on cardiovascular risk factors. *Proceedings of the Nutrition Society* 70, 215–231.

Cvetkovska, M., Hüner, N.P.A., Smith, D.R., 2017. Chilling out: The evolution and diversification of psychrophilic algae with a focus on Chlamydomonadales. *Polar Biology* 40, 1169–1184.

da Silva Vaz, B., Moreira, J.B., de Morais, M.G., Costa, J.A.V., 2016. Microalgae as a new source of bioactive compounds in food supplements. *Current Opinion in Food Science* 7, 73–77.

Draaisma, R.B., Wijffels, R.H., Slegers, P.M.E., Brentner, L.B., Roy, A., Barbosa, M.J., 2013. Food commodities from microalgae. *Current Opinion in Biotechnology* 24, 169–177.

EC 2020. European Union, https://ec.europa.eu/food/safety/novel_food_en, accessed 25 April 2020.

Ejike, C.E.C.C., Collins, S.A., Balasuriya, N., Swanson, A.K., Mason, B., Udenigwe, C.C., 2017. Prospects of microalgae proteins in producing peptide-based functional foods for promoting cardiovascular health. *Trends in Food Science and Technology* 59, 30–36.

El Sohaimy, S.A., 2012. Functional foods and nutraceuticals-modern approach to food science. *World Applied Science Journal* 20, 691–708.

Fernando, I.P.S., Kim, M., Son, K.-T., Jeong, Y., Jeon, Y.-J., 2016. Antioxidant activity of marine algal polyphenolic compounds: A mechanistic approach. *Journal Medicine and Food* 19, 615–628. doi:10.1089/jmf.2016.3706.

Fleurence, J., Morançais, M., Dumay, J., Decottignies, P., Turpin, V., Munier, M., Garcia-Bueno, N., Jaouen, P., 2012. What are the prospects for using seaweed in human nutrition and for marine animals raised through aquaculture? *Trends in Food Science and Technology* 27, 57–61.

Fraeye, I., Bruneel, C., Lemahieu, C., Buyse, J., Muylaert, K., Foubert, I., 2012. Dietary enrichment of eggs with omega-3 fatty acids: A review. *Food Research International* 48, 961–969.

FSANZ, 2020. Food Standards Australia New Zealand. (https://www.foodstandards.gov.au/industry/novel/Pages/default.aspx, accessed 25 April 2020).

Fujiwara-Arasaki, T., Mino, N., Kuroda, M., 1984. The protein value in human nutrition of edible marine algae in Japan. In: *Eleventh International Seaweed Symposium*. Springer, pp. 513–516.

Ghozzi, K., Zemzem, M., Dhiab, R. Ben, Challouf, R., Yahia, A., Omrane, H., Ouada, H.B., 2013. Screening of thermophilic microalgae and cyanobacteria from Tunisian geothermal sources. *Journal of Arid Environment* 97, 14–17.

Griffin, J.A., 2015. An Investigative Study Into the Beneficial Use of Seaweed in Bread and the Broader Food Industry.

Hsu, H.-Y., Jeyashoke, N., Yeh, C.-H., Song, Y.-J., Hua, K.-F., Chao, L.K., 2010. Immunostimulatory bioactivity of algal polysaccharides from *Chlorella pyrenoidosa* activates macrophages via Toll-like receptor 4. *Journal of Agriculture and Food Chemistry* 58, 927–936.

Jensen, G.S., 2001. Blue-green algae as an immuno-enhancer and biomodulator. *Journal of American Nutraceutical Association* 3, 24–30.

Johnson, D.B., 2012. Acidophilic algae isolated from mine-impacted environments and their roles in sustaining heterotrophic acidophiles. *Frontier Microbiology* 3, 325.

Karthikeyan, S., 2012. A critical review: Microalgae as a renewable source for biofuel production. *International Journal of Engineering Research Technology* 1. ISSN: 2278-0181.

Knudsen, A.K., Allebeck, P., Tollånes, M.C., Skogen, J.C., Iburg, K.M., McGrath, J.J., Juel, K., Agardh, E.E., Ärnlöv, J., Bjørge, T., 2019. Life expectancy and disease burden in the Nordic countries: Results from the global burden of diseases, injuries, and risk factors study 2017. *Lancet Public Heal* 4, e658–e669.

Kralovec, J.A., 2005. Chlorella composition having high molecular weight polysaccharides and polysaccharide complexes.

Kralovec, J.A., Power, M.R., Liu, F., Maydanski, E., Ewart, H.S., Watson, L.V., Barrow, C.J., Lin, T.J., 2005. An aqueous Chlorella extract inhibits IL-5 production by mast cells in vitro and reduces ovalbumin-induced eosinophil infiltration in the airway in mice in vivo. *International Immunopharmacology* 5, 689–698.

Kwak, J.H., Baek, S.H., Woo, Y., Han, J.K., Kim, B.G., Kim, O.Y., Lee, J.H., 2012. Beneficial immunostimulatory effect of short-term *Chlorella* supplementation: Enhancement of Natural Killercell activity and early inflammatory response (Randomized, double-blinded, placebo-controlled trial). *Nutrition Journal* 11, 53.

Lemahieu, C., Bruneel, C., Ryckebosch, E., Muylaert, K., Buyse, J., Foubert, I., 2015. Impact of different omega-3 polyunsaturated fatty acid (n-3 PUFA) sources (flaxseed, Isochrysis galbana, fish oil and DHA Gold) on n-3 LC-PUFA enrichment (efficiency) in the egg yolk. *Journal of Functional Foods* 19, 821–827.

Lerat, Y., Cornish, M.L., Critchley, A.T., 2018. Applications of algal biomass in global food and feed markets: From traditional usage to the potential for functional products. *Blue Biotechnology: Production and Use of Marine Molecules* 1, 143–189.

Li, M.H., Robinson, E.H., Tucker, C.S., Manning, B.B., Khoo, L., 2009. Effects of dried algae *Schizochytrium* sp., a rich source of docosahexaenoic acid, on growth, fatty acid composition, and sensory quality of channel catfish Ictalurus punctatus. *Aquaculture* 292, 232–236.

Liang, H., Gong, W.-J., Chen, Z.-L., Tian, J.-Y., Qi, L., Li, G.-B., 2009. Effect of chemical preoxidation coupled with in-line coagulation as a pretreatment to ultrafiltration for algae fouling control. *Desalination and Water Treatment* 9, 241–245.

Liang, S., Liu, X., Chen, F., Chen, Z., 2004. Current microalgal health food R & D activities in China. In: *Asian Pacific Phycology in the 21st Century: Prospects and challenges*. Springer, pp. 45–48.

Liu, C., Liu, J., Hu, S., Wang, Xin, Wang, Xuhui, Guan, Q., 2019. Isolation and identification of a halophilic and alkaliphilic microalgal strain. *PeerJ* 7, e7189.

Losses, F.A.O.G.F., Waste, F., 2011. Extent, causes and prevention. Rome Food Agric. Organ. United Nations.

Markou, G., Wang, L., Ye, J., Unc, A., 2018. Using agro-industrial wastes for the cultivation of microalgae and duckweeds: Contamination risks and biomass safety concerns. *Biotechnology Advances* 36, 1238–1254.

Mata, T.M., Martins, A.A., Caetano, N.S., 2010. Microalgae for biodiesel production and other applications: A review. *Renewable and Sustainable Energy Reviews* 14, 217–232.

Melis, A., 2009. Solar energy conversion efficiencies in photosynthesis: minimizing the chlorophyll antennae to maximize efficiency. *Plant Science* 177, 272–280.

Mendes, A., Reis, A., Vasconcelos, R., Guerra, P., da Silva, T.L., 2009. *Crypthecodinium cohnii* with emphasis on DHA production: A review. *Journal of Applied Phycology* 21, 199–214.

Mišurcová, L., Škrovánková, S., Samek, D., Ambrožová, J., Machů, L., 2012. Health benefits of algal polysaccharides in human nutrition. In: *Advances in Food and Nutrition Research*. Elsevier, pp. 75–145.

Miyashita, K., 2014. Marine antioxidants: Polyphenols and carotenoids from algae. *Antioxidants and Functioal components in Aquatic Foods*, 219–229.

Mourelle, M.L., Gómez, C.P., Legido, J.L., 2017. The potential use of marine microalgae and cyanobacteria in cosmetics and thalassotherapy. *Cosmetics* 4, 46.

Murray, M., Dordevic, A.L., Ryan, L., Bonham, M.P., 2018. An emerging trend in functional foods for the prevention of cardiovascular disease and diabetes: Marine algal polyphenols. *Critical Reviews in Food Science and Nutrition* 58, 1342–1358.

Nagy, K., Tiuca, I.-D., 2017. Importance of fatty acids in physiopathology of human body, in: Fatty Acids. IntechOpen.

Ngo, D.-H., Kim, S.-K., 2013. Sulfated polysaccharides as bioactive agents from marine algae. *International Journal of Biological Macromolecules* 62, 70–75.

Nwoba, E.G., Parlevliet, D.A., Laird, D.W., Alameh, K., Moheimani, N.R., 2020a. Outdoor phycocyanin production in a standalone thermally-insulated photobioreactor. *Bioresource Technology* 315, 123865.

Nwoba, E.G., Ogbonna, C.N., Ishika, T., Vadiveloo, A., 2020b. Microalgal pigments: A source of natural food colors. In: *Microalgae Biotechnology for Food, Health and High Value Products*. Springer, pp. 81–123.

Nwoba, E.G., Parlevliet, D.A., Laird, D.W., Alameh, K., Moheimani, N.R. 2020c. Does growing nannochloropsis sp. in innovative flat plate photobioreactors result in changes to fatty acid and protein composition? *Journal of Applied Phycology* 32, 3619–3629.

Nwoba, E.G., Parlevliet, D.A., Laird, D.W., Alameh, K., Moheimani, N.R., 2019. Light management technologies for increasing algal photobioreactor efficiency. *Algal Research* 39, 101433.

Ogbonna, I.O., Ogbonna, J.C., 2015. Isolation of microalgae species from arid environments and evaluation of their potentials for biodiesel production. *African Journal of Biotechnology* 14, 1598–1604.

Ogbonna, J.C., Ichige, E., Tanaka, H., 2002. Regulating the ratio of photoautotrophic to heterotrophic metabolic activities in photoheterotrophic culture of *Euglena gracilis* and its application to α-tocopherol production. *Biotechnology Letters* 24, 953–958.

Ogbonna, J.C., Tomiyamal, S., Tanaka, H., 1998. Heterotrophic cultivation of *Euglena gracilis* Z for efficient production of α-tocopherol. *Journal of Applied Phycology* 10, 67–74.

Ortiz, J., Uquiche, E., Robert, P., Romero, N., Quitral, V., Llantén, C., 2009. Functional and nutritional value of the Chilean seaweeds Codium fragile, *Gracilaria chilensis* and *Macrocystis pyrifera*. *European Journal of Lipid Science and Technology* 111, 320–327.

Pal, A., Kamthania, M.C., Kumar, A., 2014. Bioactive compounds and properties of seaweeds: A review. *Open Access Library Journal* 1, 1–17.

Pulz, O., Gross, W., 2004. Valuable products from biotechnology of microalgae. *Applied Microbiology and Biotechnology* 65, 635–648.

Radmer, R.J., 1996. Algal diversity and commercial algal products. *Bioscience* 46, 263–270.

Raja, R., Shanmugam, H., Ganesan, V., Carvalho, I.S., 2014. Biomass from microalgae: An overview. *Oceanography* 2, 1–7.

de Jesus Raposo, M.F., De Morais, R.M.S.C., Bernardo de Morais, A.M.M., 2013. Bioactivity and applications of sulphated polysaccharides from marine microalgae. *Marine Drugs* 11, 233–252.

Rashid, N., Ryu, A.J., Jeong, K.J., Lee, B., Chang, Y.-K., 2019. Co-cultivation of two freshwater microalgae species to improve biomass productivity and biodiesel production. *Energy Conversion and Management* 196, 640–648.

Richmond, A., 2008. *Handbook of Microalgal Culture: Biotechnology and Applied Phycology*. John Wiley & Sons, Hoboken, NJ.

Rodolfi, L., Chini Zittelli, G., Bassi, N., Padovani, G., Biondi, N., Bonini, G., Tredici, M.R., 2009. Microalgae for oil: Strain selection, induction of lipid synthesis and outdoor mass cultivation in a low-cost photobioreactor. *Biotechnology and Bioengineering* 102, 100–112.

Saarela, M. (2007). *Functional Dairy Products*. Elsevier, Amsterdam.

Sedlmeier, A., Kluttig, A., Giegling, I., Prehn, C., Adamski, J., Kastenmüller, G., Lacruz, M.E., 2018. The human metabolic profile reflects macro-and micronutrient intake distinctly according to fasting time. *Science Reports* 8, 1–8.

Seifried, H.E., Anderson, D.E., Fisher, E.I., Milner, J.A., 2007. A review of the interaction among dietary antioxidants and reactive oxygen species. *Journal of Nutrition and Biochemistry* 18, 567–579.

Sibi, G., 2018. Bioenergy production from wastes by microalgae as sustainable approach for waste management and to reduce resources depletion. *International Journal of Environmental Science and Natural Resource* 13, 77–80.

Sousa, I., Gouveia, L., Batista, A.P., Raymundo, A., Bandarra, N.M., 2008. Microalgae in novel food products. In K.N. Papadopoulos (ed.), *Food Chemistry Research Development*, pp. 75–112. Nova Science Publishers, Inc., Hauppauge, NY.

Spolaore, P., Joannis-Cassan, C., Duran, E., Isambert, A., 2006. Commercial applications of microalgae. *Journal Bioscience and Bioengineering* 101, 87–96.

Stiger-Pouvreau, V., Bourgougnon, N., Deslandes, E., 2016. Carbohydrates from seaweeds. In: J. Fleurence, I. Levine (eds.) *Seaweed in Health and Disease Prevention*. Elsevier, Amsterdam, pp. 223–274.

Suárez, E.R., Kralovec, J.A., Grindley, T.B., 2010. Isolation of phosphorylated polysaccharides from algae: the immunostimulatory principle of *Chlorella pyrenoidosa*. *Carbohydrate Research* 345, 1190–1204.

Suganya, T., Varman, M., Masjuki, H.H., Renganathan, S., 2016. Macroalgae and microalgae as a potential source for commercial applications along with biofuels production: A biorefinery approach. *Renewable Sustainable Energy Reviews* 55, 909–941.

Udenigwe, C.C., 2014. Bioinformatics approaches, prospects and challenges of food bioactive peptide research. *Trends in Food Science and Technology* 36, 137–143.

Udenigwe, C.C., Aluko, R.E., 2012. Food protein-derived bioactive peptides: production, processing, and potential health benefits. *Journal of Food Science* 77, R11–R24.

Udokanma, E.E., Emeahara, G.O., 2017. Nutrition and food proportion for maintenance of health and fitness of pregnant mothers and children. *Nigerian Journal of Healing Promotion* 10, 95–105.

van Ginneken, V.J.T., Helsper, J.P.F.G., de Visser, W., van Keulen, H., Brandenburg, W.A., 2011. Polyunsaturated fatty acids in various macroalgal species from north Atlantic and tropical seas. *Lipids and Health Disease* 10, 104.

Wells, M.L., Potin, P., Craigie, J.S., Raven, J.A., Merchant, S.S., Helliwell, K.E., Smith, A.G., Camire, M.E., Brawley, S.H., 2017. Algae as nutritional and functional food sources: Revisiting our understanding. *Journal of Applied Phycology* 29, 949–982.

Zhu, X.-G., Long, S.P., Ort, D.R., 2010. Improving photosynthetic efficiency for greater yield. *Annual Reviews in Plant Biology* 61, 235–261.

Zhu, X.-G., Long, S.P., Ort, D.R., 2008. What is the maximum efficiency with which photosynthesis can convert solar energy into biomass? *Current Opinion in Biotechnology* 19, 153–159.

# 10 Microalgae Biotechnology Research and Development Opportunities in Nigeria

*James Chukwuma Ogbonna*
University of Nigeria

*Emeka G. Nwoba and David Chuka-Ogwude*
Murdoch University

*Innocent Ogbonna*
Federal University of Agriculture

*Abosede T. Adesalu*
University of Lagos

## CONTENTS



DOI: 10.1201/9781003178378-10

## 10.1 INTRODUCTION

Microalgae are microscopic species of algae living in both freshwater and marine environments. They are basically unicellular but can live singly or in chains. The major characteristic of microalgae is their photosynthetic ability whereby they convert solar energy and inorganic carbons into chemical energy, carbohydrate, biomass, and an array of useful metabolites (Perez-Garcia et al., 2011; Priyadarshani and Rath, 2012; Raja et al., 2014). In addition to photoautotrophic growth using only light and inorganic carbons, many species can also grow heterotrophically in the absence of light, using organic carbon as both carbon and energy source (Ogbonna et al., 1998; Ogbonna and Moheimani, 2015; Ogbonna and McHenry, 2015). Furthermore, in the presence of inorganic carbon, organic carbon and light, some species can grow mixotrophically (Ogbonna et al., 2002a,b). This versatility in modes of growth enables them to grow in a wide range of environments, ranging from aquatic to terrestrial habitants. Some are free-floating and thus grow in suspended forms in both fresh and marine water bodies while others are attached to submerged substrates. Still, some grow aerially, depending on molecular carbon dioxide, humidity, and nutrients on the surface of the objects to which they are attached (Kiepper, 2013). They are also found in extreme environments such as deserts where they are exposed to hot and very cold weathers and in the Polar Regions (Bleeke et al., 2014; Raja et al., 2014). There are both prokaryotic and eukaryotic forms (Raja et al., 2014). Furthermore, due to their metabolic versatility and wide distribution, they have a wide range of applications in food and agriculture (Vigani et al., 2015; Wuang et al., 2016), as sources of useful metabolites used in medical, pharmaceutical, cosmetic, and other industries (Spolaore et al., 2006; Priyadarshani and Rath, 2012). They are also used in environmental monitoring, as well as for bioremediation of air, water, and wastewater (Ogbonna et al. 2000; Chuka-Ogwude et al., 2020; Nwoba et al., 2020a,b).

In spite of their versatility in growth habitants and applications, they remain one of the least exploited bioresource in Nigeria. This chapter explores the various research opportunities with microalgae in Nigeria, ranging from isolation and screening for the purposes of documentation and building culture collection centres, to their various industrial, medical, food, and agricultural as well as environmental applications.

## 10.2 BASIC RESEARCH OPPORTUNITIES

There are a lot of basic research on microalgae which have helped to elucidate some basic physiology and biochemistry of these species of microorganisms. A basic characteristic of microalgae is their ability to photosynthesise. Photosynthesis is a biological process that harvests light (sunlight or artificial lights), water, and $CO_2$ to produce carbon-rich compounds (e.g., carbohydrates) and $O_2$. This is one of the most important biological processes since life on earth depends on the process and products of photosynthesis. It is a model process that drives metabolite production via harvesting of solar energy and inorganic compounds and converts to chemical energy (organic products) stored as adenosine triphosphate and a reductant, nicotinamide adenine dinucleotide phosphate (NADPH). The ATP and NADPH are the drivers of biomass production. The earliest forms of photosynthetic organisms were anoxygenic photosynthetic bacteria, which formed organic compounds by extraction of electrons and protons from inorganic molecules such as hydrogen sulphide for $CO_2$ reduction (Masojıdek et al., 2004). The prokaryotic blue-green microalgae (cyanobacteria) and eukaryotic autotrophic microalgae emerged later and are the bedrocks of oxygenic photosynthesis. While the cyanobacteria are prokaryotic, with a DNA-rich nucleoplasm, and photosynthetic membranes housed by the chromoplast, the eukaryotes have the photosynthetic machinery contained in the chloroplasts. The cyanobacterial photosynthetic membranes are near the cell surface and organised in parallel. The chloroplasts of eukaryotic algae have alternate layers of thylakoids (lipoprotein membranes) and stroma (fluid phases). The eukaryotic microalgae are grouped into Chlorophyta (green algae), Rhodophyta (red algae), Chrysophyceae (golden algae), and Phaeophyceae (brown algae) based on their light-harvesting pigments (Table 10.1).

Microalgae including cyanobacteria are the best representation of solar energy transformation systems, which convert inorganic carbon into chemical molecules using photosynthesis – the only pathway for the sustainable synthesis of various complex organic compounds. Microalgae consist of two multi-subunit membrane protein complexes called photosystem I (PS1, which absorbs light wavelength of 700 nm) and photosystem II (PSII, which absorbs light wavelength of 680 nm). The photosystem comprises a light-harvesting complex (LHC) and a photochemical reaction centre, with the LHCI and LHCII associated with PSI and PSII, respectively. In a nutshell, the five main proteins of the thylakoid membranes with associated pigments are: PSI protein-pigment, PSII protein-pigment, cytochrome $b_6f$, LH chlorophylls a/b, and ATP synthase complexes (Figure 10.1a). The PSII protein-pigment complex in association with the antenna pigment, chlorophyll as well as transport proteins is found mainly in the grana lamellae. In contrast, the PSI protein pigments, in association with its antenna pigment molecules, electron transport proteins, and the enzyme, ATP synthase, are located in the stroma lamellae, which are not stacked grana. The PSI and PSII are connected by the cytochrome

**TABLE 10.1**
**Classification of Microalgae Based on Pigment Composition (Nwoba et al., 2019; Carvalho et al., 2011)**

| Light-harvesting Pigments Class | Major Pigment Constituents | Absorption Spectrum (nm) | Behaviour in Solvent | Pigment Colour | Algal Division |
|---|---|---|---|---|---|
| Chlorophylls | $a$, $b$, $c_1$, $c_2$, $d$, $f$ | 450–475, 630–680, 700–750 | Hydrophobic | Green | Cyanophyta, Prochlorophyta, Glaucophyta, Rhodophyta, Cryptophyta, Heterokontophyta, Haptophyta, Dinophyta, Euglenophyta, Chlorarachniophyta, Chlorophyta |
| Phycobilins | C-phycocyanin, Phycoerythrin, Allophycocyanin | 500–650 | Hydrophilic | Red, blue | Cyanophyta, Glaucophyta, Rhodophyta, Cryptophyta |
| Carotenoids | α-, β-, & ε-carotene, Lutein, Astaxanthin, Violaxanthin, Fucoxanthin, Zeaxanthin | 400–550 | Hydrophobic | Red, yellow, orange | Cyanophyta, Prochlorophyta, Glaucophyta, Rhodophyta, Cryptophyta, Heterokontophyta, Haptophyta, Dinophyta, Euglenophyta, Chlorarachniophyta, Chlorophyta |

$b_6f$ (Figure 10.1a and b). The light-harvesting pigments of microalgae and cyanobacteria are chlorophylls, carotenoids, and phycobilins (Table 10.1). Chlorophylls are subdivided into *a*, *b*, *c*, *d*, and *f*, with chlorophyll *a* as the most abundant and present in all oxygenic photoautotrophs (Masojıdek et al., 2004). Chlorophylls absorb in the blue, blue-green region (400–475 nm) and red (630–750 nm) regions of visible light wavelengths (Figure 10.1c) (Nwoba et al., 2019). While chlorophyll *a* forms the core of the LH protein complexes and involved in direct light-energy capture, other pigments which constitute accessory pigments (chlorophylls *b*, *c*, *d*, *f*, carotenoids, and phycobilins) expand the wavelength absorption of algae and transfer excitation energy to chlorophyll *a* (Carvalho et al., 2011). This excitation energy is used for photosynthetic splitting of water molecules into protons, electrons, and oxygen (Figure 10.1a and b).

During photosynthesis, the LHC proteins absorb light and channel the excitation energy to PSI/PSII. The excess energy which is not used for biomass formation (photochemistry) is dissipated as heat (or fluorescence) to avoid PSII photodamage through non-photochemical quenching reactions. Under high irradiance,

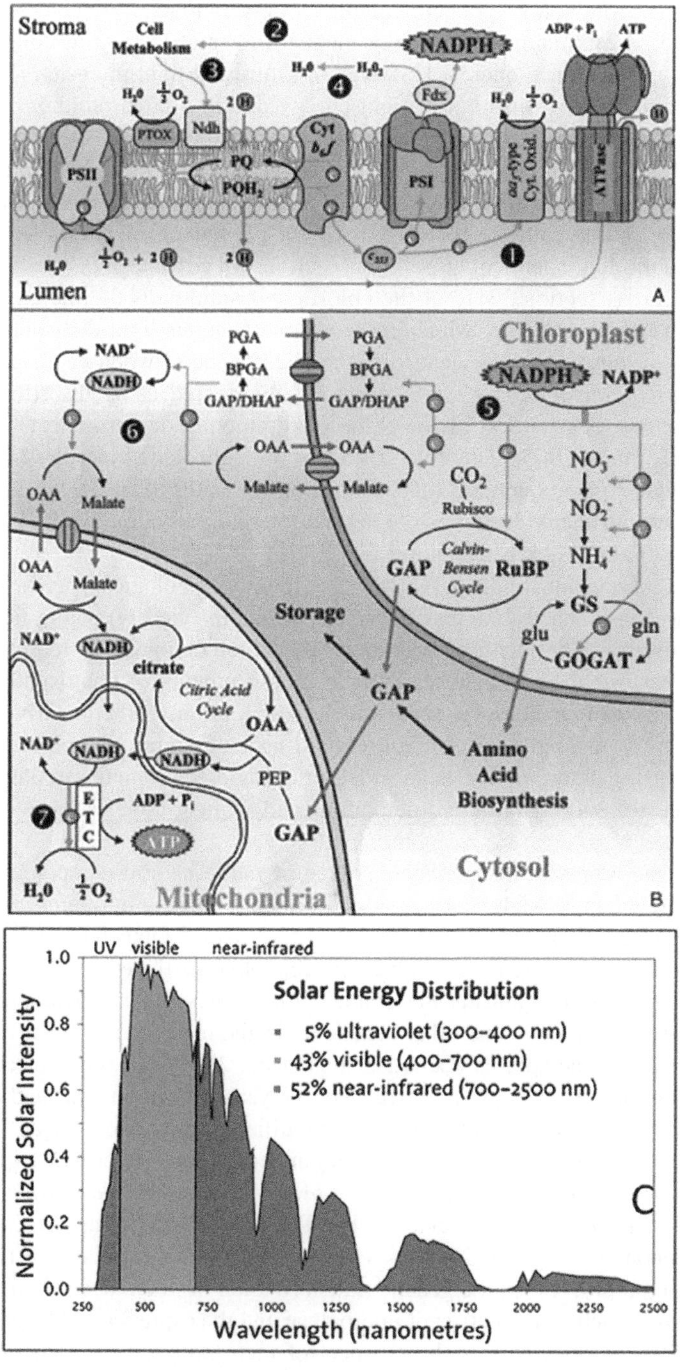


**FIGURE 10.1** Photosynthetic and metabolic electron transport pathways. Cyanobacterial type thylakoid membrane (a), downstream metabolic pathways (Behrenfeld et al., 2004) (b), and solar energy distribution (c).

energy losses from non-photochemical quenching can reach as high as 90% (Polle et al., 2003). Minimising these losses means higher efficiency in conversion of light energy to chemical energy. However, at saturating light intensity, microalgae are inefficient in the utilisation of light energy due to a large number of antenna pigments. To moderate the light absorption rate of microalgae, the concentration of the LHC pigments can be reduced by decreasing the cross-sectional size of the antenna molecule. Under this scenario, microalgae would capture the amount of light energy that can be efficiently used for photochemistry, and significantly minimise non-photochemical quenching reaction. An advantage of selective reduction of the cross-sectional size of the chlorophyll antenna is the increase in light penetration into the culture, which ensures homogeneous light distribution in the cultivation systems and tolerance to higher irradiance (Nwoba et al., 2019). For instance, a *Chlamydomonas reinhardtii* was engineered using insertional mutagenesis technique by truncating the chlorophyll antenna size proteins (Kirst et al., 2012b; Polle et al., 2003). The mutant (tla1) with truncated chlorophyll molecules attained higher photosynthetic light saturation of 2,500 $\mu E/m^2/s$, while the wild type with an untruncated antenna could only tolerate a light intensity of 1,000 $\mu E/m^2/s$ (Kirst et al., 2012a, b; Perrine et al., 2012).

The measurement of photosynthesis or photophysiological performance of microalgae can be used to assess the stress or health (vitality) or physical fitness of algal cultures. The techniques used in algal biotechnology to monitor photosynthetic performance of cultures include measurements of photosynthetic oxygen evolution using a Clark-type oxygen electrode, photosynthetic carbon fixation using infrared gas analysis or $^{14}C$ radio-labelling techniques, and chlorophyll fluorescence using fluorometers (e.g., pulse amplitude fluorometers) (Baker, 2008; Beardall et al., 2003; Bolhàr-Nordenkampf and Öquist, 1993; Krause and Weis, 1991).

Microalgae represent a sustainable source of many natural compounds such as polyunsaturated fatty acids, carotenoids, food dyes, phycobiliproteins, antitumour chemicals, etc., which have been harnessed for various biotechnological applications. A powerful driving force in algae biotechnology is the enticing option to use genetically improved organisms. Selectable marker genes, reporter genes, promoters, transformation techniques and other genetic tools and methods are already available for several few algae species (León-Bañares et al., 2004) and this molecular toolbox is becoming increasingly powerful. Unlike yeasts, bacteria, and higher plants, which have a large number of genetically modified species, microalgae have very few species that have been successfully engineered. Model microalgal species are *Chlamydomonas reinhardtii*, *Thalassiosira pseudomonas*, *Phaeodactylum tricornutum*, and *Synechocystis* (Ajjawi et al., 2017). The lack of genetic engineering toolbox for many economic microalgal species and transcriptional silencing of transgenes have hampered progress in the genetic transformation of microalgae (León-Bañares et al., 2004). Genetic engineering of microalgae and the expression of heterogeneous genes in these species creates the prospect of increasing the productivity of conventional microalgal products. Novel bioactive compounds can be produced through metabolic engineering for industrial and pharmaceutical uses. The use of microalgae as a model eukaryotic host is attractive, especially specific genes such as those connected to photosynthesis, photoreception, and functioning of flagella, which cannot be expressed in other hosts such as yeasts (Fernández et al., 1989). The production of high-quality human proteins such as edible vaccines, antibodies, and hormones could be done through genetic engineering of microalgae (Rosales-Mendoza et al., 2016). Transgenic microalgae have the ability to be grown photoautotrophically and heterotrophically in photobioreactors, which are suitable for control of culture conditions (Richmond, 2008; Zaslavskaia et al., 2001).

The commercial application of transgenic microalgae for metabolite production is yet to be documented; however, many studies have reported the prospects of genetically modified microalgae for biotechnological applications. A genetically modified microalga with a greater capacity to chelate heavy metals by expressing a foreign metallothionein gene or mothbean pyrroline-5-carboxylate synthetase genes, which stimulates the overproduction of proline in *C. reinhardtii* has been documented (Cai et al., 1999; Dunahay et al., 1996). A glucose transporter gene from human red blood cells was expressed in the transgenic diatom, *Phaeodactylum tricornutum* and conferred the alga the ability to grow as an obligate photoautotroph that could utilise glucose as a sole carbon source in the dark (Zaslavskaia et al., 2001). The photo-biological production of $H_2$ has been reported in a transgenic *C. reinhardtii*, in which the chloroplastic sulphate permease gene was downregulated through the insertion of anti-sense CrepSulf gene (León-Bañares et al., 2004). Recently, the lipid productivity of a transgenic *Nannochloropsis gaditana*, a non-model microalgal species, was doubled by the downregulation or attenuation of $Zn(II)_2Cys_6$ transcription regulator for lipid production using CRISPR-Cas9 reverse-genetics pipeline-based insertional mutagenesis (Ajjawi et al., 2017). Transgenic microalgae have the potential to be used as biofactories for the production of edible vaccines, recombinant vaccines, and value-added bioactive compounds (Table 10.2).

Although a lot of basic research on microalgae are going on, a lot are yet unknown about this group of microorganisms. For example, although it is generally believed that the photosynthetic efficiencies of microalgae are higher than those of higher plants, photosynthetic efficiencies in large-scale culture systems are still much lower than the potential values. This is partly due to the problem of light shading and light distribution in such culture systems and partly due to lack of knowledge on physiological response of microalgae to various environmental stresses such as temperature, pH, salinity, and nutrient deficiencies. A lot of work is therefore needed in this direction. Furthermore, microalgae are good model organisms to study physiology of higher plants. Although it has been shown that various species of microalgae have the potentials for production of various metabolites, genetic modification can be used to improve their productivities while they can be used as hosts for expression of various useful genes. Yet in comparison with other groups of microorganisms, there are still very few genetically modified microalgae. Thus, a lot can still be done on their genetic modification for various purposes.

**TABLE 10.2**
**Transgenic Microalgae for Disease Treatment (Dauvillée et al., 2010; Dreesen et al., 2010; Feng et al., 2014; Gregory et al., 2012; Rosales-Mendoza et al., 2016)**

| Microalgae Species | Infectious Agent | Target Disease | Antigen Description | Immunogenic Properties |
|---|---|---|---|---|
| *Chlamydomonas reinhardtii* | *Plasmodium falciparum* | Malaria | Chimeric protein containing Pfs25 antigen of *Plasmodium* surface protein ligated to cholera toxin B (CTB) subunit expressed in *Chlamydomonas* chloroplast genome | Mucosal IgA response to both CTB and Pfs25, and systemic IgG response to CTB in mice |
| *Dunaliella salina* | White spot syndrome | White spot syndrome virus | *D. salina* expressed VP28 protein of the white spot syndrome virus | Crayfish vaccinated with *Dunaliella*-VP28 showed 41% (significantly higher) survival rate compared to the controls. |
| *Chlamydomonas reinhardtii* | *Plasmodium falciparum*, *Plasmodium berghei* | Malaria | Chimeric protein consisting of *P. falciparum* MSP1 antigen, *P. berghei* MSP1, and *P. berghei* AMA1, C-terminal domains only, fused to granule-bound starch synthase. Antigens are expressed in *C. reinhardtii* nuclear genome, targeted to the chloroplast starch granules | Systemic IgG production and protection against *P. berghei* challenge in mice model immune sera and purified IgG specific to starch-bound PfMSP1-19 antigen blocked red blood cell entry by *P. falciparum* in vitro. |
| *Chlamydomonas reinhardtii* | *Staphylococcus aureus* | Staphylococcosis | Chimeric protein consisting of D2 fibronectin-binding domain of *S. aureus* fused to cholera toxin B (CTB) subunit, expressed in *C. reinhardtii* chloroplast genome. | Mucosal IgA and systemic IgG responses, and protection (80% survival rate) against lethal *S. aureus* challenge in mice. |

## 10.3 RESEARCH IN MICROALGAE BIORESOURCES CONSERVATION

It has been estimated that there are several million species of microalgae in nature (Kim et al., 2010; Bleeke et al., 2014) but only about thirty thousand to forty thousand species have been identified and described (Hargreaves and Tucker, 2004; Scharff, 2015). Out of these, only a few thousand strains are kept in various culture collections all over the world. Unfortunately, only a very few of these culture collections are in African continent and there is yet no functional microalgae culture collection in Nigeria. There are many studies on distribution of phytoplankton in Nigeria (Kadiri, 1999, 2002, 2011; Kadiri et al., 2020; Adeniyi and Akinwole, 2017; Essien et al., 2008; Adesalu et al., 2014; Adesalu, 2017; Chia et al., 2011, 2014a,b, 2015). Through these studies, both useful and harmful species within Baccillariophyta, Chlorophyta, Euglenophyta, Cyanophyta, Dinophyta, and Cyanophyta have been documented. However, studies on isolation and preservation of useful species are lagging in Nigeria. Furthermore, although about 15 species are currently cultivated industrially (Gouveia et al., 2008; Scharff, 2015), there is yet no microalgae-based industry in Nigeria.

Microalgae are still one of the least understood groups of microorganisms (Guedes and Malcata, 2012). Nigeria is endowed with a variety of habitats, ranging from humid swap forests, forests, savannah, and deserts all supporting the growth of thousands of species of microalgae. Many species have been harvested from Lake Chad for food for decades but these species have not yet been commercially cultivated in Nigeria. There has been no large project aimed at screening for, isolation and characterisation of the various species of microalgae in Nigeria. Although there have been a handful of small projects aimed at isolating specific strains of microalgae for various applications (e.g., Ogbonna and Ogbonna, 2015; Idenyi et al., 2020), the scope of such work has been limited and there is a need for a large-scale research projects on isolation, documentation, and preservation of microalgae. Presently, most of the isolated strains are discarded after the project because of lack of culture collection. Microalgae are very important bioresource in Nigeria and we need to do more in terms of isolation, characterisation, documentation, and preservation of useful strains.

## 10.4 RESEARCH ON MEDICAL APPLICATIONS OF MICROALGAE

Many species of microalgae have been used as food and food supplements because they are a good source of macro- and micro-nutrients such as microminerals and vitamins. According to Becker (2004), microalgae are an abundant source of vitamins (vitamins A, B1, B2, C, and E) and minerals (nicotinate, biotin, folic acid, pantothenic acid, niacin, iodine, potassium, iron, magnesium, and calcium) all of which can be used to alleviate many diseases. Edible algae are recognised as complete foods that provide correct balance of proteins, carbohydrates, vitamins, and minerals (Pooja, 2014). For example, many species of cyanobacteria (blue-green algae) are photosynthetic prokaryotes used as food by humans and recognised as an excellent source of vitamins and proteins and found in health food stores throughout the world (Becker, 2004; Spolaore et al., 2006; Del Campo et al., 2000; Sawraj et al., 2008). They have the potential for large-scale production of many pharmaceutical ingredients because their photosynthetic efficiencies have been reported to be ten times higher than those of the terrestrial plants (Sathasivam et al., 2019).

Aside from their nutritive values, they are also rich in various bioactive molecules such as polysaccharides, fibres, antioxidants, and omega-3 fatty acids which may have a potential in promoting health in humans. Many algae naturally accumulate high amount of carotenoids such as β-carotene, astaxanthin, and canthaxanthin. Recently, Walsh et al. (2018) reported that Coccolithophores, unicellular marine phytoplankton, are an excellent candidate as an additive to composite materials used in bone repair. One important characteristic of microalgae is that the nature and concentrations of secondary metabolites produced by microalgae can be controlled by the environmental (culture) conditions (Blaine and Pyne, 1986). In view of these various bioactive compounds found in microalgae, they have been used for their antioxidant, antimicrobial (Bouhlal et al., 2011), antiviral (Kim and Karadeniz, 2011), antifungal (De Felício, 2010), anti-allergic (Na et al., 2005), anticoagulant (Dayong et al., 2008), anticancer (Kim et al., 2011), antifouling (Devi et al., 2011), hypocholesterolemic and hypolipidemic (Lamela et al., 1989; Panlasigui et al., 2003; Nishide and Uchida, 2003), antitumour, antithrombic, anti-inflammatory, and many other therapeutic properties such as their beneficial effects against metabolic disorders such as obesity, diabetes, and hyperlipidaemia (Keshri, 2012; Chu and Phang, 2019). Thalasso-therapy is a new therapy emerging based on macroalgal use. In this therapy, algae pastes are made by cold-grinding or freeze-crushing, applied to the person's body and then warmed under infrared radiation. This treatment, in conjunction with seawater hydrotherapy, is said to provide relief for rheumatism and osteoporosis. The therapy is growing popular in France (Barsanti and Gualtieri, 2006). Similar investigation can also be conducted for several microalgae species.

Nigerian weather condition is very suitable for cultivation of microalgae all-round the year (Ali and Ogbonna, 2011; Nwoba et al., 2020a). In Nigeria, potentials of many strains of microalgae for production of vitamins (Afiukwa and Ogbonna, 2007), as well as protein and bioactive compounds (Ozioko et al., 2015; Eze et al., 2017a,b; Nwoba, 2017) have been investigated. As outlined above, the potential applications of microalgae in healthcare sector are huge but only a few species of microalgae have been investigated in Nigeria. There is, therefore, an urgent need to explore these potentials with the various local species of microalgae in Nigeria.

## 10.5 RESEARCH ON VARIOUS INDUSTRIAL APPLICATIONS OF MICROALGAE

Industrial applications of microalgae transverse essentially all the areas of biotechnology. Microalgal research has recently been a considerable global issue because of algae's extensive application potentials in renewable energy, biopharmaceutical, nutraceutical industries, and other environment-related concerns. Microalgae and their products are renewable and sustainable. They are economical sources of materials useful to man including biofuels, bioactive medicinal materials, and useful sources of food ingredients.

Microalgal metabolites could be either primary or secondary depending on the phase of growth of its production and whether the metabolite is reusable by the producing strain or not. The primary microalgal metabolites include protein (produced during active multiplication of the algal cells) and carbohydrates and lipids which are produced in most algae as energy-storage molecules during nutrient deficiency or light conditioning periods (Vuppaladadiyam et al., 2018). Cellular compositions of different primary metabolites of some species of microalgae are varied. Protein contents of many species of algae are high, and the range reported varying along species line. *Spirulina maxima*, *Spirulina platensis*, *Synechococcus* sp., and *Scenedesmus obliquus* have very high protein contents as high, whereas some other strains such as *Spirogyra* sp. contain proteins as low as 6%. Microalgae rich in protein (high $N_2$ content) are not very ideal for biofuel production but could be very good for other applications like in food and feed.

### 10.5.1 Lipids

Microalgal lipids are classified into neutral lipid which is composed of triacylglycerol (TAG) seen in the cytosol of cells and the membrane lipids. This is a polar amphipathic phospho- and glycolipids. In microalgae, neutral and polar lipids are interconvertible. During stress, membrane lipids can be converted to TAGs. Lipophilic compounds of microalgae consist of both primary and secondary metabolites. Some microalgae have large amounts of long-chain polyunsaturated fatty acids (PUFA). This unsaturated fatty acid is very important as a nutritional supplement for the prevention or treatment of certain diseases. Eicosapentaenoic acid (EPA 20:5 ω-3), docosahexaenoic acid (DHA 22:6 ω-3), α-linolenic acid (ALA 18:3 ω-3), γ-linolenic acid (GLA 18:3 ω-6), and arachidonic acid (ARA 20:6 ω-6) are some of the very important polyunsaturated fatty acids produced by some microalgae (Vuppaladadiyam et al., 2018). Microalgae also produce sterols. Sterols are bioactive compounds produced by eucaryotes. Sterols produced by microalgae include phytosterols, fucosterols, and stenols. These compounds are important in food and health applications. Phytosterols are believed to lower blood cholesterol levels. Phytosterols are also reported to be anti-inflammatory, anti-cancerogenic, anti-oxidative, immunomodulatory, anti-hypercholesterolemic, and hepatoprotective. In addition, they are precursor metabolites for other bioactive molecules.

### 10.5.2 Carbohydrates

Most microalgae have rigid cellulose cell wall like terrestrial plants. These cellulose components form structural carbohydrates but some have additional biopolymers. Besides this structural carbohydrate, some microalgae also accumulate large quantities of energy-storage carbohydrates, like starch in the dinoflagellates and brown algae, alginate, laminaran, fucocidin, or mannitol in some brown algae and diatoms (Vuppaladadiyam et al., 2018). Cyanobacteria, the blue-green algae, store energy as glycogen. There are many structural and biological applications of microalgal carbohydrates or their derivatives. Some algal carbohydrate products are important in cosmetics and skin care products. Some are emulsifiers in food industries and in medicine. Sulphated polysaccharides in particular are important in medicine.

### 10.5.3 Proteins

Microalgae can synthesise all the essential amino acids with comparable qualities with those of other plants and animals. Microalgal proteins have both functional and structural roles. They contribute to cell growth and are major constituents of the photosynthetic apparatus in addition to contributing to the carbon dioxide fixation pathway of the cells. Protein contents of microalgae are variable. Microalgae with high protein contents are very good sources of functional foods, nutraceuticals, food and feed additives and conversely disfavouring algal biofuel production. Some microalgal proteins have good foam and gel properties. Some form biofilms and some have antimicrobial potentials.

### 10.5.4 Secondary Metabolites

In addition to primary metabolites of industrial importance, microalgae also produce many other secondary metabolites. These include antioxidants, bioactive compounds, carotenoids, hydrocarbons, ketones, lectins, mycosporine, phycocolloids, pigments, polyketide, sterol, vitamins, and waxes. Some of the microalgae that have been used to produce various secondary metabolites are shown in Table 10.3.

It is important to note that one species of microalgae produces various secondary metabolites, depending on the growth condition, and by implication, one metabolite is produced by many species of microalgae. Production of these microalgal metabolites is enhanced by their exposure to stress conditions including but not limited to light intensity, nutrient deprivation, pH, salinity, and temperature (Khan et al., 2018). Apart from stress, cultivation of many microalgae under optimal condition had led to the improvement in the productivities of many useful metabolites. It is therefore very important that we evaluate various species of microalgae for production of various metabolites.

Research in industrial applications of microalgae has gone a long way but for full industrialisation of microalgae biotechnology, much still needs to be done. In some cases, these information gap could have arisen from lack of awareness about the product, scarce incentive to produce, and the under exploitation of the microalgae. For example, light utilisation mechanisms of microalgae are not fully understood and needs to be exploited as well as the need for more information about the metabolic pathway of microalgae. Similarly, there is gap in information on the response of carbon partitioning to different primary metabolites (carbohydrates, proteins, and lipids). There is under exploitation of microalgae for production of human dietary foods. Research lacuna exists in the use of microalgae for different formulations as functional food ingredients. There is also a need for reliable information on costs for individual processes which depend on the process route, the algal species, costing methodology, and the by-products generated. The challenge of producing materials that do not compete with fossil fuel–derived products and terrestrial products needs to be addressed while the technology to process algal-based biofuel has not been fully developed. There is still a lot to be done on optimisation of processes for metabolites production by microalgae while full understanding of the influence of stress conditions on metabolite production is necessary to enable their commercial production.

**TABLE 10.3**
**Secondary Metabolites of Industrial Importance Produced by Microalgae**

| Species | Secondary Metabolite | Physiological Roles | Applications | Reference |
|---|---|---|---|---|
| *Chlorococcum sp.* | Astaxanthin | Antioxidant | In food industry, as natural colour enhancer | Ma and Chen (2001) |
| *Chlorella zofingiensis* | Astaxanthin, lutein | Antioxidant | In food industry, as natural colour enhancer | Del Campo et al. (2004) |
| *Dunaliella salina* | β-Carotene, carotenoids, zeaxanthin, lutein, trans-β-carotene | Protects membrane from peroxidation, antioxidants | In food industry, as natural colour enhancer. Critical role in vision and the immune system | Lamers et al. (2010), Fu et al. (2013), Khan et al. (2018) |
| *Dunaliella salina* | Lutein | Anti-aging, anti-cancerogenic, anti-inflammatory and anti-oxidative | Feed additive and food colourant | Lamerset al. (2010), Fu et al. (2013), Khan et al. (2018) |
| *Eustimatos cf. polyhem* | β-Carotene | | In food industry, as natural colour enhancer | Li et al. (2012) |
| *Haematococcus pluvialis* | Astaxanthin | Strong antioxidant. Multifunctional carotenoid important in reduction of oxidative stress diseases | In food industry, as natural colour enhancer | Cheng et al. (2016), Khan et al. (2018) |
| *Spirulina platensis* | C-phycobiliprotein | Light-energy harvesters and antioxidant to repair the photosynthetic apparatus | Pigment used in microscopy as fluorescent agents | Chen et al. (2013), Khan et al. (2018) |
| *Tetraselmis suecica* | Carotenoids | Light-energy harvesters, Antioxidants | Colour enhancer | Ahmed et al. (2015) |
| *Euglena gracilis* | Tocopherols | | | |
| | Phycobiliproteins | Light-energy harvesters and antioxidant to repair the photosynthetic apparatus | | Vuppaladadiyam et al. (2018) |
| | Phycobilin (phycoerythrin and phycocyanin) | Efficient photosynthetic pigments with unique protein-binding sites | Cosmetics, dairy, and as food colourants | Vuppaladadiyam et al. (2018) |

(*Continued*)

**TABLE 10.3 (*Continued*)**
**Secondary Metabolites of Industrial Importance Produced by Microalgae**

| Species | Secondary Metabolite | Physiological Roles | Applications | Reference |
|---|---|---|---|---|
| *Haslea ostrearia* | Vitamins | Antioxidant | | Vuppaladadiyam et al. (2018) |
| *Spirulina maxima, Chlorella vulgaris* | Phenolic compounds | Antioxidant | | Vuppaladadiyam et al. (2018) |
| | Phycobilin | Antioxidant | | Vuppaladadiyam et al. (2018) |
| | Polyketides | Antioxidant | High value-added compound used in pharmaceutical industry | Vuppaladadiyam et al. (2018) |
| | Polyhydroxyalkanoates | Antioxidant | | Vuppaladadiyam et al. (2018) |

## 10.6 RESEARCH ON POTENTIAL APPLICATIONS OF MICROALGAE IN FOOD AND AGRICULTURE

The ubiquity, phenotypic plasticity, and ability to utilise inorganic carbon via solar-powered redox reactions put microalgae at a distinctive advantage in the microbial world. There is also a huge variety of species with an extensive range of derived products and applications. Their microscopic nature and high growth rates have placed them on the top of the list of third-generation biomass choices and their use will significantly reduce concerns regarding competition with food crops for arable land (Chuka-ogwude et al., 2020). Though water consumption in microalgae cultivation has been a cause for concern, it is noteworthy to know that water requirements for algae cultivation is similar to that required for cotton and wheat cultivation per unit area, but less than required for the cultivation of corn (Hannon et al., 2010). There are a lot of derived products from microalgae biomass and this section discusses those related to food and agriculture.

### 10.6.1 Single Cell Proteins

The term "single cell proteins" (SCP) was originally introduced in the 1966 by Carol L. Wilson and used to describe protein-rich foods derived from yeast and served as dietary supplements for human and livestock (Doelle, 1994). The term has since evolved to include various microorganisms including fungi, bacteria, and algae grown from cheap carbon sources. The term "single cell protein" is derived from the fact that the whole cell of the organism is consumed after the cells have been harvested and dried. SCPs may be enriched with high contents of protein (primary focus), carbohydrates, fats, nucleic acids, and vitamins, and proportions of these in the cells are all malleable depending on the cultivation conditions under which the cells were grown. For an SCP to be considered useful it must possess the following properties: (1) it must be safe to eat, (2) it must have high nutritional values, especially amino acids, (3) it must have the functionality found in common staple foods, (4) production must be economically sustainable, and (5) it must be socially acceptable (Nangul and Bhatia, 2019). From a production stand point, some other criteria are necessary for the successful application of any microbe as an SCP including: genetic and metabolic plasticity, and high growth rate/cell doubling time (Scrimshaw and Dillen, 1977). These are all properties that microalgae possess. In the last few years, the growth of the food supplement industry has piqued interest in research on bioactive compounds derived from algae especially bioactive compounds that fall into the category of "essential nutrients". These are called high-value products because of their higher market values in comparison to other microalgae-derived products. They include polyunsaturated fatty acids (PUFAs) and carotenoids. Various products of SCPs under different brand names have been made over the years in either powder, capsules, tablets, and liquid forms, mostly as food supplements and nutraceuticals. The most prominent microalgae species to be sold as SCPs are *Chlorella* and *Spirulina* species. As at 2006, there were more than 70 companies in the world actively engaged in the production of *Chlorella* sp. with Taiwan Manufacturing and Co, being the largest producer with an annual biomass turnaround of 400 tonnes of *Chlorella* biomass (Milledge, 2011). *Chlorella* is rich in protein (51%–58% dry weight), carotenoids, and various vitamins and has been sold as a food source for dietary supplement (Borowitzka, 1995), especially as it is said to have numerous health benefits. Suggested health benefits of *Chlorella* consumption include preventive action against hyper-cholesterol, gastric ulcers, and antitumour activities. The most important substance in *Chlorella*, regards to food and food supplements, is β-1, 3-glucan, which is an active immunostimulatory and performs other functions such as free radical scavenging and reduction of blood lipids (Iwamoto, 2004). Another prominent microalgae species used as SCP is *Spirulina platensis.* This cyanobacterium can be found in a variety of environments, especially in alkaline, saline, and brackish waters, frequently forming blooms. The first production facility for *Spirulina* production was established in the late 1970s and now produced all over the world with most production facilities in Asiatic regions (Hu, 2004). *Spirulina* has been used for human diet for at least 700 years on the continents of America and Africa (Hu, 2004). *Spirulina* contains high percentage of proteins (up to 65% dry weight), significant amounts of essential fatty acids (γ-linoleic acids, GLA), carotenoids, polysaccharides, phycobiliproteins, vitamins, and minerals (Hu, 2004).

### 10.6.2 Carotenoids

Microalgae contain carotenoids, accessory pigments in the LHC (antenna complex) of microalgae, and usually yellow, orange, or red pigments (Nwoba et al., 2020b). The most important use of carotenoids is as food supplements and as food colourants. There are over 400 known carotenoids and some of the most important and commercially available ones are discussed next. Previously, carotenoids were sourced from terrestrial plants and chemical synthesis, but recently algae are considered more attractive because of safety, higher growth rates and lower land mass requirement for cultivation. β-Carotene is one of the most prominent carotenoids. It is an orange-red pigments which is converted to vitamin A in the human and consequently used as a supplement for vitamin A in nutraceuticals. It is also used as a food colourant and as supplements for animal feed. The main source of natural β-carotene is carrots, but algae are considered a viable alternative (Nwoba et al., 2020b). The most common microalgae to produce β-carotene are *Dunaliella salina*, *Scenedesmus almeriensis*, and *Dunaliella bardawil*. The average percentage content of β-carotene in most algae is around 0.1 to 2% of dry weight, but *Dunaliella* species can produce up to 14% β-carotene if grown under the right conditions of high light intensity, high salinity, and nutrient limitation (Borowitzka, 1995).

Astaxanthin is another important carotenoid belonging to the xanthophyll family. It is a reddish pigment that can be gotten from microalgae biomass and considered as high-value product in the market. Its major usage is as a dietary supplement for fish farming, it causes the distinct reddish colour observed in aquatic foods such as shrimps, lobsters, crabs, and salmons, and as a powerful antioxidant in food supplements for humans. It is also considered a good protectant against sunburn when applied topically. The main choice to produce astaxanthin is the freshwater algae *Haematococcus pluvialis* which can produce up to 3% dry weight astaxanthin given the right cultivation conditions (Radmer and Parker, 1994; Hata et al., 2001). *Chlorella zofingiensis* has also been successfully applied for the commercial production of astaxanthin (Guedes et al., 2011).

Zeaxanthin are yellow-coloured carotenoid pigments commonly found in corn, orange, berries, egg yolk and marigold flowers (Bhalamurugan et al., 2018). Their major application is in the cosmetic, pharmaceutical, and nutraceutical industries. As food supplements, their primary purpose is as antioxidants against free radical and are considered to be useful in protecting the eyes from free radical damage. *Scenedesmus almeriensis* and *Nannochloropsis oculata* are the microalgae species commonly employed for the commercial production of zeaxanthin (Granado-Lorencio et al., 2009). Lutein, like zeaxanthin, is another yellow-coloured carotenoid pigment (orange-red in high concentrations). It is commonly found in almost all fruits and vegetables. The most prominent sources for human consumption are maize and egg yolk. Like zeaxanthin, their dietary importance, is tied to their antioxidant properties and are also considered to be instrumental in protecting the eyes from free radical damage. The microalgae species commonly applied for production of lutein are *Muriellopsis* sp., *Scenedesmus almeriensis*, *Chlorella protothecoides*, *Chlorella zofingiensis*, *Chlorococcum citriforme*, and *Neospongiococcus gelatinosum* (Fernández-Sevilla et al., 2009). Percentage composition of lutein in microalgae biomass is dependent on culture conditions (Nwoba et al., 2020b).

### 10.6.3 Fatty Acids

PUFAs are an important group of essential nutrients to humans since we lack the prerequisite enzymes for their synthesis and must obtain them from external sources. The major sources of PUFAs for human consumption like omega-3 fatty acids are oily fish and fish oils but concerns regarding accumulation of toxins in fish due to pollution of water bodies have heightened the push towards alternative sources

(Radmer and Parker, 1994). Docosahexaenoic acid (DHA), an omega-3 fatty acid, is used as dietary supplements in infant formula as it is essential for the development of the nervous system (Bhalamurugan et al., 2018). Other omega-3 fatty acids like eicosapentaenoic acid (EPA) are also used as dietary supplements and food enrichments. *Nannochloropsis oculata* is rich in omega-3 fatty acids like EPA and DHA and a process to produce EPA from *Phaeodactylum tricornutum* has been developed in Spain. The Cyanobacterium *Aphanizomenon flos aquae* has also been revealed to produce the alpha-linolenic acid (ALA). Research is ongoing on the production of omega-6 fatty acids (Jensen et al., 2001); linolenic acid from *Spirulina* and arachidonic acid from *Porphyridium* sp. However, the only commercially produced PUFA from microalgae is obtained from the dinoflagellate *Crypthecodinium cohnii.*

### 10.6.4 Animal Feed

The use of microalgae as animal feed has grown rapidly in the last few decades and biomass has been used as feedstock for fish and farm animals. About 30% of algal biomass produced in the world is sold as animal feed (Radmer and Parker, 1994). This is most likely because no significant downstream process is required for the biomass when it is used as animal feed and there it tends to be a more economical approach for the utilisation of the biomass. The commonly used microalgae species for animal feed are *Chlorella, Isochrysis, Phaeodactylum, Chaetoceros, Nannochloropsis, Tetraselmis, Dunaliella, Scenedesmus, Thalassiosira*, and *Skeletonema* spp. *Haematococcus* has been successfully employed as fish feed for its astaxanthin content and has proven to be non-toxic on consumption (Dore and Cysewski, 2003). *Haematococcus pluvalis,* which is rich in astaxanthin, is commonly used in aquaculture especially as a colouring agent in salmonid feeds. *Chlorella,* which is rich in protein and having a high growth rate, could be used as an alternative source of protein in animal feed since it is less expensive (Bhalamurugan et al., 2018). *Scenedesmus* species are considered as a source of monounsaturated, polyunsaturated, and saturated fatty acids in animal and fish feeds (Ishaq et al., 2016). *Spirulina platensis* which is rich in proteins, carbohydrates, and vitamins are also considered valuable feed for livestock. β-Carotene obtained from *Dunaliella* species is used as a colouring agent and vitamin A for aquaculture. *Nannochloropsis* species being rich in EPA have been used as a nutritional supplement in the aquaculture industry (Mohammad, 2014). *Spirulina platensis* has been shown to have significant positive impact on chicks when introduced into broiler feeds in poultry farming (Evans et al., 2015). *Chlorella vulgaris* supplemented into poultry feed have also shown to improve the growth, intestinal microflora, and immune system in chicks (Kang et al., 2013). Fermented *Chlorella vulgaris* biomass incorporated into poultry has been studied for its effects on egg production, egg quality, liver lipids, and intestinal microflora in laying hens. However, there are concerns for digestibility of microalgae in feeds, except for *Spirulina platensis,* due the high content of cellulose in the cell wall of the organisms and often requires some pre-treatment. This, however, is much less of a concern in ruminants, which are capable of digesting cellulosic organics. Microalgae can also be used for efficient pH bio-stabilisation and bio-purification of aquaculture water (Ahamefule et al., 2018).

### 10.6.5 Biofertilisers

Fertilisers contain nutrients required for plant growth and the most prominent of these nutrients added in fertiliser composition are nitrogen and phosphorus. The biomass of microalgae contains nitrogen and phosphorous, alongside other essential nutrients and trace elements that can be harnessed by plants. As organic fertiliser, biomass of microalgae can be added as dry mass or wet mass to soil and has been shown to improve crop yield. For instance, using *Chlorella vulgaris* biomass as a biofertiliser has been shown to improve growth in lettuce plants. Dry biomass of *Chlorella vulgaris* has also been shown to enhance germination of lettuce (*Lactuca sativa*) seeds in comparison with unfertilised controls indicating improvements in soil fertility. It also significantly increased the amount of pigments (chlorophyll *a*, chlorophyll *b*, and carotenoids) in the lettuce seedlings (Faheed and Fattah, 2008). There is also evidence of increased growth and crop yield of up to 51% in *Zea mays* grown on soil treated with a combination of *Spirulina platensis* and cow dung manure (Dineshkumar et al., 2017). Cyanobacteria of the genera *Aulosira*, *Anabaena*, *Nostoc*, and *Tolypothrix* can fix atmospheric nitrogen and are used as biofertilisers to fix atmospheric nitrogen in the cultivation of paddy crops such as rice (Rizwan et al., 2018). In addition to the nitrogen fixation properties of cyanobacteria, studies have shown that species like *Westiellopsis prolifica* and *Anabaena variabilis* are capable of high levels of extracellular phosphate solubilisation by activity of alkaline phosphatase (Natesan and Shanmugasundaram, 1989) and phthalic acids (Yandigeri et al., 2011) and can be used to improve phosphate absorption of inorganic phosphate fertilisers, and can be exploited for the efficient utilisation of low cost, low grade rock phosphate fertilisers.

## 10.7 RESEARCH ON ENVIRONMENTAL BIOTECHNOLOGY

Environment as the total surrounding of a place needs to be conserved, protected from over-exploitation, degradation, and pollution. Environmental biotechnology is involved in the control of environmental degradation, pollution, and bioremediation. It is also involved in monitoring and conversion of many environmental wastes to wealth and production of many useful materials. Production of biofuels, namely, biodiesel, bioethanol, biogas, and biohydrogen, is central in environmental biotechnology. This is because biofuels are alternative, renewable, and environmentally friendly alternatives to the fossil fuels whose drilling, transportation, and use have done a lot of harm to the environment – ranging from soil pollution, greenhouse gas emission to acidic rains. Other aspects of environmental biotechnology involving microalgae include the use of microalgae for environmental monitoring, carbon dioxide fixation by microalgae, remediation of acid and other harmful gases, water and wastewater treatment, and other bioremediation processes with microalgae. Some of these environmental applications of microalgae are highlighted below.

### 10.7.1 Bioenergy

Bioenergy research is very important for the economic stability and energy security of any nation, and significantly contributes to reduction in global greenhouse gases (GHGs) emission. Bioenergy is energy derived from biological materials including plants, animal materials as well as organic wastes. There are many reasons for the global intensive research on development of bioenergies as alternative to fossil fuels. These include, but are not limited to, (1) energy demand arising from over-population and industrialisation, (2) rising fossil fuel prices coupled with the instability in prices, (3) depletion in the fossil fuel resources, and (4) GHGs release leading to global warming, acid rain, climate change, and other environmental problems. Microalgae have been extensively used for production of various forms of bioenergy such as biodiesel, bioethanol, biogas, and biohydrogen. Other forms of biofuels include biomethanol and bioethers (biodimethyl ether, biomethylbutyl ether, bioethyltetrabutyl ether) (Khan et al., 2018). Out of these, bioethanol and biodiesel are the major and most researched. The use of microalgae as a biofuel feedstock comes with the following advantages: (1) microalgae have high growth rate/short generation time, (2) they have high lipid and carbohydrate contents, (3) they have high photosynthetic capacity, and (4) they can be optimised for high lipid productivity and low carbon footprint.

As a result of the above advantages, microalgae are considered perfect substitute to fossil fuels with respect to renewability, sustainability, costs, and environmental impacts. Some of the types of bioenergy with great potential to supplement or replace fossil fuels include the following.

#### 10.7.1.1 Biodiesel

This is the most common liquid fuel from algae. It is a fatty acid methyl ester (FAME) produced by transesterification between triglycerides and alcohol preferably, methanol at between 60°C and 70°C in the presence of alkaline or acidic catalyst at atmospheric pressure. Biodiesel is a renewable fuel produced from oils (plant, animal, or waste cooking oils or oils from microorganisms). Microalgal oils are the ideal oil for use because of lack of competition with food and feed like the other oil crops. Biodiesel are biodegradable and non-toxic, and it is essentially free of sulphur and aromatics and thus produces lower exhaust emissions than conventional diesel fuels. Chemically, biodiesel is defined as mono alkyl esters of long-chain fatty acids derivative of biological lipids. Biodiesel is typically produced through the reaction of a fat or oil, which contains triglycerides, with an alcohol, in presence of a catalyst to yield methyl esters (biodiesel) and glycerine. The resulting biodiesel, after its purification, is quite similar to conventional diesel fuels in its main features, and it can be blended with conventional diesel fuel. Blends are designated by BXX, where XX is the proportion of biodiesel and conventional diesel fuel. For example, B20 means 20% of biodiesel and 80% of petro-diesel. Biodiesel are preferable to petrol derived diesel because of the following advantages: (1) it is produced from biological sources which makes it sustainable and renewable, (2) biodiesel is ecologically friendly, (3) it releases no sulphur and less CO, HC, particulate matter, and aromatic compounds to the environment, (4) it could be a source of income in rural communities, (5) the fuel properties of biodiesel are similar to the conventional fuel, (6) it can used in existing unmodified diesel engines, (7) production of biodiesel locally will significantly reduce oil importation, and thus save foreign reserves, and (8) biodiesel is non-toxic, biodegradable and safe to handle.

Biodiesel is produced by transesterification of large, branched triglycerides into smaller, straight chain molecules of methyl esters, using an alkali, acid, or enzyme as catalyst. There are three stepwise reactions with intermediate formation of diglycerides and monoglycerides resulting in the production of three moles of methyl esters and one mole of glycerol from triglycerides. The overall reaction is: alcohols such as methanol, ethanol, propanol, butanol, and amyl alcohol are used in the transesterification process. Methanol and ethanol are used most frequently, especially methanol because of its low cost, and physical and chemical advantages. They can quickly react with triglycerides and sodium hydroxide is easily dissolved in these alcohols. Microalgal species used or experimented for the production of biodiesel are *Botryococcus braunii*, *Chlorella* sp., *Chlorella pyrenoidosa*, *Dunaliella tertiolecta*, *Nannochloropsis oculata*, *Spirulina* sp. *Schizochytrium limacinum*, *Dictyochloropsis splendida*, *Desmodesmus quadricautus*, and *Oscillatoria* sp.

A lot of work have been done on biodiesel oil production by various species of microalgae in Nigeria (Ogbonna et al., 2015; Eze et al., 2017a; Ogbonna and Ogbonna, 2018a,b; Asoiro et al., 2019; Eze et al., 2020; Idenyi et al., 2020). With the suitable weather conditions, more intensive screening for oleaginous species and technology development can lead to economically viable commercial biodiesel oil production in Nigeria.

#### 10.7.1.2 Bioethanol

Bioethanol is ethanol produced by the fermentation of sugars from biological materials. Microalgae having starch contents up to 50% are ideal feedstock for the production of bioethanol. Although microalgae have high potentials for production of bioethanol, microalgal bioethanol production is still at the preliminary stage of research. In this process, the carbohydrate components of algal cellular materials are converted into bioethanol. However, microalgae contain different forms of carbohydrates including glycogen, cellulose, starch, and agar with varying sugar yields on hydrolysis. In some cases, enzymatic hydrolysis of algal residues after lipid extraction leads to the production of sugars which are then converted to bioethanol. There is a need for intensive research on isolation of carbohydrate-rich strains of microalgae as well as optimisation of culture conditions for maximum carbohydrate accumulation. Advantages of microalgal bioethanol include: (1) algal bioethanol yield is higher than the first and second generations bioethanol feedstocks due to high growth rates, (2) algal biomass does not require severe pre-treatment before it is converted into ethanol as is the case with other feedstocks with complex lignin and hemicellulose-based compounds, and (3) use of microalgae for bioethanol production can reduce the overall cost since microalgal fermentation can circumvent the need for concentrated harvesting (Tiwari and Kiran, 2018). Microalgal species used in the production of bioethanol include *Chlamydomonas reinhardtii*, *Chlorococcum* sp., *Scenedesmus dimorphus*, *Chlorella vulgaris*, *Porphyridium cruentum*, and *Dunaliella salina*.

#### 10.7.1.3 Biogas

Biogas otherwise called biomethane is produced from the anaerobic digestion of organic matter which leads to the formation of gas with a methane content of 55%–75% and a carbon dioxide content of 25%–45% (Herold et al., 2020). Biomethane is expected to play an essential role in the circular bioeconomy because it is a clean, cheap, and versatile fuel. It is produced from the anaerobic digestion of all types of biomasses and consists essentially of methane and carbon dioxide. Anaerobic digestion of high-water containing biomass (80%–90%) is ideal for conversion into biogas which could have up to 75% methane. Microalgal biogas production is relatively very low probably because of low carbon: nitrogen ratio. Microalgal species used or experimented for the production of biogas are *Chlorella vulgaris, Spirulina* sp., *Scenedesmus obliquus, Spirulina maxima*, and *Euglena gracilis*. The most interesting aspect of microalgae biogas is that the biomass can be used after extraction of high-value metabolites.

Algal biogas production is perceived of as an anaerobic microbiological digestion processes summarised in three phases. The first step is the hydrolysis of macromolecules, proteins, carbohydrates, and lipids into soluble sugars. This is followed by the fermentation of the soluble sugars into volatile fatty acids, alcohol, hydrogen, and carbon dioxide. Prior to this, algal biomasses are first disrupted. The final stage is methanogenesis where methane and carbon dioxide are formed (Patel et al., 2012).

#### 10.7.1.4 Biohydrogen

Microalgae are photosynthetic microorganisms that convert solar energy into chemical energy which can be redirected physiologically to produce hydrogen ($H_2$). Microalgae have intrinsic potential to produce hydrogen by splitting water molecule using solar energy. This capacity is possible because microalgae have hydrogenase enzyme capable of converting hydrogen ion during photosynthesis into molecular hydrogen. The splitting of the water molecule in this process (photolysis of water) also leads to the generation of oxygen, and oxygen inhibits the dehydrogenase enzyme. This means that hydrogen production is self-limiting. Therefore, to produce hydrogen optimally, removal of oxygen is very crucial. To remove oxygen, normal growth of the algae is necessary, subsequently, allowing sulphur deprivation to introduce anaerobic condition. This can sustain hydrogen gas production for up to seven days. There are short-comings regarding large-scale hydrogen production for fuel, although this production system could give the advantages of not releasing GHGs and release of water as by-product. Microalgal species used or experimented for the production of biohydrogen include *Chlorella* sp., *Anabaena cylindrica, Mastigocladus laninosus*, and *Chlamydomonas reinhardtii*.

### 10.7.2 Other Environmental Applications of Microalgae

#### 10.7.2.1 Use of Microalgae for Environmental Monitoring

Environmental problems including air and water pollution, over fertilisation, wastewater treatment, and ecological disasters could be monitored using algal growth. For example, algal blooms are good indication of heavy pollution of rivers and other water bodies. Many species of microalgae have very high growth rates and responds sharply to some environmental stress. Thus, their growth and characteristics (size, growth rates, pigmentations, shapes, etc.) can be used as indications of pollution or absence of certain substances in the environment. For example, the presence of some heavy metals in water affects the growth and pigmentation of some species of microalgae. Biomonitoring involves programmes and activities needed to characterise the quality of an environment. Thus, information on algal growth and photosynthesis, ATP formation, radioactive carbon assimilation by algae, oxygen evolution, and algal fluorescence induction phenomenon are all methodologies that could be applied in monitoring activity.

#### 10.7.2.2 Use of Microalgae for Carbon Dioxide Fixation

Fossil fuel resources have been variously criticised because of their high release of GHGs and consequently increasing in global warning and climate change. There have been predictions of increase in the global $CO_2$ emissions giving the current anthropogenic activities especially burning of fossil fuels. Microalgae are photosynthetically active group of solar energy harvesters capable of fixing $CO_2$ from different sources to form sugars. The use of microalgae in carbon dioxide fixation is called algal $CO_2$ sequestration. Many algal species can fix $CO_2$ more efficiently than the terrestrial (C4) plants because of their rapid growth rates and higher photosynthetic efficiencies (Ravindran et al., 2016). Microalgal-based carbon capture has some advantages. For instance, removal of $CO_2$ from the atmosphere reduces global warming. In addition, the $CO_2$ captured by the algae is used in the useful metabolite formation. The ability of some microalgae to grow in environments rich in $CO_2$ concentration (e.g., flue gas) and assimilate $CO_2$ is a very important attribute in pollution control. Some microalgae like *Chlorella* can sequester carbon dioxide from industrial flue gas even if the concentration is 100%. Some microalgal species used in the $CO_2$ sequestration include *Euglena*, *Botryococcus*, *Scenedesmus*, *Chlorella*, *Dunaliella*, and *Chlorococcum* spp.

#### 10.7.2.3 Use of Microalgae for Remediation of Acid and Other Harmful Gases

Microalgae can be applied in the remediation of acid and other harmful gases. Various natural effects lead to the formation of acids in the environment. For instance, acidic solutions can be formed by natural decomposition of sulphide materials through the processes in wastewater. Leaching of heavy metal contaminants from sources can also result in high acidity. Microalgae can be used as decontaminating agents and are necessary in the clean-up of some of these contaminants. Use of microalgae to perform this feat has several advantages. The process is low cost and easy to manipulate. Algal growth can lead to other product formation. Different gases including ammonia, carbon dioxide, hydrogen sulphide, methane, nitrogen, oxygen, and sulphur (iv) oxide are some of the gases that can be assimilated by microalgae. One kilogram of some species of microalgae can fix 1.83 kg of carbon dioxide. In addition, some species of microalgae use $SO_x$ and $NO_x$ as nutrient along with $CO_2$. In this way, microalgae can contribute in the remediation of wastewater containing such gases.

#### 10.7.2.4 Use of Microalgae for Water and Wastewater Treatment

Wastewater is raw untreated spent water with some impurities that is potentially a pollutant (Dar et al., 2019). Wastewater contains some nutrient elements which can easily be assimilated, absorbed, or adsorbed by microalgae. The use of microalgae for wastewater treatment is called phycoremediation. Microalgae require essentially nitrogen, phosphorous, and some other mineral elements dissolved in water for their growth. Most wastewater including domestic, industrial, and agricultural wastewater contain some of these mineral elements. An interesting practice in microalgal wastewater treatment is the integration of algal production system with wastewater treatment. For instance, it is becoming a common practice to integrate algal biofuel production with wastewater treatment. Why the integration is necessary because the global requirement for freshwater is very high. The amount of nitrogen, phosphorous, and other mineral elements required for mass cultivation of microalgae is also very high, and this increases the cost of production. By using wastewater, the competition for freshwater is reduced as well as the cost of getting the desired nutrients and in addition, treating the water. Over the years, nitrogen, phosphorous, and other nutrients have been removed during wastewater treatment using other biological and chemical means. These methods are costly and treatment of the resulting sludge is a challenge. The use of microalgae for wastewater treatment reduces cost and sludge formation since algae can effectively utilise all the nutrients in the wastewater. Microalgae-mediated wastewater treatment also offers the advantage of better pathogen removal and reduction of greenhouse gas emission (Ravindran et al., 2016). Microalgae are important in removing heavy metals and nutrients from water. Microalgal removal of heavy metals from wastewater is a good attribute because the accumulation of heavy metals by algae is useful in biomonitoring and bioremediation. Technically, some advantages of the use of algae in this feat include low cost of raw material, high adsorptive capacity, and no production of toxic by-products. Phycoremediation of wastewater is usually a secondary or tertiary treatment method for industrial wastewater (Ravindran et al., 2016).

In Nigeria, potential applications of microalgae in wastewater treatment (Anaga and Abu, 1996; Nwuche et al., 2014; Ogbonna et al., 2018; Okpozu et al., 2019) and metal accumulation (Abirhire and Kadiri, 2011; Chia et al., 2014a, 2015) have been investigated. However, more studies using various local isolates are required to establish the potentials of phycoremediation in Nigeria.

## 10.8 GENERAL CONCLUSION

From the above discussion, it is obvious that microalgae have a wide range of applications which if commercially exploited will contribute significantly to sustainable development of Nigeria. Although research on biodiversity, isolation, and cultivation of various microalgae species are ongoing in Nigeria, the scope is still limited and there is yet no commercial cultivation of microalgae in Nigeria. Our knowledge of species diversity in Nigeria is still very limited, we have done little on identifying indigenous species with potential applications to our local problems, we have done little on optimisation of the culture conditions, while little has been done on development of cultivations systems and photobioreactors suitable to our climatic and weather conditions. Microalgae research is therefore a fertile area that needs to be exploited by all those interested in applied phycology. There is also urgent need to establish microalgae collection centres in Nigeria for preservation and maintenance of local isolates/strains with potential valuable applications.

## REFERENCES

Abirhire, O., and Kadiri, M. O. (2011). Bioaccumulation of heavy metals using microalgae. *Asian Journal of Microbiology, Biotechnology and Environmental Science*, 13: 91–94.

Adeniyi, A. O., and Akinwole, A. O. (2017). Phytoplankton composition and physico-chemical parameters of lower river Niger, Agenebode, Edo state Nigeria. *International Journal of Fisheries and Aquatic Studies*, 5(3): 256–260.

Adesalu, T. A., Adesanya, T., and Ogwuzor, C. J. (2014). Phytoplankton composition and water chemistry of a tidal creek (Ipa-Itako) part of Lagos Lagoon. *Journal of Ecology and the Natural Environment*, 6(11): 373–388. doi:10.5897/jene2014.0473.

Adesalu, T. A. (2017). Freshwater diatoms diversity of national parks in Nigeria I: Okomu national park, South-South, Nigeria. *Ife Journal of Science*, 19(2): 269. doi:10.4314/ijs.v19i2.7.

Anaga, A., and Abu, G. O. (1996). A laboratory-scale cultivation of *Chlorella* and *Spirulina* using waste effluent from a fertilizer company in Nigeria. *Bioresource Technology*, 58(1): 93–95.

Asoiro, F. U., Okonkwo, W. I., and Nweze, N. O. (2019). Studies on the growth rate, oil yield and properties of some indigenous freshwater microalgae species. *Journal of Environmental Science and Technology*, 12(4): 164–176.

Afiukwa, C. A., and Ogbonna, J. C. (2007). Effects of mixed substrates on growth and vitamin production by *Euglena gracilis*. *African Journal of Biotechnology*, 6(22): 2612–2615.

Ahamefule, C. S., Ogbonna, J. C., Moneke, A. N., and Ossai, N. I. (2018). Application of photosynthetic microalgae as efficient pH bio-stabilizers and bio-purifiers in sustainable aquaculture of *Clarias gariepinus* (African catfish) fry. *Nigerian Journal of Biotechnology*, 35(2): 23–26.

Ahmed, F., Fanning, K., Netzel, M., Schenk, P. M. (2015). Induced carotenoid accumulation in *Dunaliella salina* and *Tetraselmis suecica* by plant hormones and UV-C radiation. *Applied Microbiology and Biotechnology*, 99(22): 9407–9416.

Ajjawi, I., Verruto, J., Aqui, M., Soriaga, L. B., Coppersmith, J., Kwok, K., Peach, L., Orchard, E., Kalb, R., and Xu, W. (2017). Lipid production in Nannochloropsis gaditana is doubled by decreasing expression of a single transcriptional regulator. *Nature Biotechnology*, 35(7): 647–656.

Ali, F. U., and Ogbonna, J. C. (2011). Suitability of Nigerian weather conditions for cultivation of microalgae. *Nigerian Journal of Biotechnology*, 22: 60–65.

Baker, N. R. (2008). Chlorophyll fluorescence: A probe of photosynthesis in vivo. *Annual Review in Plant Biology*, 59: 89–113.

Barsanti, L., and Gualtieri, P. (2006). *Algae: Anatomy, Biochemistry, and Biotechnology*. Taylor and Francis, Boca Raton, FL. 301pp.

Beardall, J., Quigg, A., and Raven, J. A. (2003). Oxygen consumption: Photorespiration and chlororespiration. In: Larkum, A. W. D., Douglas, S. E., Raven, J. A. (eds.) *Photosynthesis in Algae*, Springer, Dordrecht, pp. 157–181.

Becker, W. (2004). Microalgae in human and animal nutrition. In: Richmond, A. (Ed.) *Microalgal Culture, Handbook*, Blackwell, Oxford, pp. 312–351.

Prasil, M. J., O., Babin, M., and Bruyant, F. (2004). In search of a physiological basis for covariations in light-limited and light-saturated photosynthesis 1. *Journal of Phycology*, 40(1): 4–25.

Bhalamurugan, G. L., Valerie, O., and Mark, L. (2018). Valuable bioproducts obtained from microalgal biomass and their commercial applications: A review. *Environmental Engineering Research*, 23(3): 229–241. doi:10.4491/eer.2017.220.

Blaine, M., and Pyne, J. W. (1986). Biologically active compounds from microalgae. *Enzyme and Microbial Technology*, 8(7): 386–394.

Bleeke, F., Rwehumbiza, V. M., Winckelmann, D., and Klöck, G. (2014). Isolation and characterization of new temperature tolerant microalgal strains for biomass production. *Energies*, 7(12): 7847–7856. doi:10.3390/en7127847.

Bolhàr-Nordenkampf, H., and Öquist, G. (1993). Chlorophyll fluorescence as a tool in photosynthesis research. In: Hall, D. O., Scurlock, J. M. O., Bolhàr-Nordenkampf, H. R., Leegood, R. C., Long, S. P. (eds.) *Photosynthesis and Production in a Changing Environment*, Springer, Dordrecht, pp. 193–206.

Borowitzka, M. A. (1995). Microalgae as sources of pharmaceuticals and other biologically active compounds Michael. *Journal of Applied Phycology*, 7: 3–15. doi:10.5005/jp/books/14244_69.

Bouhlal, R., Haslin, C., Chermann, J. C., Colliec-Jouault, S., Sinquin, C., Simon, G., Cerantola, S., Riadi, H., and Bourgougnon, N. (2011). Antiviral activities of sulfated polysaccharides isolated from *Sphaerococcus coronopifolius* (Rhodophytha, Gigartinales) and *Boergeseniella thuyoides* (Rhodophyta, Ceramiales). *Marine Drugs*, 9: 1187–1209.

Cai, X.-H., Brown, C., Adhiya, J., Traina, S. J., and Sayre, R. T. (1999). Growth and heavy metal binding properties of transgenic Chlamydomonas expressing a foreign metallothionein gene. *International Journal of Phytoremediation*, 1(1): 53–65.

Carvalho, A. P., Silva, S. O., Baptista, J. M., and Malcata, F. X. (2011). Light requirements in microalgal photobioreactors: An overview of biophotonic aspects. *Applied Microbiology and Biotechnology*, 89(5): 1275–1288.

Chen, C. Y., Kao, C., Tsai, C. J., Lee, D. J., and Chang, J. S. (2013). Engineering strategies for simultaneous enhancement of C-phycocyanin production and $CO_2$ fixation with *Spirulina platensis*. *Bioresource Technology*, 145: 307–312.

Cheng, J., Li, K., Yang, Z. Zhou, J., and Cen, K. (2016). Enhancing the growth rate and astaxanthin yield of *Haematococcus pulvalis* by nuclear irradiation and high concentration of carbon dioxide stress. *Bioresource Technology*, 204: 49–54.

Chia, M. A., Bako, S. P., Alonge, S. O., and Adamu, A. K. (2011). Green algal interactions with physicochemical parameters of some manmade ponds in Zaria, northern Nigeria. *Revista Brasileira de Botânica*, 34(3): 285–295. doi:10.1590/s0100-84042011000300004.

Chia, M. A., Chimdirim, P. K., and Japhet, W. S. (2014a). Lead induced antioxidant response and phenotypic plasticity of *Scenedesmus quadricauda* (Turp.) de Brébisson under different nitrogen concentrations. *Journal of Applied Phycology*, 27(1): 293–302. doi:10.1007/s10811-014-0312-8.

Chia, M. A., Galadima, S. Y., and Japhet, W. S. (2015). Combined effect of atrazine and copper on the growth, biomass production, morphology and antioxidant response of *Scenedesmus quadricauda*. *Phycologia*, 54(2): 109–117. doi:10.2216/14-71.1.

Chia, M. A., Odoh, O. A., and Ladan, Z. (2014b). The Indigo blue dye decolorization potential of immobilized *Scenedesmus quadricauda*. *Water, Air, & Soil Pollution*, 225(4). doi:10.1007/s11270-014-1920-2.

Chu, W. L., and Phang, S. M. (2019). Bioactive compounds from microalgae and their potential applications as pharmaceuticals and nutraceuticals. In: Hallmann, A., Rampelotto, P. (eds.) *Grand Challenges in Algae Biotechnology. Grand Challenges in Biology and Biotechnology*. Springer, Cham, pp. 429–469.

Chuka-ogwude, D., Ogbonna, J., and Moheimani, N. R. (2020). A review on microalgal culture to treat anaerobic digestate food waste effluent. *Algal Research*, 47: 101841. https://doi.org/10.1016/j.algal.2020.101841.

Dar, R. A., Sharma, N., Kaur, K., and Phutela, U. P. (2019). Feasibility of microalgal technologies in pathogen removal from wastewater. In: Gupta, S. K., Bux, F. (eds.), *Application of Microalgae in Wastewater Treatment*, pp. 237–268. https://doi.org/10.1007/978-3-030-13913-1_12.

Dauvillée, D., Delhaye, S., Gruyer, S., Slomianny, C., Moretz, S. E., d'Hulst, C., Long, C. A., Ball, S. G., and Tomavo, S. (2010). Engineering the chloroplast targeted malarial vaccine antigens in Chlamydomonas starch granules. *PloS One*, 5(12): e15424.

Dayong, S., Jing, L., Shuju, G., and Lijun, H. (2008). Antithrombotic effect of romophenol, the alga-derived thrombin inhibitor. *Journal Biotechnology*, 136: 577–588.

De Felício, R., De Albuquerque, S., Young, M. C. M., Yokoya, N. S., and Debonsi, H. M. (2010). Trypanocidal, leishmanicidal and antifungal potential from marine red alga Bostrychia tenella J. Agardh (Rhodomelaceae, Ceramiales). *Journal of Pharmaceutical and Biomedical Annals*, 52: 763–769.

Del Campo, J. A., Moreno, J., Rodríguez, H., Vargas, M. A., Rivas, J., and Guerrero, M. G. (2000). Carotenoid content of chlorophycean microalgae: factors determining lutein accumulation in *Muriellopsis* sp. (Chlorophyta). *Journal of Biotechnology*, 76:51–59.

Del Campo, J. A., Rodriguez, H., Moreno, J., Vargas, M. A. Rivas, J., and Guerrero, M. G. (2004). Accumulation of astaxanthin and lutein in *Chlorella zofingiensis* (Chlorophyta). *Applied Microbiology and Biotechnology*, 64 (6): 848–854.

Devi, G. K., Manivannan, K., Thirumaran, G., Rajathi, F. A. A., and Anantharaman, P. (2011). In vitro antioxidant activities of selected seaweeds from Southeast coast of India. Asian. Pacific. *Journal of Tropical Medicine*, 4: 205–211.

Dineshkumar, R., Subramanian, J., Gopalsamy, J., Jayasingam, P., Arumugam, A., Kannadasan, S., and Sellaiyan, K. (2017). The impact of using microalgae as biofertilizer in Maize (Zea mays L.). *Waste and Biomass Valorization*, 10(5): 1101–1110. https://doi.org/10.1007/s12649-017-0123-7.

Doelle, H. W. (1994). Microbial Process Development. Retrieved from https://books.google.com.au/books?id=_qDabLa-0ukC

Dore, J. E., and Cysewski, G. R. (2003). *Haematococcus algae meal as a source of natural astaxanthin for aquaculture feeds.* Cyanotech Corporation, Kailua, HI, USA.

Dreesen, I. A., Charpin-El Hamri, G., and Fussenegger, M. (2010). Heat-stable oral alga-based vaccine protects mice from Staphylococcus aureus infection. *Journal of Biotechnology*, 145(3): 273–280.

Dunahay, T. G., Jarvis, E. E., Dais, S. S., and Roessler, P. G. (1996). Manipulation of microalgal lipid production using genetic engineering. *Applied Biochemistry and Biotechnology*, 57(1): 223–230.

Essien, J. P., Antai, S. P., and Benson, N. U. (2008). Microalgae biodiversity and biomass status in Qua Iboe Estuary mangrove swamp, Nigeria. *Aquatic Ecology*, 42(1): 71–81.

Evans, A. M., Smith, D. L., and Moritz, J. S. (2015). Effects of algae incorporation into broiler starter diet formulations on nutrient digestibility and 3 to 21 d bird performance. *Journal of Applied Poultry Research*, 24(2): 206–214. https://doi.org/10.3382/japr/pfv027.

Eze, C. N., Aoyagi, H., and Ogbonna, J. C. (2020). Simultaneous accumulation of lipid and carotenoid in freshwater green microalgae *Desmodesmus subspicatus* LC172266 by nutrient replete strategy under mixotrophic condition. *Korean Journal of Chemical Engineering*, 37(9): 1522–1529.

Eze, C. N., Ogbonna, J. C., Ogbonna, I. O., and Aoyagi, H. (2017). A novel flat plate air-lift photobioreactor with inclined reflective broth circulation guide for improved biomass and lipid productivity by *Desmodesmus subspicatus* LC172266. *Journal of Applied Phycology*, 29: 2745–2754. https://doi.org/10.1007/s10811-017-1153-z.

Eze, C. N., Ogbonna, J. C., Ndu, O. O., and Ochiogu, S. I. (2017). Evaluation of some biological activities of *Euglena gracilis* biomass produced by a fed-batch culture with some crop fertilizers. *African Journal of Biotechnology*, 16(8): 337–345.

Faheed, F., and Fattah, Z. (2008). Effect of chlorella vulgaris as bio-fertilizer on growth parameters and metabolic aspects of lettuce plant. *Journal of Agriculture and Social Science*, 4: 165–169.

Feng, S., Feng, W., Zhao, L., Gu, H., Li, Q., Shi, K., Guo, S., and Zhang, N. (2014). Preparation of transgenic *Dunaliella salina* for immunization against white spot syndrome virus in crayfish. *Archives of Virology*, 159(3): 519–525.

Fernández, E., Schnell, R., Ranum, L., Hussey, S. C., Silflow, C. D., and Lefebvre, P. A. (1989). Isolation and characterization of the nitrate reductase structural gene of Chlamydomonas reinhardtii. *Proceedings of the National Academy of Sciences*, 86(17): 6449–6453.

Fernández-Sevilla, J. M., Fernández, F. G. A., and Grima, E. M. (2009). Biotechnological production of lutein and its applications. *Applied Microbiology and Biotechnology*, 86: 27–40.

Fu, W., Gudmundsson, Ó., Paglia, G., Herjólfsson, G., Andrésson, Ó. S., and Palsson, B. Ø. (2013). Enhancement of carotenoid biosynthesis in the green microalga *Dunalellia salina* with light emitting diodes and adaptive laboratory evolution. *Applied Microbiology and Biotechnology*, 97(6): 2395–2403.

Gouveia, A., Batista, A. P., Sousa, I., Raymundo, A., and Bandarra, N. M. (2008). Microalgae in novel food products. In: Papadopoulos, K. N. (ed.). *Food Chemistry Research Developments*. Nova Science Publishers, Inc., Hauppauge, NY, pp. 1–37.

Granado-Lorencio, F., Herrero-Barbudo, C., Acien, G., Molina-Grima, E., Fernandez-Sevilla, J. M., Pérez-Sacristán, B., and Blanco-Navarro, I. (2009). In vitro bioaccesibility of lutein and zeaxanthin from the microalgae *Scenedesmus almeriensis*. *Food Chemistry*, 114: 747–752. https://doi.org/10.1016/j.foodchem.2008.10.058.

Gregory, J. A., Li, F., Tomosada, L. M., Cox, C. J., Topol, A. B., Vinetz, J. M., and Mayfield, S. (2012). Algae-produced Pfs25 elicits antibodies that inhibit malaria transmission. *PloS One*, 7(5): e37179.

Guedes, A. C., and Malcata, F. X. (2012). Nutritional value and uses of microalgae in aquaculture. In: Muchlisin, Z. (ed.). *Aquaculture*. InTech, Available from: http://www.intechopen.com/books/aquaculture/nutritionalvalue-and-uses-of-microalgae-in-aquaculture.

Guedes, A. C., Amaro, H. M., and Malcata, F. X. (2011). Microalgae as sources of carotenoids. *Marine Drugs*, 9(4): 625–644. https://doi.org/10.3390/md9040625

Hannon, M., Gimpel, J., Tran, M., Rasala, B., and Mayfield, S. (2010). Biofuels from algae: Challenges and potential. *Biofuels*, 1(5): 763–784. https://doi.org/10.4155/bfs.10.44.

Hargreaves, J. A., and Tucker, C. S. (2004). Managing ammonia in fish ponds. Southern Regional Aquaculture Center, Publication No. 4603.

Hata, N., Ogbonna, J. C., Taroda, H., and Tanaka, H. (2001). Production of astaxanthin by *Haematococcus pluvialis* in a sequential heterotrophic-photoautotrophic culture. *Journal of Applied Phycology*, 13: 395–402.

Herold, C., Ishika, T., Nwoba, E. G., Tait, S., Ward, A., and Moheimani, N. R. (2020). Biomass production of marine microalga *Tetraselmis suecica* using biogas and wastewater as nutrients. *Biomass and Bioenergy*, 145, 105945.

Hu, Q. (2004). Industrial production of microalgal cell-mass and secondary products – major industrial species: Arthrospira (Spirulina) platensis. In: Richmond, A. (ed.), *Handbook of Microalgal Culture: Biotechnology and Applied Phycology*. Blackwell Publishing Ltd, Hoboken, NJ, pp. 264–272.

Idenyi, J. N., Eya, J. C., Ogbonna, J. C., Chia, M. A., Alam, M. A., and Ubi, B. E. (2021). Characterization of strains of *Chlorella* from Abakaliki, Nigeria, for the production of high-value products under variable temperatures. *Journal of Applied Phycology*, 33 (1): 275–285.

Ishaq, A., Peralta, H. M., and Basri, H. (2016). Bioactive compounds from green microalga – scenedesmus and its potential applications: A brief review. *Pertanika Journal of Tropical Agricultural Science*, 39: 1–16.

Iwamoto, H. (2004). Industrial production of microalgal cell-mass and secondary products – major industrial species: Chlorella. In: Richmond, A. (ed.), *Handbook of Microalgal Culture: Biotechnology and Applied Phycology*. Blackwell Publishing Ltd., Hoboken, NJ, pp. 225–264.

Jensen, G., Ginsberg, D. I., and Drapeau, C. (2001). Blue-green algae as an immuno-enhancer and biomodulator. *JANA*, 3: 24–30.

Kadiri, M. (2011). Notes on harmful algae from Nigerian coastal waters. *Acta Botanica Hungarica*, 53(1–2): 137–143.

Kadiri, M. O. (1999). Phytoplankton distribution in some coastal waters of Nigeria. *Nigerian Journal of Botany*, 12: 51–62.

Kadiri, M. O. (2002). A spectrum of phytoplankton flora along salinity gradient in the eastern Niger Delta area of Nigeria. *Acta Botanica Hungarica*, 44(1–2): 75–83.

Kadiri, M. O., Isagba, S., Ogbebor, J. U., Omoruyi, O. A., Unusiotame-Owolagba, T. E., Lorenzi, A. S., and Chia, M. A. (2020). The presence of microcystins in the coastal waters of Nigeria, from the Bights of Bonny and Benin, Gulf of Guinea. *Environmental Science and Pollution Research*, 27(28): 35284–35293.

Kang, H. K., Salim, H., Akter, N., Kim, D.-S., Kim, J. H., Bang, H. T., and Suh, O. (2013). Effect of various forms of dietary Chlorella supplementation on growth performance, immune characteristics, and intestinal microflora population of broiler chickens. *The Journal of Applied Poultry Research*, 22: 100–108. https://doi.org/10.3382/japr.2012-00622.

Keshri, J. P. (2012). Algae in medicine. In: Keshri, J. P., and Mukhopadhyay, R. (eds.) *Medicinal Plants: Various Perspectives*. Pub. Department of Botany & Publication Unit, The University of Burdwan. pp. 31–50.

Khan, M. I. Shin, J. H., and Kim, J. D. (2018). The promising future of microalgae: current status, challenges, and optimization of a sustainable and renewable industry for biofuels, feed, and other products. *Microbial Cell Factory*, 17: 36.

Kiepper, B. H. (2013). Microalgae utilization in wastewater treatment [Bulletin 1419]. Georgia, USA: University of Georgia Cooperative Extension.

Kim, J., Lingaraju, B. P., Rheaume, R., Lee, J., and Siddiqui, K. F. (2010). Removal of ammonia from wastewater effluent by *Chlorella vulgaris*. *Tsinghua Science and Technology*, 15(4): 391–396. https://doi.org/10.1016/S1007-0214(10)70078-X.

Kim, S. K., and Karadeniz, F. (2011). Anti- HIV activity of extracts and compounds from marine algae. *Advances in Food and Nutrition Research*, 64: 255–265.

Kim, S. K., Thomas, N. V., and Li, X. (2011). Anticancer compounds from marine macroalgae and their application as medicinal foods. *Advances in Food and Nutrition Research*, 64: 213–224.

Kirst, H., García-Cerdán, J. G., Zurbriggen, A., and Melis, A. (2012a). Assembly of the light-harvesting chlorophyll antenna in the green alga Chlamydomonas reinhardtii requires expression of the TLA2-CpFTSY gene. *Plant Physiology*, 158(2): 930–945.

Kirst, H., Garcia-Cerdan, J. G., Zurbriggen, A., Ruehle, T., and Melis, A. (2012b). Truncated photosystem chlorophyll antenna size in the green microalga Chlamydomonas reinhardtii upon deletion of the TLA3-CpSRP43 gene. *Plant Physiology*, 160(4): 2251–2260.

Krause, G. H., and Weis, E. (1991). Chlorophyll fluorescence and photosynthesis: The basics. *Annual Review of Plant Physiology and Plant Molecular Biology*, 42(1): 313–349.

Lamela, M., Anca, J., Villar, R., Otero, J., and Calleja, J. M. (1989). Hypoglycemic activity of several seaweed extracts. *Journal Ethnopharmacology*, 27: 35–43.

Lamers, P. P., van de Laak, C. C., Kaasenbrood, P. S., Lorier, J., Janssen, M., De Vos, R. C., et al. (2010). Carotenoid and fatty acid metabolism in light-stressed *Dunaliella salina*. *Biotechnology and Bioengineering*, 106 (4): 638–648.

León-Bañares, R., González-Ballester, D., Galván, A., and Fernández, E. (2004). Transgenic microalgae as green cell-factories. *Trends in Biotechnology*, 22(1): 45–52.

Li, Z., Ma, X., Li, A., and Zhang, C. (2012). A. novel potential source of carotene: *Eustigmatoscf.* polyphem (Eustigmatophyceae) and pilot B- carotene production in bubble column and flat panel photobioreactors. *Bioresource Technology*, 117: 257–263.

Ma, R. Y.-N., and Chen, F. (2001). Enhanced production of free transastaxanthin by oxidative stress in the culture of *Chlorococcum* sp. *Process Biochemistry*, 36(12): 1175–1179.

Masojıdek, J., Koblızek, M., and Torzillo, G. (2004). Photosynthesis in microalgae. In: Richmond, A. (ed.), *Handbook of Microalgal Culture: Biotechnology and Applied Phycology*. Wiley-Blackwell, Hoboken, NJ, pp. 20–39.

Milledge, J. J. (2011). Commercial application of microalgae other than as biofuels: A brief review. *Reviews in Environmental Science and Biotechnology*, 10(1): 31–41. https://doi.org/10.1007/s11157-010-9214-7.

Mohammad, M. (2014). An overview: biomolecules from microalgae for animal feed and aquaculture. *Journal of Biological Research-Thessaloniki*, 21(6): 2241–5793.

Na, H. J., Moon, P. D., Lee, H. J., Kim, H. R., Chae, H. J., Shin, T., Seo, Y., Hong, S. H., and Kim, H. M. (2005). Regulatory effect of atopic allergic reaction by Carpopeltis affinis. *Journal of Ethnopharmacology*, 101: 43–48.

Nangul, A., and Bhatia, R. (2019). Microorganisms: A marvelous source of single cell proteins. *Journal of Microbiology, Biotechnology and Food Sciences*, 9(1): 15–18.

Natesan, R., and Shanmugasundaram, S. (1989). Extracellular phosphate solubilization by the cyanobacterium Anabaena ARM310. *Journal of Biosciences*, 14(3): 203–208. https://doi.org/10.1007/BF02716680.

Nishide, E., and Uchida, H. (2003). Effects of Ulva powder on the ingestion and excretion of cholesterol in rats. In: Chapman, A. R. O., Anderson, R. J., Vreeland, V. J., and Davison, I. R. (eds.) *Proceedings of the 17th International Seaweed Symposium*, Oxford University Press, Oxford, pp. 165–168.

Nwoba, E. G., Vadiveloo, A., Ogbonna, C. N., Ubi, B. E., Ogbonna, J. C., and Moheimani, N. R. (2020a). Algal cultivation for treating wastewater in African developing countries: A review. *Clean – Soil, Air, Water.* doi:10.1002/clen.202000052.

Nwoba, E. G., Ogbonna, C. N., Ishika, T., and Vadiveloo, A. (2020b). Microalgal pigments: A source of natural food colors. In: Alam, M., Xu, J. L., and Wang, Z. (eds.) *Microalgae Biotechnology for Food, Health and High Value Products.* Springer, Singapore, pp. 81–123.

Nwoba, E. G. (2017). Effect of selected light spectra on the growth of chlorella spp. (Chlorophyta). *Nigerian Journal of Biotechnology*, 32(1): 69–76.

Nwoba, E. G., Parlevliet, D. A., Laird, D. W., Alameh, K., and Moheimani, N. R. (2019). Light management technologies for increasing algal photobioreactor efficiency. *Algal Research*, 39: 101433.

Nwuche, C. O., Ekpo, D. C., Eze, C. N., Aoyagi, H., and Ogbonna, J. C. (2014). Use of palm oil mill effluent as medium for cultivation of *Chlorella sorokiniana. British Biotechnology Journal*, 4(3): 305–316.

Ogbonna, I. O., Moheimani, N. R., and Ogbonna, J. C. (2015). Potentials of microalgae biodiesel production in Nigeria. *Nigerian Journal of Biotechnology*, 29(1): 45–61.

Ogbonna, I. O., and Ogbonna, J. C. (2018a) Effects of carbon source on growth characteristics and lipid accumulation by microalaga *Dictyosphaerium* sp with potential for biodiesel production. *Energy and Power Engineering*, 10(2): 29.

Ogbonna, I. O., and Ogbonna, J. C. (2018b). Evaluation of oil producing potential of a new isolate—*Chlorella lewinii* SUB3545914 for biodiesel production under heterotrophic cultivation. *Journal of Sustainable Bioenergy Systems*, 8: 67–81.

Ogbonna, I. O., and Ogbonna, J. C. (2015). Isolation of microalgae species from arid environments and evaluation of their potentials for biodiesel production. *African Journal of Biotechnology*, 14(8): 1596–1604.

Ogbonna, I. O., Okpozu, O. O., Ikwebe, J., and Ogbonna, J. C. (2018). Utilization of *Desmodesmus subspicatus* LC 172266 for simultaneous remediation of cassava wastewater and accumulation of lipids for biodiesel production. *Biofuels*, 1–8. doi:10.1080/17597269.2018.1426164.

Ogbonna, J. C., Ichige, E., and Tanaka, H. (2002a). Interactions between photoautotrophic and heterotrophic metaboilism in photoheterotrophic culture of *Euglena gracilis. Applied Microbiology and Biotechnology*, 58: 532–538.

Ogbonna, J. C., Ichige, E., and Tanaka, H. (2002b). Regulating the ratio of photoautotrophic to heterotrophic metabolic activities in photoheterotrophic culture of *Euglena gracilis* and its application to α-tocopherol production. *Biotechnology Letters*, 24: 953–958.

Ogbonna, J. C., and McHenry, M. P. (2015). Culture systems incorporating heterotrophic metabolism for biodiesel oil production by microalgae. In: Moheimani, N. R., McHenry, M. P., Boe, K., and Bahri, P. A. (eds.) *Biomass and Biofuels from Microalgae. Advances in Engineering and Biology.* Springer International Publishing, Cham, pp. 63–74.

Ogbonna, J. C., and Moheimani, N. R. (2015). Potentials of exploiting heterotrophic metabolism for biodiesel oil production by microalgae. In: Moheimani, N. R., McHenry, M. P., Boe, K., and Bahri, P. A. (eds.) *Biomass and Biofuels from Microalgae. Advances in Engineering and Biology*. Springer International Publishing, Cham, pp. 45–61.

Ogbonna, J. C., Tomiyama, S., and Tanaka, H. (1998). Hetetrophic cultivation of *Euglena gracilis* Z for efficient production of α-tocopherol. *Journal of Applied Phycology*, 10: 67–74.

Ogbonna, J. C., Yoshizawa, H., and Tanka, H. (2000). Treatment of high strength organic wastewater by a mixed culture of phosynthetic microorganisms. *Journal of Applied Phycology*, 12(3–5):277–284. https://doi.org/10.1023/A:1008188311681.

Okpozu, O. O., Ogbonna, I. O., Ikwebe, J., and Ogbonna, J. C. (2019). Phycoremediation of cassava wastewater by *Desmodesmus armatus* and the concomitant accumulation of lipids for biodiesel production. *Bioresource Technology Reports*, 7: 1000256.

Ozioko, F. U., Chiejina, N. V., and Ogbonna, J. C. (2015). Effect of some phytohormones on growth characteristics of *Chlorella sorokiniana* IAM C212 under photoautotrophic conditions. *African Journal of Biotechnology*, 14(30): 2367–2376.

Panlasigui, L. N., Baello, O. Q., Dimatangal, J. M., and Dumelod, B. D. (2003). Blood cholesterol and lipid-lowering effects of carrageenan on human volunteers. *Asia-Pacific Journal of Clinical Nutrition*, 12: 209–214.

Patel, B., Tamburic, B., Zemichael, F. W. Dechatiwongse, P., and Hellgardt, K. (2012). Algal biofuels: A credible prospective? *International Scholarly Research Network ISRN Renewable Energy*, 2012: 14. doi:10.5402/2012/631574.

Perez-Garcia, O., Escalante, F. M. E., de-Bashan, L. E., and Bashan, Y. (2011). Heterotrophic cultures of microalgae: Metabolism and potential products. *Water Research*, 45(1): 11–36. https://doi.org/10.1016/j.watres.2010.08.037.

Perrine, Z., Negi, S., and Sayre, R. T. (2012). Optimization of photosynthetic light energy utilization by microalgae. *Algal Research*, 1(2): 134–142.

Polle, J. E., Kanakagiri, S.-D., and Melis, A. (2003). tla1, a DNA insertional transformant of the green alga Chlamydomonas reinhardtii with a truncated light-harvesting chlorophyll antenna size. *Planta*, 217(1): 49–59.

Pooja, S. (2014). Algae used as medicine and food-a short review. *Journal of Pharmaceutical Science Research*, 6(1): 33–35.

Priyadarshani, I., and Rath, B. (2012). Commercial and industrial applications of microalgae. *Journal of Algal Biomass Utilization*, 3(4): 89–100.

Radmer, R. J., and Parker, B. C. (1994). Commercial applications of algae: Opportunities and constraints. *Journal of Applied Phycology*, 6(2): 93–98. https://doi.org/10.1007/BF02186062.

Raja, R., Shanmugam, H., Ganesan, V., and Carvalho, I. S. (2014). Biomass from microalgae: An overview. *Journal of Oceanography and Marine Research*, 2(1): 1–7. doi:10.4172/2332-2632.1000118.

Ravindran, B., Gupta, S. K., Cho, W., Kim, J. K., Lee, S. R., Jeong, K., Lee, D. J., and Choi, H. (2016). Microalgae potential and multiple roles—Current progress and future prospects—An Overview. *Sustainability*, 8: 1215. doi:10.3390/su8121215.

Richmond, A. (2008). *Handbook of Microalgal Culture: Biotechnology and Applied Phycology*. John Wiley & Sons, Hoboken, NJ.

Rizwan, M., Mujtaba, G., Memon, S. A., Lee, K., and Rashid, N. (2018). Exploring the potential of microalgae for new biotechnology apicplations and beyond: A review. *Renewable and Sustainable Energy Reviews*, 92: 394–404. https://doi.org/10.1016/j.rser.2018.04.034.

Rosales-Mendoza, S., Angulo, C., and Meza, B. (2016). Food-grade organisms as vaccine biofactories and oral delivery vehicles. *Trends in Biotechnology*, 34(2): 124–136.

Sathasivam, S., Radhakrishnan, R., Hashem, A., and Abd_Allah, E. F. (2019). Microalgae metabolites: A rich source for food and medicine. *Saudi Journal of Biological Sciences*, 26 (4): 709–722.

Sawraj, S., Kate, B. N., and Banerjee U. C. (2008). Bioactive compounds from cyanobacteria and microalgae: An overview. *Critical Reviews in Biotechnology*, 25(3): 73–95. doi:10.1080/07388550500248498.

Scharff, C. (2015). Use of microalgae as renewable resources. *Journal of Central European Green Innovation*, 3(3): 149–156.

Scrimshaw, N. S., and Dillen, J. C. (1977). Single cell protein as food and feed. In: Garattini, S., Paglialunga, S., and Scrimshaw, N. (eds.), *Single Cell Protein-safety for Animal and Human Feeding*. Surrey University Press, Oxford, UK, pp. 171–173.

Spolaore, P., Joannis-cassan, C., Duran, E., and Isambert, A. (2006). Commercial applications of microalgae. *Journal of Bioscience and Bioengineering*, 101: 87–96.

Tiwari, A., and Kiran, T. (2018). Advances in Biofuels and Bioenergy, pp. 239–249. http://dx.doi.org/10.5772/intechopen.73012.

Vigani, M., Parisi, C., Rodrıguez-Cerezo, E., Barbosa, M. J., Sijtsma, L., Ploeg, M., et al. (2015). Food and feed products from microalgae: Market opportunities and challenges for the EU. *Trends in Food Science & Technology*, 42(1): 81–92. https://doi.org/10.1016/j.tifs.2014.12.004.

Vuppaladadiyam, A. K., Prinsen, P., Raheem, A., and Luque, R. (2018). Microalgae cultivation and metabolites production: A comprehensive review. *Biofuels, Bioproducts and Biorefinery*, 12: 304–324. doi:1002/bbb.1864.

Walsh, P. J., Fee, K., Clarke, S. A., Julius, M., and Buchanan, F. J. (2018). Blueprints for the next generation of bioinspired and biomimetic mineralised composites for bone regeneration. *Marine Drugs*, 16(8): 288.

Wuang, S. C., Khin, M. C., Chua, P. Q. D., and Luo, Y. D. (2016). Use of *Spirulina* biomass produced from treatment of aquaculture wastewater as agricultural fertilizers. *Algal Research*, 15: 59–64. https://doi.org/10.1016/j.algal.2016.02.009.

Yandigeri, M. S., Yadav, A. K., Srinivasan, R., Kashyap, S., and Pabbi, S. (2011). Studies on mineral phosphate solubilization by cyanobacteria Westiellopsis and Anabaena. *Microbiology*, 80(4): 558–565. https://doi.org/10.1134/S0026261711040229.

Zaslavskaia, L., Lippmeier, J., Shih, C., Ehrhardt, D., Grossman, A., and Apt, K. (2001). Trophic conversion of an obligate photoautotrophic organism through metabolic engineering. *Science*, 292(5524): 2073–2075.

# 11 Emerging Eco- and Bio-technologies in the Use of Algal Biomass for Biofuels and High-Value Products

*Taofikat A. Adesalu*
University of Lagos

*Mathias A. Chia*
Ahmadu Bello University

*John N. Idenyi*
Ebonyi State University

*Emeka G. Nwoba*
Murdoch University

**CONTENTS**



DOI: 10.1201/9781003178378-11

## 11.1 INTRODUCTION

Ecotechnology is the use of technological means for ecosystem management based on a deep understanding of the principles on which natural ecological systems are built and the transfer of such principles into ecosystem management in a way to minimize the costs of the technologies and their harm to the global environment (Straskraba, 1993). Examples of such processes are pretreatment and saccharification of algal biomass, fermentation, gasification, pyrolysis, hydrothermal liquefaction, and anaerobic digestion for the production of biohydrogen, bio-oils, biomethane, biochar, and various bio-based products (Kumar et al., 2020). Algae have the potential to play an important role as a renewable raw material for different applications. Some other characteristics of eco-friendly technologies are recycled or reused materials and reduction of greenhouse gas emissions and pollutants.

Biotechnology research goals therefore include finding ways to increase the reproductive rate, improve metabolism of inputs and enhance the production of desired oils, fuel-grade alcohols, or proteins in useful species. Biotechnology is already employed in sequencing and annotating the genomes of algal species which aids researchers in understanding the metabolic processes through which algae convert carbon and nutrients into lipids or carbohydrates. Genetic engineering techniques currently used in plant and microbial biotechnology, including synthetic biology and metabolic engineering, are then employed to enable algae more predictably produce desired lipids for biofuels, alcohols, proteins, enzymes and other molecules, or carbohydrate-rich biomass for bioprocessing.

Algae (grouped into microalgae and macroalgae/seaweed) are a diverse group of aquatic organisms that can be unicellular and/or simple multicellular that have high growth rates with photosynthetic efficiencies due to their simple structures. While the phylogenetic relationships continue to be resolved among groups, algae according to Barkia et al. (2019) are assigned to 11 major phyla: Cyanophyta, Chlorophyta, Rhodophyta, Glaucophyta, Euglenophyta, Chlorarachniophyta, Charophyta, Cryptophyta, Haptophyta, Heterokontophyta, and Dinophyta. They represent a largely under-exploited group of natural resources with a tremendous potential to produce high-value natural products and are a promising new source of biomass for production of food, feed, fuel, or chemicals (Vandamme et al. 2013). They are presented as new model organisms for a wide range of biotechnological applications, including biodiesel production (Demirbas and Demirbas, 2011), wastewater bioremediation (Craggs et al., 1996; Nwoba et al., 2020b), and dietary supplements for animal and human nutrition (Enzing et al., 2014; Leu and Boussiba, 2014). Microalgal extracts are established commercial sources of high-value chemicals such as β-carotene, astaxanthin, docosahexanoic acid, eicosahexanoic acid, phycobilin pigments, proteins, lipids, carbohydrates, vitamins, and antioxidants, with applications in cosmetics, nutritional, nutraceuticals, functional foods, and pharmaceutical industries (Adesalu et al., 2016; Borowitzka, 2013; Kit et al., 2017; Chia et al., 2015; Kunrunmi et al., 2017). According to Carvalho and Malcata (2000), the cellular content of each fraction varies according to the specific strain of algae and their physiological responses to biotic and abiotic factors, e.g., light intensity, photoperiod, temperature, nutrients, and growth phase. Biofuel generally refers to liquid gas, and solid fuels predominantly produced from biomass, it includes bioethanol, biomethanol, vegetable oils, biodiesel, biogas, bio-synthetic gas (bio-syngas), bio-oil, biochar, Fischer-Tropsch liquids, and biohydrogen (Demirbas, 2008). The author projected that the biofuel economy will grow rapidly during the 21st century and its economic development. According to Demirbas (2008), by the year 2050, the most biomass-intensive scenario and modernized biomass energy would contribute about one half of total energy demand in developing countries based on agricultural production. The high growth rate of microalgae makes it possible to satisfy the massive demand on biofuels using limited land resources without causing potential biomass deficit. Microalgal cultivation consumes less water than land crops, and the tolerance of microalgae to high $CO_2$ content in gas streams allows high-efficiency $CO_2$ mitigation. This chapter presents the emerging eco- and bio-technologies in the use of algal biomass for biofuels and high-value products. Third-generation feedstock for biofuels, generation of algal biomass using ecotechnological and biotechnological approaches, phycoremediation, bioproducts, and health and safety considerations for the utilization of algal biomass are discussed.

## 11.2 MICROALGAE AS THIRD-GENERATION FEEDSTOCK FOR BIOFUEL PRODUCTION

Fossil fuels such as coal, natural gas, and petroleum for many decades have been world first-class energy source, and there are growing worries on future energy reserves and security. For instance, International Energy Agency World Energy Outlook (Khatib, 2012) projected global energy consumption to reach 53% between 2008 and 2035, representing an increase of 1.6% per year. Furthermore, increased combustion of fossil fuels strongly contributes to the emission of greenhouse gases and harmful gaseous pollutants such as $CO_2$, $NO_2$, $SO_2$, and CO, which promote global warming and significant environmental challenges. A restraint on the use of fossil fuels has been recommended due to their dangerous environmental effects as well as the uncertainty from the global supply, and this has triggered a critical search for renewable and eco-friendly energy resources. Besides the photovoltaic, wind, and hydro renewable energy technologies, biomass-based energy sources are desirable because they target the larger energy sector that is primarily dependent on liquid fuel use (Ooms et al., 2016; Nwoba et al., 2019; Ogbonna and Nwoba, 2021). Renewable biomass from rapeseed, soya bean, canola, algae, and lignocellulosic plants can be converted to biofuels, and this is attractive from the climate change standpoint (Jez et al., 2017; Ogbonna and Nwoba, 2021). The 1G (first-generation) biofuel feedstocks are produced from edible food crops such as canola, soybean, and corn and are associated with sustainability challenges. Significant among these are the food vs. fuel crisis, land-use change, and environmental impacts such as loss of biodiversity, ecotoxicity, and eutrophication (Mamo et al., 2013). These challenges have triggered the development of 2G (second-generation) biofuel sources, which are generated from non-edible crops, such as poplar, miscanthus, lignocellulosic agro-residues, and agro-industrial wastes. The 2G bioenergy sources are deemed more sustainable because they circumvent food vs. fuel crisis and land-use change, but the use of agricultural fertilizers to grow these crops could increase the price of food crops in the market, as well as result in environmental burdens such as eutrophication and loss of biodiversity. The lack of technology maturation in the conversion of lignocellulosic materials to biofuels has made this pathway for the production of renewable energy technically and economically unviable (Sims et al., 2010) (Figure 11.1). These bottlenecks have led to the development of 3G (third-generation) biofuel solution (Figure 11.2), which are primarily from algal biomass. Currently, algal biofuels are touted as one of the best sources of carbon-neutral, high-energy-density liquid fuels (Stephens et al., 2010). Microalgae use biochemical, physiological, and ecological systems to provide abundant bioenergy resources such as biodiesel, biogas (biomethane, biohydrogen), bio-oil, bioethanol, and other kinds of fuels (Chaudry et al., 2015), and these fuel products can be obtained using conversion technologies such as transesterification, anaerobic digestion, liquefaction, and pyrolysis. As a renewable energy candidate, certain narratives have elevated microalgae over traditional crops for bioenergy production, and these include: (1) fast growth rate and high photosynthetic conversion competence with some, doubling their numbers in one day. Due to this, they have higher net biomass yield and areal productivity compared with dedicated energy crops. For instance, while annual biomass productivity of sugarcane was estimated at 61–95 ton/ha/year of fresh weight, cultivation of brown microalgae for seven months yielded

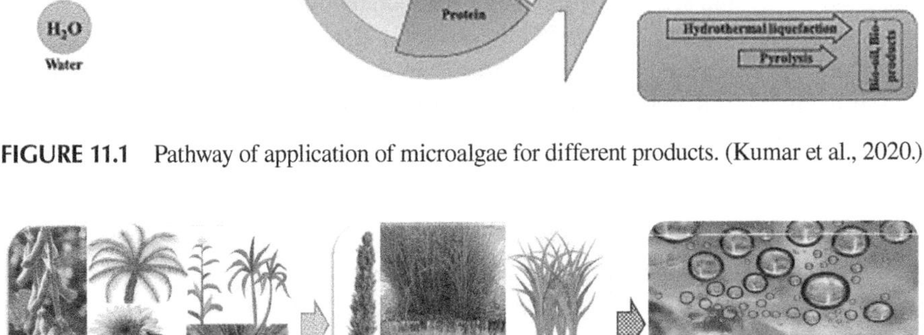

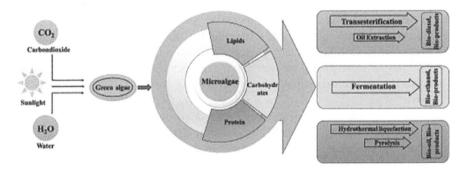


**FIGURE 11.1** Pathway of application of microalgae for different products. (Kumar et al., 2020.)

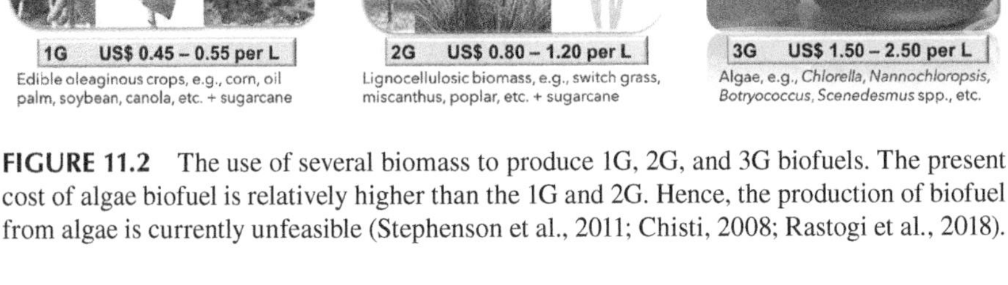


**FIGURE 11.2** The use of several biomass to produce 1G, 2G, and 3G biofuels. The present cost of algae biofuel is relatively higher than the 1G and 2G. Hence, the production of biofuel from algae is currently unfeasible (Stephenson et al., 2011; Chisti, 2008; Rastogi et al., 2018).

**TABLE 11.1**
**Biodiesel Production Efficiencies of Bioenergy Crops (Schenk et al., 2008; Chisti 2007; Rastogi et al., 2018; Atabani et al., 2012)**

| Bioenergy Crop | Oil Yield (L/ha) |
|---|---|
| Mustard seed | 572 |
| Soya bean | 446 |
| Cotton | 325 |
| Palm oil | 5 950 |
| Jatropha | 1 892 |
| Rapeseed/ canola | 1 190 |
| Sunflower | 952 |
| Algae | 50 000 – 136 000 |

Mustard seed

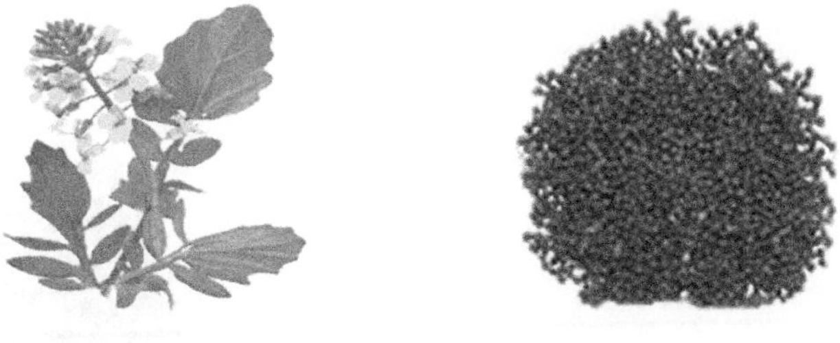

Soya bean

Cotton

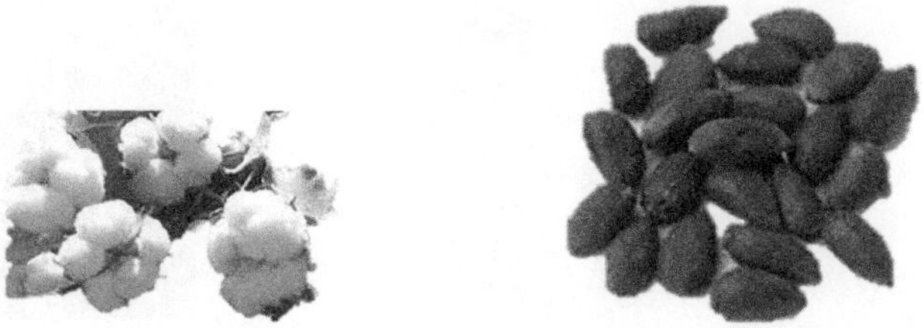

Palm oil

Jatropha

Rapeseed/ canola

Sunflower

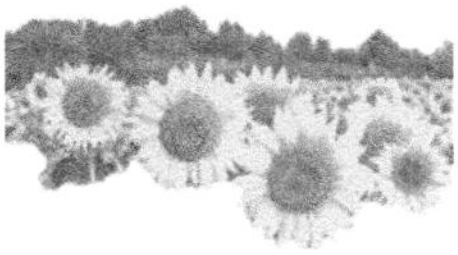

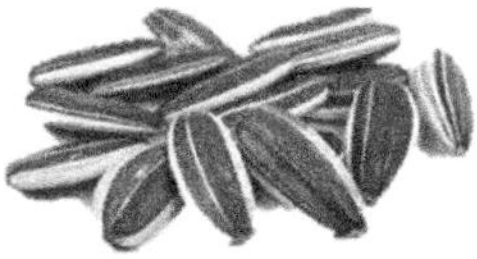

Algae

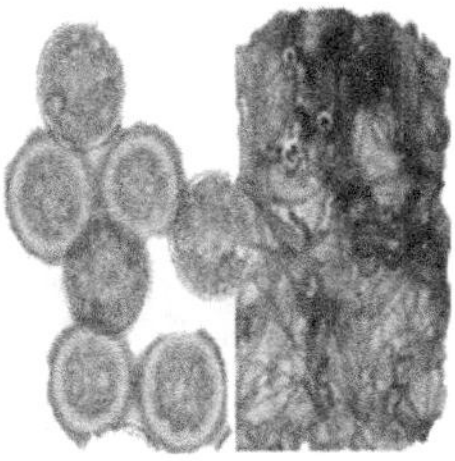

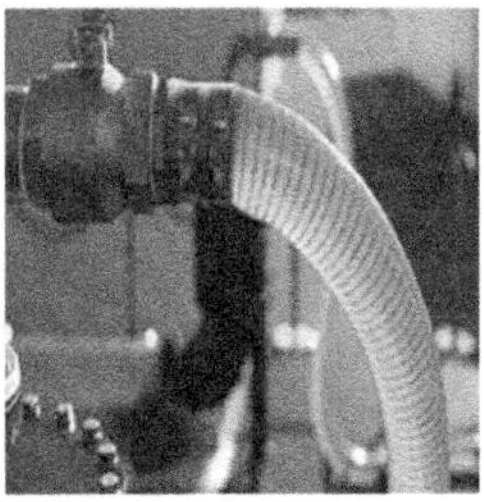

131 ton/ha of dry weight biomass, which is even 27.5%–53.4% higher than that obtained with a fresh weight of sugarcane (Kraan, 2013). As a photosynthetic-based crop, microalgae have higher efficiency of converting solar energy to chemical energy that is stored in the biomass. The theoretical maximum photosynthetic efficiency of microalgae is well over 10% compared to the 6% for best C4 plants according to Zhu et al. (2008). (2) High lipid yields of up to 60 $m^3$/ha vs. 2 $m^3$/ha for *Jatropha*, and 0.2 $m^3$/ha for corn (Chisti, 2007) (Table 11.1). Certain species of microalgae have lipid contents that are more than 50% dry weight of biomass (Jez et al., 2017) (Table 11.2). About 200 barrels of oil can be obtained by farming microalgae, which is several hundred-fold more than dedicated biodiesel crops such as soybean (Hu et al., 2008) (Table 11.1). (3) High affinity for $CO_2$ sequestration, due to the faster growth rates, with a high affinity to fix $CO_2$ into biomass (Voloshin et al., 2016). (4) Non-reliance of microalgaculture on agricultural land, hence the ability to be cultivated on marginal arable land hereby reducing competition with food-based crops. (5) Avoidance of dependence on fertilizer and the limiting freshwater, microalgae can be grown in flexible culture systems such as open or closed system, utilizing seawater and wastewater (nutrients source) for their growth (Ishika et al., 2017). (6) Higher lipid yield per hectare (158 tons) of microalgae than macroalgae (60–100 tons), which is a more favorable biomaterial for lipid-based biorefineries (Chisti, 2007). However, macroalgae have higher carbohydrate content as well as simple harvesting procedures, with a gasification-driven net energy production of 11 GJ/ton dry biomass than microalgae (9.5 GJ/ton dry biomass) (Aitken et al., 2014).

**TABLE 11.2**
**Oil Contents of Selected Microalgae (Chisti 2007)**

| Microalgae | Oil Content (% Dry Weight) |
|---|---|
| *Botryococcus braunii* | 25–75 |
| *Chlorella* sp. | 28–32 |
| *Crypthecodinium cohnii* | 20 |
| *Cylindrotheca* sp. | 16–37 |
| *Dunaliella primolecta* | 23 |
| *Isochrysis* sp. | 25–33 |
| *Monallanthus salina* | >20 |
| *Nannochloris* sp. | 20–35 |
| *Nannochloropsis* sp. | 31–68 |
| *Neochlorisoleo abundans* | 35–54 |
| *Nitzschia* sp. | 45–47 |
| *Phaeodactylum tricornutum* | 20–30 |
| *Schizochytrium* sp. | 50–77 |
| *Tetraselmis suecia* | 15–23 |

## 11.3 PHYCOREMEDIATION OF WASTEWATER, CARBON DIOXIDE BIOSEQUESTRATION, AND SIMULTANEOUS GENERATION OF BIOPRODUCTS

Phycoremediation is a green technology with vast environmental benefit and more so as no hazardous secondary by-products are produced. It is the use of macroalgae or microalgae for the removal or biotransformation of pollutants, including nutrients like organic/inorganic carbon, nitrogen, phosphorous, sulfates, heavy metals and xenobiotics, from wastewater and biosequestration of air pollutants such as carbon dioxide (Olguín, 2003; Richards and Mullins 2013; Renuka et al., 2015). During phycoremediation process, microalgae use carbon, nitrogen, phosphorus, and other salts from the wastewater necessary for the growth and production of microalgae (Nwoba et al., 2020b, 2016). Among these nutrients, the assimilation of nitrogen and phosphorus is vital for the growth of microalgal species. Nitrogen in wastewaters can be utilized in the form of $NO_3^-$ (nitrate), $NO_2^-$ (nitrite), or $NH_4^+$ (ammonium) by microalgae through conversion to organic nitrogen (Hadiyanto et al., 2013). As shown in Figure 11.3, the translocation of inorganic nitrogen across the plasma membrane of the microalgae cells follows an enzymatic reduction to nitrate and further to nitrite by nitrate reductase. Nitrate reductase uses the reduced form of nicotinamide adenine dinucleotide (NADH) as a co-enzyme to transfer two electrons that fuel the conversion of nitrate into nitrite. Nitrite is enzymatically converted into ammonium by nitrite reductase using ferredoxin (Fd), which transfers six electrons in the reaction. Finally, ammonium is incorporated into glutamine by glutamine synthetase using glutamate (Glu) and adenosine triphosphate (ATP) to drive the reaction (Cai et al., 2013). On the other hand, the phosphorus content in the wastewater is absorbed by microalgae to build-up the lipids, nucleic acids, proteins, and other intermediate products and mediate energy metabolism of the cells. Accordingly, assimilation of the nitrate and phosphate ions for algal growth and production of biomass leads to the decrease

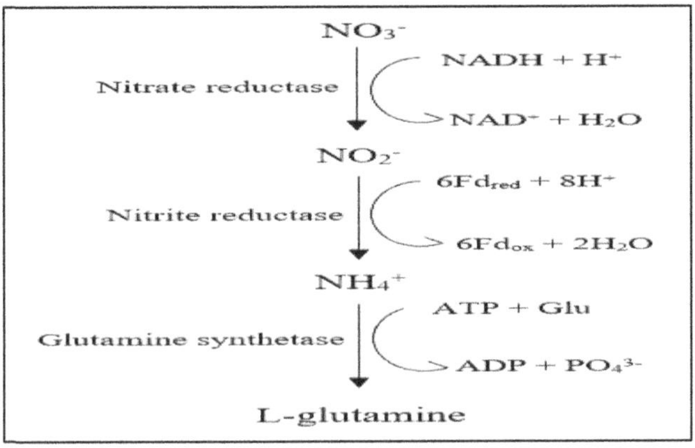


**FIGURE 11.3** Conversion of inorganic nitrogen into organic form through nutrient uptake by microalgae.

of their concentrations in wastewater thereby enhancing the quality of effluents discharged (Zeng et al., 2012).

The utilization of wastewaters as nutrient source for algae production can simultaneously reduce greenhouse gas emission from point sources and significantly decrease the cost of production of bulk commodities (e.g. biofuels) at large scale. Hence, combining wastewater remediation with the production of bioproducts is attracting increasing research attention due to its sustainability and cost benefits. During photosynthesis, algae require $CO_2$ as a carbon source in the presence of light and water to generate commercial products. In this, the addition of $CO_2$ to wastewater treatment systems would enhance the productivity of algae, resulting in the removal of all nutrients (especially N and P). This will improve the quality of treated wastewater, meeting a regulatory requirement for environmental discharge while producing a significant amount of biomass with remarkable lipid content. The $CO_2$ for the algal growth can be sourced from power plants such as flue gas or derived from the organic matter in the waste streams. It has been reported that algae-based treatment of wastewater (domestic, industrial, and animal manure), when integrated with the production of biofuel, can decrease greenhouse gas emissions by 1% (Benemann, 2013; Blackburn and Lee-Chang, 2018). Therefore, the algae-based wastewater treatment technologies could be developed for nutrients and water recycling, significantly maximizing biofuel outputs. For instance, in the case of microalgae, high-rate algal pond (HRAPs) is a cheap but efficient wastewater treatment system with high treatment performance combined with high microalgal productivity. The biomass produced by HRAPs is adequate for conversion to biofuel and can be further valorized to high-value bioproducts in the biorefinery context (Batten et al., 2013). In terms of nutrient biofiltration and $CO_2$ biomitigation, microalgae are considered the best renewable biosolution agents (Nwoba et al., 2017). Many microalgal species, such as *Chlorella, Spirulina, Chlamydomonas,* and *Scenedesmus* spp. (with *Chlorella* and *Scenedesmus* spp. being the best-studied), have great potential for nutrient recovery from different wastewaters (Ji et al., 2013). The harvested biomass can be used as a feedstock for food, feed, bioenergy, or fine chemical production (Figure 11.4). However, the harvesting of the microalgae is both cost-ineffective and energy-intensive, resulting in high processing costs and making bulk commodity production from microalgae economically unfeasible. In contrast, macroalgae are easy to harvest, using cheap procedures such as straining or scraping (Nwoba et al., 2017). Macroalgae such as *Ulva, Rhizoclonium, Cladophora,* and *Oedogonium* spp. have been applied in the bioremediation of various effluents such as aquaculture, dairy, animal manure, and municipal effluents (de Paula Silva et al., 2012; Al-Hafedh et al., 2012; Neori et al., 2004; Nwoba et al., 2017). The potential bioproducts from the application of algal biotechnology in wastewater bioremediation are shown in Figure 11.4.

## 11.4 BIOPRODUCTS FROM ALGAE

Bioproducts are natural products or compounds obtained from renewable and biodegradable resources (Mathimani and Pugazhendhi, 2019); they are usually non-toxic to consumers and the environment, and they can be derived from micro- and macro-organisms that include plants, algae, cyanobacteria, bacteria, and animals. Algae have

**FIGURE 11.4** Potential bioproducts from algal biomass when integrated with wastewater treatment (Nwoba et al., 2019).

diverse elemental constituents (hydrogen, carbon, nitrogen, phosphorus) and at the stationary growth phase, their biochemical composition is generally made up of 30%–40% proteins, 10%–20% lipids, and 5%–15% carbohydrates. Changes in biochemical composition make algae excellent sources of non-fuel bioproducts such as proteins, lipids, carbohydrates, antioxidants, and biofertilizers with applications in cosmetics, food, agricultural, pharmaceutical, and nutraceutical industries (Table 11.3).

### 11.4.1 Lipids

Lipids are major constituents of algal biomass that have been generally employed in the generation of biofuels, but they have several important non-bioenergy applications. Algal lipids may comprise about 15%–90% of total biomass depending on species type and culture conditions (Chisti, 2007; Metting, 1996). Lipids are divided into different classes ranging from steryl esters to acetone mobile polar lipids (Chia et al., 2013), with the basic building blocks of lipids, fatty acids, which are carboxylic acids having a hydrocarbon chain and a terminal carboxyl group (Figure 11.5). Fatty

**TABLE 11.3**
**Summary of Microalgae Genera, Their Bioproducts, and Applications**

| Algae | Algal Bioproducts | Applications | References |
|---|---|---|---|
| *Chlorella* | Lipids (PUFAs, fatty acids, omega-3, EPA, sterols, oleic acid) pigments (carotenoids, astaxanthin, zeaxanthin, violaxanthin, lutein, canthaxanthin); proteins (peptide); carbohydrates (sulfated polysaccharides) | Food, animal feeds, pharmaceuticals, cosmetic, nutraceuticals | Ip and Chen (2005), Liu et al. (2008), Saygideger and Deniz (2008), Bajguz and Piotrowska-Niczyporuk (2013), Chen et al. (2009), de Souza et al. (2018), Yu et al. (2015) |
| *Spirulina* (*Arthrospira*) | Vitamins (B12, C, K), pigments (allophycocyanin, phycocyanins); lipid (diacylglycerols, linolenic acid, oleic acid, lauric acid); polysaccharides; proteins | Food, cosmetics, nutraceuticals, pharmaceuticals | Liu and Chen (2014), Safi et al. (2014), Sonani et al. (2016), Sousa et al. (2008) |
| *Porphyridium* | Carbohydrates (sulfated polysaccharides, ß-1,3-glucan), (phycoerythrobilin, phycocyanin) | Pharmaceuticals, cosmetic, nutraceuticals | Koller et al. (2014), Spolaore et al. (2006) |
| *Haematococcus* | Pigments (astaxanthin, β-carotene, zeaxanthin, canthaxanthin, lutein, oleic acid | Food, animal feeds, pharmaceuticals, cosmetic, nutraceuticals | Koller et al. (2014), Pulz and Gross (2004), Spolaore et al. (2006) |
| *Nostoc* | Lipids, carotenoid, polysaccharides, ß-1,3-glucan, phenolic compounds | Food, nutraceuticals | de Morais et al. (2015), Nakashima et al. (2009) |
| *Dunaliella* | EPA, DHA, sterols, carotenoid, β carotene, β-cryptoxanthin, vitamin E (α-tocopherol), protein, diacylglycerol | Nutraceuticals, pharmaceuticals, food | Hamed (2016), Luo et al. (2015), Tang and Suter (2011) |
| *Schizochytrium* | Lipids (DHA, EPA) | Nutraceuticals | Pulz and Gross (2004) |
| *Synechocystis* | Pigments (β-carotene, zeaxanthin); protein (SOD) | | Pulz and Gross (2004), Sun et al. (2016) |
| *Aphanizomenon* | Chlorophyll *a* | Nutraceuticals, pharmaceuticals | Bishop and Zubeck (2012) |
| *Anabaena* | Phenolic compounds, lycopene, superoxide dismutase (SOD) | Cosmetics, pharmaceuticals | de Morais et al. (2015); Singh et al. (2017) |
| *Nannochloropsis* | Lipids (EPA, fatty acids); pigments (canthaxanthin, β-carotene); proteins | Pharmaceuticals, nutrition, cosmetics, food and animal feeds | Borowitzka (1995), Koller et al. (2014), Pulz and Gross (2004), Spolaore et al. (2006) |

acids can be saturated or unsaturated, a state that determines their applications as non-biofuel products. For example, *Diacronemavl kianum* and *Isochrysis galbana* have a high fat content of 18% and 24%, respectively, and both species contain substantial amounts of polyunsaturated fatty acids, (ω-3) omega-3 (Batista et al., 2013). The high PUFA content of other algae such as *Tetraselmis suecica*, *Porphyridium cruentum*, *Isochrysis galbana*, *Skeletonema costatum*, *Chaetoceros calcitrans*,

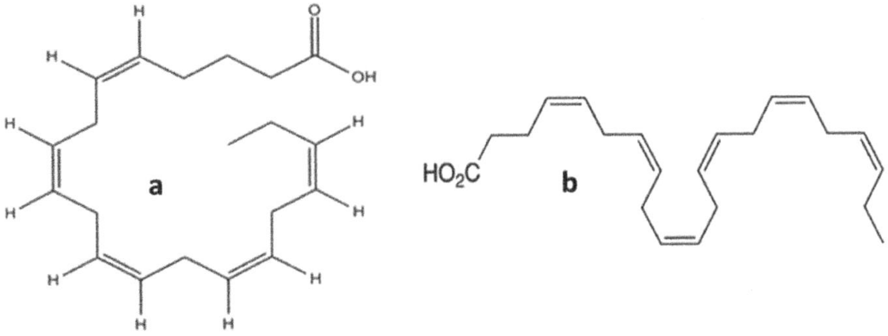


**FIGURE 11.5** Structural examples of omega-3 fatty acids produced by microalgae. (a) Eicosapentaenoic acid and (b) docosahexaenoic acid. (Molecule of the week archive, American Chemical Society.)

and *Nannochloropsis* sp. makes them valuable biomaterials for cosmetic upgrading (Servel et al., 1994). Saturated and polyunsaturated fatty acids can coexist in copious amounts in the lipid content of microalgae as observed in the report of Kunrunmi et al. (2017) on Epe lagoon algal species (Figure 11.6) and Adesalu et al. (2016). Algae that contain high polyunsaturated fatty acids (PUFAs) are suitable for application in nutraceuticals, human food, and fish oil replacement (Gosch et al., 2012). Important human health lipids such as eicosapentaenoic (EPA) and docosahexaenoic (DHA) are synthesized in high concentrations by several algae and are very important candidates in the prevention and control of various cardiovascular diseases (Sillman, 2016). *Spirulina* is rich in γ-linolenic acid that is given to patients with high cholesterol to reduce low-density lipoprotein (Li et al., 2019; Mazokopakis et al., 2014).

### 11.4.2 Proteins

Some microalgal biomass contain approximately 50%–70% proteins which are macromolecules consisting of amino acid residues (Bleakley and Hayes, 2017). Depending on the species, algae have a wide range of proteins that vary in their proportions with macroalgae having an average proportion of 20% total proteins, which is higher than those of other crop residues such as wheat straw, sugarcane leaf, and corn stover (Lammens et al., 2012). Unlike macroalgae, microalgae such as *Chlorella vulgaris, Diacronemavl kianum, Isochrysis galbana*, and *Spirulina maxima* have a high protein content of 38%, 38%, 40%, and 44%, respectively (Batista et al., 2013) which implies that these species are excellent candidates for application in the food industry and as animal feed supplements as the bioaccumulation of proteins is dependent on the nutrients composition and environmental factors to which microalgae are exposed (Chia et al., 2015).

When proteins are fragmented with acids, alkali, and enzymes, the product is a protein hydrolysate, which has the same nutritional qualities as the original protein material per constituent amino acids (Figure 11.7). These preparations are especially useful as special diets for patients that are unable to consume conventional food proteins. Also, the building blocks of proteins, peptides, have excellent antioxidant capacities that rely on amino acid composition, peptide molecular weight, molecular and surface hydrophobicity, and

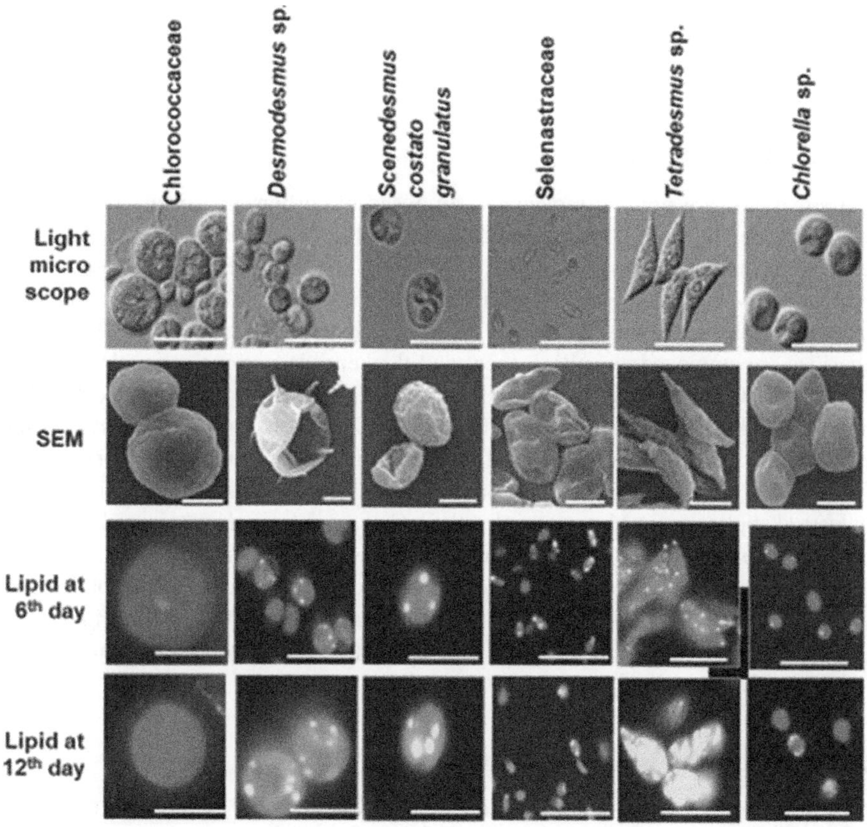


**FIGURE 11.6** Light microscope images (scale bars: 10 µm) of isolated green microalgae (first row). Lipid accumulation study using Nile red dye (second and third rows). Algal cells with lipid droplets shown in golden color. Nile red stained cells were at 100× using fluorescence microscope (scale bars: 10 µm) via UV light with excitation and emission at 530 and 57 nm, respectively. (Kunrunmi et al. 2017.)

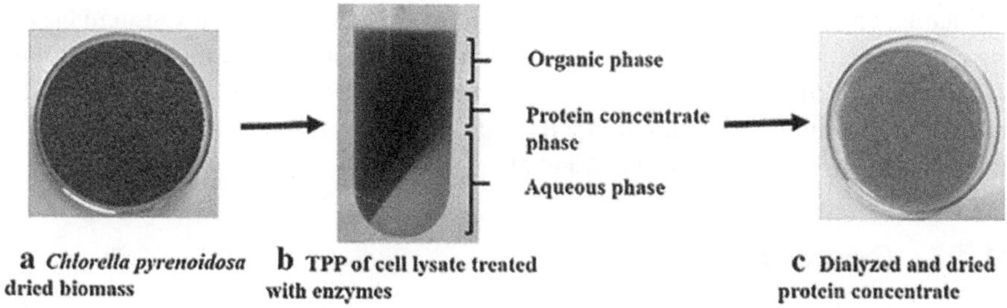


**FIGURE 11.7** Protein hydrolysate obtained from *Chlorella pyrenoidosa*. (Wagner and Adrian 2009.)

the type of reactive oxygen species (ROS) or free radical. Bioactive peptides scavenge free radicals, absorb oxygen radicals, and inhibit lipid peroxidation (Del Campo et al., 2007) and together, these characteristics are critical in disease control.

### 11.4.3 Carbohydrates

Carbohydrates are organic molecules composed of carbon, hydrogen, and oxygen, and they come in various forms, starch, sugars, polysaccharides, and cellulose. Algae are a rich source of carbohydrate (Chew et al., 2017) as starch and glucose are the principal raw materials used in producing bioethanol or biohydrogen; cellulose, a structural polysaccharide of the cell wall, is used to produce agar. Algal polysaccharides are employed as gelling agent and thickener in various cosmetic formulas (Jain et al., 2005), while alginate, fucoidans, agar, and carrageenan (polysaccharides) help in the distribution of water in the skin, serving as a good moisturizer (Wang et al., 2013). *Chlorella* contains β-1–3-glucan (active immunostimulatory), while *Kappaphycu salvarezii* is used to produce carrageenan and generate glucose for industrial applications (Masarin et al., 2016). Carrageenan is also an ingredient in the production of toothpaste, while alginate is utilized in the pharmaceutical, medical, textile, printing, biocatalyst, paper industries, and a carrier for cell immobilization (Roesijadi et al., 2010 Bixler and Porse, 2011; Kraan, 2013). Agar is used in preparing solid culture media for microbial studies and as a laxative in the pharmaceutical industry (McHugh, 2003).

### 11.4.4 Pigments as Antioxidants

Antioxidants are molecules that inhibit the production of oxidative or free radicals; these molecules help to reduce or prevent oxidative damage caused by ROS on cells. Algae are a source of unexplored natural antioxidants (Li et al., 2007) with *Anabaena* sp., *Porphyridium purpureum*, *Isochrysis galbana*, *Phaeodactylum tricornutum*, and *Synechocystis* sp. being valuable sources of antioxidants (de Souza et al., 2018). Carotenoid, phycocyanin, neoxanthin, phycoerythrin, zeaxanthin, fucoxanthin, astaxanthin, and violaxanthin are compounds with antioxidant properties that are used in nutraceutical as additives and food colorants (Pulz and Gross, 2004; Radmer, 1996). For instance, *Dunaliella* is a rich source of carotenes, lutein, neoxanthin, zeaxanthin, and violaxanthin, while the pigment astaxanthin has strong anti-aging, anti-inflammatory, anti-obesity, sun-proofing, and immune system boosting characteristics, which are beneficial in food, feed, and nutraceutical applications (Cheng et al., 2016). Also, β-carotene, phycobiliproteins, and astaxanthin are used as supplements in the production of ice creams, noodles, biscuits, tea, drinks, bread, and beers (Liang et al., 2004; Nwoba et al., 2020a).

### 11.4.5 Organic Fertilizers

Inorganic fertilizers have adverse effects on plants, soil, animals, consumers, and the general environment, to mitigate these problems, and organic fertilizers have increasingly been used in the agricultural sector. Uysal et al. (2015) revealed a positive effect of organic fertilizer obtained from *Chlorella vulgaris* on maize and wheat with the microalga improving the amount of organic matter and water holding capacity of the soil, leading to a higher germination rate and plant height.

Most cyanobacteria are capable of fixing atmospheric nitrogen, which makes them efficient biofertilizers (Vaishampayan et al., 2001; Adesalu and Olugbemi, 2015). When combined with members of other phytoplankton groups, some microalgae perform excellently as biofertilizers, for instance, *Chlorella vulgaris* and *Spirulina platensis* are good biofertilizers for rice cultivation because they increase rice yield by 7%–20.9% (Dineshkumar et al., 2018) and significantly decrease the total sulfides and ferrous iron content of the soil (Aiyer, 1965). Also, *Spirulina*-based fertilizer has been consistently used to enhance the growth of leafy vegetables such as *Eruca sativa*, *Amaranthus gangeticus*, and *Brassica rapa* (Wuang et al., 2016). Treatment of soil with the marine macroalgal powders as an organic fertilizer improved the vegetative, structural characteristics, and yield of *Vicia faba* plants than those treated with and without inorganic fertilizer (Sabh and Shallan, 2008).

## REFERENCES


Adesalu T.A., Temenu T. O. and Julius M.L. (2016). Molecular characterization, lipid analysis and GC-MS determination of bioactive compounds identified in a West African strain of the green alga *Oedogonium* (Chlorophyta). *Journal of Pharmacognosy and Phytochemistry* 5(6):01–06.

Adesalu, T.A. and Olugbemi, O.M. (2015). Soil algae: A case study of two vegetable farmlands in Lagos and Ogun States, Southwest, Nigeria. *Ife Journal of Science* 17(3):765–772.

Aitken D., Bulboa C., Godoy-Faundez A., Turrion-Gomez J.L. and Antizar-Ladislao B. (2014). Life cycle assessment of macroalgae cultivation and processing for biofuel production. *Journal of Cleaner Production* 75:45–56.

Aiyer R.S. (1965). Comparative algological studies in rice fields in Kerala State.

Al-Hafedh Y.S., Alam A., Buschmann A.H. and Fitzsimmons K.M. (2012). Experiments on an integrated aquaculture system (seaweeds and marine fish) on the Red Sea coast of Saudi Arabia: Efficiency comparison of two local seaweed species for nutrient biofiltration and production. *Reviews in Aquaculture* 4(1):21–31.

Atabani A.E., Silitonga A.S., Badruddin I.A., Mahlia T., Masjuki H. and Mekhilef S. (2012). A comprehensive review on biodiesel as an alternative energy resource and its characteristics. *Renewable and Sustainable Energy Reviews* 16(4):2070–2093.

Bajguz A. and Piotrowska-Niczyporuk A. (2013). Synergistic effect of auxins and brassinosteroids on the growth and regulation of metabolite content in the green alga *Chlorella vulgaris* (Trebouxiophyceae). *Plant Physiology and Biochemistry* 71:290–297.

Barkia I., Saari N. and Manning S. R. (2019). Microalgae for high-value products towards human health and nutrition. *Marine Drugs* 17(5):304.

Batista A.P., Gouveia L., Bandarra N.M., Franco J.M. and Raymundo A. (2013). Comparison of microalgal biomass profiles as novel functional ingredient for food products. *Algal Research* 2: 164–173.

Batten D., Beer T., Freischmidt G., Grant T., Liffman K., Paterson D., Priestley T., Rye L. and Threlfall G. (2013). Using wastewater and high-rate algal ponds for nutrient removal and the production of bioenergy and biofuels. *Water Science and Technology* 67 (4):915–924.

Benemann J. (2013). Microalgae for biofuels and animal feeds. *Energies* 6 (11):5869–5886

Bishop W.M. and Zubeck H.M. (2012). Evaluation of microalgae for use as nutraceuticals and nutritional supplements. *Journal of Nutrition and Food Science* 2:1–6.

Bixler H.J. and Porse H. (2011). A decade of change in the seaweed hydrocolloids industry. *Journal of Applied Phycology* 23:321–335.

Blackburn S.I. and Lee-Chang K.J. (2018). Microalgae: A renewable resource for food and fuels and more. *Blue Biotechnology: Production and Use of Marine Molecules* 1:1–32.

Bleakley S. and Hayes M. (2017). Algal proteins: Extraction, application, and challenges concerning production. *Foods* 6:33.

Borowitzka M.A. (1995). Microalgae as sources of pharmaceuticals and other biologically active compounds. *Journal of Applied Phycology* 7:3–15.

Borowitzka M.A. (2013). High-value products from microalgae—Their development and commercialisation. *Journal of Applied Phycology* 25:743–756. doi: 10.1007/s10811-013-9983.

Cai T., Park S.Y. and Li Y. (2013). Nutrient recovery from wastewater streams by microalgae: status and prospects. *Renewable and Sustainable Energy Reviews* 19:360–369.

Carvalho A.P. and Malcata F.X. (2000). Effect of culture media on production of polyunsaturated fatty acids by *Pavlovalutheri. Cryptogamie Algol*ogy 21:59–71. doi:10.1016/S0181-(00)00101-X.

Chaudry S., Bahri P.A. and Moheimani N.R. (2015). Pathways of processing of wet microalgae for liquid fuel production: A critical review. *Renewable and Sustainable Energy Reviews* 52:1240–1250.

Chen T., Wei D., Chen G., Wang Y. and Chen F. (2009). Employment of organic acids to enhance astaxanthin formation in heterotrophic *Chlorella zofingiensis. Journal of food Processing and Preservation* 33:271–284.

Cheng J., Li K., Yang Z., Zhou J. and Cen K. (2016). Enhancing the growth rate and astaxanthin yield of *Haematococcus pluvialis* by nuclear irradiation and high concentration of carbon dioxide stress. *Bioresource Technology* 204: 49–54.

Chew K.W., Yap J.Y., Show P.L., Suan N.H., Juan J.C., Ling T.C., Lee D.-J. and Chang J.-S. (2017). Microalgae biorefinery: High value products perspectives. *Bioresource Technology* 229: 53–62.

Chia M.A., Lombardi A.T., da Graça Gama Melão M. and Parrish C.C. (2015). Combined nitrogen limitation and cadmium stress stimulate total carbohydrates, lipids, protein and amino acidaccumulation in *Chlorella vulgaris* (Trebouxiophyceae). *Aquatic Toxicology* 160. doi:10.1016/j.aquatox.2015.01.002.

Chia M.A., Lombardi A.T. and Melão M.G.G. (2013). Growth and biochemical composition of *Chlorella vulgaris* in different growth media. *Anais da Academia Brasileira de Ciências* 85. doi:10.1590/0001-3765201393312.

Chisti Y. (2007). Biodiesel from microalgae. *Biotechnology Advances* 25(3):294–306.

Chisti Y. (2008). Biodiesel from microalgae beats bioethanol. *Trends in Biotechnology* 26(3):126–131.

Craggs R.J., Adey W.H., Jenson K.R., St John M.S., Green F.B. and Oswald W.J. (1996). Phosphorus removal from wastewater using an algal turf scrubber. *Water Science and Technology* 33:191–198. doi: 10.2166/wst.1996.0138.

Demirbas A. (2008). Biofuels sources, biofuel policy, biofuel economy and global biofuel projections. *Energy Conversion and Management* 49(8): 2106–2116.

Demirbas, A. and Demirbas, M.F. (2011). Importance of algae oil as a source of biodiesel. *Energy Conversion and Management* 52(1):163–170.

Del Campo, J.A., García-González, M. and Guerrero, M.G. (2007). Outdoor cultivation of microalgae for carotenoid production: Current state and perspectives. *Applied Microbiology and Biotechnology* 74(6):1163–1174.

de Morais M.G., da Silva Vaz B., de Morais E.G. and Costa J.A.V. (2015). Biologically active metabolites synthesized by microalgae. *Biomedical Research International* 2015:835761.

de Paula Silva P.H., De Nys R. and Paul N.A. (2012). Seasonal growth dynamics and resilience of the green tide alga *Cladophora coelothrix* in high-nutrient tropical aquaculture. *Aquaculture Environment Interactions* 2(3):253–266.

de Souza M.P., Hoeltz M., Gressler P.D., Benitez L.B. and Schneider R.C.S. (2018). Potential of microalgal bioproducts: General perspectives and main challenges. *Waste and Biomass Valorization* 10:1–18.

Dineshkumar R., Kumaravel R., Gopalsamy J., Sikder M.N.A. and Sampathkumar P. (2018). Microalgae as bio-fertilizers for rice growth and seed yield productivity. *Waste and biomass valorization* 9:793–800.

Enzing C., Ploeg M., Barbosa M. and Sijtsma L. (2014). Microalgae-based products for the food and feed sector: An outlook for Europe. *JRC Scientific and Policy Report* 19–37. doi: 10.2791/3339.

Gosch B.J., Magnusson M., Paul N.A. and de Nys R. (2012). Total lipid and fatty acid composition of seaweeds for the selection of species for oil-based biofuel and bioproducts. *Gcb Bioenergy* 4:919–930.

Hadiyanto M.C., Soetrisnanto D. and Christwardhana M. (2013). Phytoremediations of palm oil mill effluent (POME) by using aquatic plants and microalgae for biomass production. *Journal of Environmental Science and Technology* 6(2):79–90.

Hamed I. (2016). The evolution and versatility of microalgal biotechnology: A review. *Comprehensive Review of Food Science and Food Safety* 15:1104–1123.

Hu Q., Sommerfeld M., Jarvis E., Ghirardi M., Posewitz M., Seibert M. and Darzins A. (2008). Microalgal triacylglycerols as feedstocks for biofuel production: Perspectives and advances. *The plant Journal* 54 (4):621–639.

International Energy Agency World Energy Outlook (IEA, 2011). Flagship Report, November 2011.

Ishika T, Moheimani N.R. and Bahri P.A. (2017). Sustainable saline microalgae co- cultivation for biofuel production: A critical review. *Renewable and Sustainable Energy Reviews* 78:356–368.

Ip P.F. and Chen, F. (2005). Employment of reactive oxygen species to enhance astaxanthin formation in *Chlorella zofingiensis* in heterotrophic culture. *Process Biochemistry* 40:3491–3496.

Jain R., Raghukumar S., Tharanathan R. and Bhosle N.B. (2005). Extracellular polysaccharide production by thraustochytrid protists. *Marine Biotechnology* 7:184–192.

Jez S., Spinelli D., Fierro A., Dibenedetto A., Aresta M., Busi E. and Basosi R. (2017). Comparative life cycle assessment study on environmental impact of oil production from microalgae and terrestrial oilseed crops. *Bioresource Technology* 239:266–275.

Ji M.K., Kim H.C., Sapireddy V.R., Yun H.S., Abou-Shanab R.A., Choi J., Lee W., Timmes T.C. and Jeon B.H. (2013). Simultaneous nutrient removal and lipid production from pretreated piggerywastewater by Chlorella vulgaris YSW-04. *Applied Microbiology and Biotechnology* 97(6):2701–2710.

Khatib, H. (2012). IEA world energy outlook 2011—A comment. *Energy policy* 48:737–743.

Kit W.C., Jing Y.Y., Pau L.S., Ng H.S., Joon C.J., Tau C.L., Duu-Jong L. and Jo-Shu C. (2017). Microalgae biorefinery: High value products perspectives. *Bioresource Technology* 229:53–62

Kraan S. (2013). Mass-cultivation of carbohydrate rich macroalgae, a possible solution for sustainable biofuel production. *Mitigation and Adaptation Strategies for Global Change* 18(1):27–46.

Koller M., Muhr A. and Braunegg G. (2014). Microalgae as versatile cellular factories for valued products. *Algal Research* 6:52–63.

Kumar M., Sun Y., Rathour R., Pandey A., Thakur S. and Tsang, D.C.N. (2020). Algae as potential feedstock for the production of biofuels and value-added products: Opportunities and challenges. *Science of the Total Environment* 716:137116.

Kunrunmi O. A., Adesalu T.A. and Shashi K. (2017). Genetic identification of new microalgal species from Epe Lagoon of West Africa accumulating high lipids. *Algal Research* 22:68–78.

Lammens T.M., Franssen M.C.R., Scott E.L. and Sanders J.P.M. (2012). Availability of protein- derived amino acids as feedstock for the production of bio-based chemicals. *Biomass and Bioenergy* 44:168–181.

Leu S. and Boussiba S. (2014). Advances in the production of high-value products by microalgae. *Industrial Biotechnology* 10:169–183.doi: 10.1089/ind.2013.0039.

Li H.B., Cheng K.W., Wong C.C., Fan K.W., Chen F. and Jiang Y. (2007). Evaluation of antioxidant capacity and total phenolic content of different fractions of selected microalgae. *Food Chemistry* 102:771–776.

Li T.T., Tong A.J., Liu Y.Y., Huang Z.R., Wan X.Z., Pan Y.Y., Jia R.B., Liu B., Chen X.H. and Zhao C. (2019). Polyunsaturated fatty acids from microalgae *Spirulina platensis* modulates lipid metabolism disorders and gut microbiota in high-fat diet rats. *Food Chemistry and Toxicology* 131:110558.

Liang S., Liu X., Chen F. and Chen Z. (2004). Current microalgal health food R & D activities in China. In *Asian Pacific Phycology in the 21st Century: Prospects and Challenges.* Springer, pp. 45–48.

Liu J. and Chen F. (2014). Biology and industrial applications of *Chlorella*: Advances and prospects. In: *Microalgae Biotechnology*. Springer, Berlin, pp. 1–35.

Liu Z.Y., Wang G.C. and Zhou B.C. (2008). Effect of iron on growth and lipid accumulation in *Chlorella vulgaris. Bioresource Technology* 99:4717–4722.

Luo X., Su P. and Zhang W. (2015). Advances in microalgae-derived phytosterols for functional food and pharmaceutical applications. *Marine Drugs* 13:4231–4254.

Mamo G, Faryar R., and Karlsson E.N. (2013). Microbial glycoside hydrolases for biomass utilization in biofuels applications. In: *Biofuel Technologies*. Springer, Berlin, pp. 171–188.

Masarin F., Cedeno F.R.P., Chavez E.G.S., De Oliveira L.E., Gelli V.C. and Monti R. (2016). Chemical analysis and biorefinery of red algae Kappaphycus alvarezii for efficient production of glucose from residue of carrageenan extraction process. *Biotechnology and Biofuels* 9:122.

Mathimani T. and Pugazhendhi A. (2019). Utilization of algae for biofuel, bio-products and bio- remediation. *Biocatalysis and Agricultural Biotechnology* 17:326–330.

Mazokopakis E.E., Papadomanolaki M.G., Fousteris A.A., Kotsiris D.A., Lampadakis I.M. and Ganotakis E.S. (2014). The hepatoprotective and hypolipidemic effects of *Spirulina (Arthrospira platensis)* supplementation in a Cretan population with non-alcoholic fatty liver disease: A prospective pilot study. *Annals of Gastroenterology: Quarterly Publication of the Hellenic Society of Gastroenterology* 27, 387.

McHugh D.J. (2003). A guide to the seaweed industry FAO Fisheries Technical Paper 441. Food and Agricultural Organisation, United Nations, Rome.

Metting F.B. (1996). Biodiversity and application of microalgae. *Journal of Industrial Microbiology* 17:477–489.

Nakashima Y., Ohsawa I., Konishi F., Hasegawa T., Kumamoto S., Suzuki Y. and Ohta S. (2009). Preventive effects of *Chlorella* on cognitive decline in age-dependent dementia model mice. *Neuroscience Letters* 464:193–198.

Neori A., Chopin T., Troell M., Buschmann A.H., Kraemer G.P., Halling C., Shpigel M. and Yarish C. (2004). Integrated aquaculture: rationale, evolution and state of the art emphasizing seaweed biofiltration in modern mariculture. *Aquaculture* 231(1):361–391.

Nwoba E.G., Ayre J.M., Moheimani N.R., Ubi B.E. and Ogbonna J.C. (2016). Growth comparison of microalgae in tubular photobioreactor and open pond for treating anaerobic digestion piggery effluent. *Algal Research* 17:268–276.

Nwoba E.G., Moheimani N.R., Ubi B.E., Ogbonna J.C., Vadiveloo A., Pluske J.R. and Huisman J.M. (2017). Macroalgae culture to treat anaerobic digestion piggery effluent (ADPE). *Bioresource Technology* 227:15–23.

Nwoba E.G., Parlevliet D.A., Laird D.W., Alameh K. and Moheimani N.R. (2019). Light management technologies for increasing algal photobioreactor efficiency. *Algal Research* 39:101433.

Nwoba E. G., Ogbonna C. N., Ishika T. and Vadiveloo A. (2020a). Microalgal pigments: A source of natural food colors. In *Microalgae Biotechnology for Food, Health and High-Value Products*. Springer, Singapore, pp. 81–123.

Nwoba E.G., Vadiveloo A., Ogbonna C.N., Ubi B.E., Ogbonna J.C. and Moheimani N.R. (2020b). Algal cultivation for treating wastewater in African developing countries. A review. *CLEAN–Soil, Air, Water* 48(3):2000052.

Ogbonna C. N. and Nwoba E. G. (2021). Bio-based flocculants for sustainable harvesting of microalgae for biofuel production. A review. *Renewable and Sustainable Energy Reviews* 139:110690.

Olguín E. J. (2003). Phycoremediation: Key issues for cost-effective nutrient removal processes. *Biotechnology Advances* 22(1–2):81–91.

Ooms M.D., Dinh C.T., Sargent E.H. and Sinton D. (2016). Photon management for augmented photosynthesis. *Nature communications* 7:12699.

Pulz O., and Gross W (2004). Valuable products from biotechnology of microalgae. *Applied Microbiology and Biotechnology* 65, 635–648.

Radmer R.J. (1996). Algal diversity and commercial algal products. *Bioscience* 46:263–270.

Rastogi R.P., Pandey, A., Larroche, C. and Madamwar, D. (2018). Algal Green Energy–R&D and technological perspectives for biodiesel production. *Renewable and Sustainable Energy Reviews* 82:2946–2969.

Renuka, N., Sood, A., Prasanna, R. and Ahluwalia, A.S. (2015). Phycoremediation of wastewaters: A synergistic approach using microalgae for bioremediation and biomass generation. *International Journal of Environmental Science and Technology* 12(4):1443–1460.

Richards, R.G. and Mullins, B.J. (2013). Using microalgae for combined lipid production and heavy metal removal from leachate. *Ecological modelling* 249:59–67.

Roesijadi G., Jones S.B., Snowden-Swan L.J. and Zhu Y. (2010). Macroalgae as a biomass feedstock: A preliminary analysis. Pacific Northwest National Lab.(PNNL), Richland, WA (United States).

Sabh A.Z. and Shallan M.A. (2008). Effect of organic fertilization of broad bean (*Vicia faba* L.) by using different marine macroalgae in relation to the morphological, anatomical characteristics and chemical constituents of the plant. *Australasian Journal of Basic and Applied Science* 2:1076–1091.

Safi C., Ursu A.V., Laroche C., Zebib B., Merah O., Pontalier P.Y. and Vaca-Garcia C. (2014). Aqueous extraction of proteins from microalgae: Effect of different cell disruption methods. *Algal Research* 3:61–65.

Saygideger S. and Deniz F. (2008). Effect of 24-epibrassinolide on biomass, growth and free proline concentration in *Spirulina platensis* (Cyanophyta) under NaCl stress. *Plant Growth Regulation* 56:219–223.

Schenk P.M., Thomas-Hall S.R., Stephens E., Marx U.C., Mussgnug J.H., Posten C., Kruse O., and Hankamer B. (2008). Second generation biofuels: High-efficiency microalgae for biodiesel production. *Bioenergy Research* 1(1):20–43.

Servel M.O., Claire C., Derrien A., Coiffard L. and De Roeck-Holtzhauer Y. (1994). Fatty acid composition of some marine microalgae. *Phytochemistry* 36:691–693.

Sillman J. (2016). Sustainability of protein production by bioreactor processes using wind and solar power as energy sources.

Sims R.E., Mabee W., Saddler J.N. and Taylor M. (2010). An overview of second generation biofuel technologies. *Bioresource Technology* 101(6):1570–1580.

Singh A.K., Sharma L., Mallick N. and Mala J. (2017). Progress and challenges in producing polyhydroxyalkanoate biopolymers from cyanobacteria. *Journal of Applied Phycology* 29:1213–1232.

Sonani R.R., Rastogi R.P., Patel R. and Madamwar D. (2016). Recent advances in production, purification and applications of phycobiliproteins. *World Journal of Biological Chemistry* 7:100.

Sousa I., Gouveia L., Batista A.P., Raymundo A. and Bandarra N.M. (2008). Microalgae in novel food products. *Food Chemistry Research Developments* 75–112.

Spolaore P., Joannis-Cassan C., Duran E. and Isambert A. (2006). Commercial applications of microalgae. *Journal of Bioscience and Bioengineering* 101:87–96.

Straskraba, M. (1993). Ecotechnology as a new means for environmental management. *Ecological Engineering* 2(4):311–331.

Stephens E., Ross I.L., Mussgnug J.H., Wagner L.D., Borowitzka M.A., Posten C., Kruse O. and Hankamer B. (2010). Future prospects of microalgal biofuel production systems. *Trends in Plant Science* 15(10):554–564.

Stephenson P.G., Moore C.M., Terry M.J., Zubkov M.V. and Bibby T.S. (2011). Improving photosynthesis for algal biofuels: Toward a green revolution. *Trends in Biotechnology* 29 (12):615–623.

Sun Y., Huang Y., Liao Q., Fu Q. and Zhu X. (2016). Enhancement of microalgae production by embedding hollow light guides to a flat-plate photobioreactor. *Bioresource Technology* 207:31–38.

Tang G. and Suter P.M. (2011). Vitamin A, nutrition, and health values of algae: *Spirulina, Chlorella,* and *Dunaliella. Journal of Pharmacy and Nutrition Science* 1:111–118.

Uysal O., Uysal F.O. and Ekinci K. (2015). Evaluation of microalgae as microbial fertilizer. *European Journal of Sustainable Development* 4:77.

Vaishampayan A., Sinha R.P., Hader D.P., Dey T., Gupta A.K., Bhan U. and Rao A.L. (2001). Cyanobacterial biofertilizers in rice agriculture. *Botany Reviews* 67:453–516.

Vandamme D., Foubert, I. and Muylaer K. (2013). Flocculation as a low-cost method for harvesting microalgae for bulk biomass production. *Trends in Biotechnology* 31(4):233–239.

Voloshin R.A., Rodionova M.V., Zharmukhamedov S.K., Veziroglu T.N. and Allakhverdiev S.I. (2016). Biofuel production from plant and algal biomass. *International Journal of Hydrogen Energy* 41(39):17257–17273.

Wagner C. and Adrian R. (2009). Cyanobacteria dominance: Quantifying the effects of climate change. *Limnology and Oceanog*raphy 54:2460–2468. doi:10.4319/lo.2009.54.6_part_2.2460.

Wang J., Jin W., Hou Y., Niu X., Zhang H. and Zhang Q. (2013). Chemical composition and moisture-absorption/retention ability of polysaccharides extracted from five algae. *International Journal of Biological Macromolecules* 57:26–29.

Wuang S.C., Khin M.C., Chua P.Q.D. and Luo Y.D. (2016). Use of *Spirulina* biomass produced from treatment of aquaculture wastewater as agricultural fertilizers. *Algal Research* 15:59–64.

Yu X., Chen L. and Zhang W. (2015). Chemicals to enhance microalgal growth and accumulation of high-value bioproducts. *Frontier Microbiology* 6:56.

Zeng X., Danquah M.K., Zheng C., Potumarthi R., Chen X.D. and Lu Y. (2012). NaCS–PDMDAAC immobilized autotrophic cultivation of Chlorella sp. for wastewater nitrogen and phosphate removal. *Chemical Engineering Journal* 187:185–192.

Zhu X.G., Long S.P., and Ort D.R. (2008).What is the maximum efficiency with which photosynthesis can convert solar energy into biomass? *Current opinion in Biotechnology* 19 (2):153–159.

# 12 African Mushrooms as Functional Foods and Nutraceuticals

*Charles Oluwaseun Adetunji and Olugbemi Tope Olaniyan*
Edo State University Uzairue

*Juliana Bunmi Adetunji*
Osun State University

*Osarenkhoe O. Osemwegie*
Landmark University

*Benjamin Ewa Ubi*
Ebonyi State University
Tottori University

## CONTENTS



## 12.1 INTRODUCTION

Mushrooms have been proven to provide energy and curative properties for soldiers during wars, and the Romans observed them as the "Food of the Gods from the time of the ancient Greeks." For decades, the Chinese culture has valued mushrooms as a natural food for the well-being of mankind, hence an "elixir of life." Without underrating their religious and ecological implications in the existence of humanity, they have existed as part of human cultures for many years and have gained substantial

DOI: 10.1201/9781003178378-12

attention in different cultures of the world because of their culinary, medicinal, and sensory features (Yasin et al., 2019). Today, they are referred to as nutritional foods because they offer essential nutrients, such as potassium, selenium, niacin, riboflavin, proteins, vitamin D, and fiber. Numerous bioactivity and nutraceuticals have been ascribed to mushrooms such as their role in the treatment or prevention of Alzheimer, Parkinson, stroke and hypertension, and cancer (Valverde et al., 2015). Additionally, they are revered as immune system boosters, antibacterial, and antihypercholesterolemic in many oriental countries (María et al., 2015).

Several species of mushroom are known to be innately endowed with tremendous amount of bioactive constituents that afford a variety of health promoting benefits. Mushrooms based on the benefits that they afford humanity may be categorized as edible, medicinal, nutraceutical, probiotics, and pharmaceutics. Medicinal mushrooms, which may not be edible, have recently, attracted attention beyond their popular use as therapeutics to becoming a huge business in the global market and preference in mushroom cultivation industry. Numerous mushroom and other beneficial microorganisms have been reported to possess several attributes that could help in resolving diverse global challenges (Adetunji and Anani, 2021a,b,c; Adetunji et al., 2021; Dwivedi et al., 2021; Olaniyan and Adetunji, 2021, Mishra et al., 2020; Thangadurai et al., 2021; Anani et al., 2020; Islam et al 2021). Since its discovery, medicinal mushrooms have been cultivated throughout the world with several strains being developed or biopharmed to potentiate bioactive molecules and polysaccharides for improved pharmaceutical benefits (Llauradó et al., 2015). Mushrooms that are rich in $\beta$-glucans and immunomodulating polysaccharides have been shown to have anti-infection, anti-inflammatory, antitumor, and antiallergic/asthmatic properties (Jong et al., 1991; Sadler and Saltmarsh, 1998; Osemwegie et al., 2002; Karaman et al., 2003; Isikhuemhen and LeBauer, 2004; Gezar et al., 2006; Lekgari, 2010; Thatoi and Singdevsachan, 2014; Khatua et al., 2017).

Therefore, this book chapter intends to provide comprehensive perspectives on myco-chemistry, nutritional values, and therapeutics of African mushrooms for the benefits of humanity.

## 12.2 ANTIMICROBIAL AND OTHER MEDICINAL VALUES OF MUSHROOMS

It has been documented that several microorganisms have developed higher level of drug resistance to numerous synthetic and semisynthetic drugs. Therefore, there is a need to search for several alternative drugs or bioactive constituents of biogenetic resources that could be applied for effective management of these drug-resistant strains. Moreover, the biodiversity of tropical fungi has become a recognized biogenetic resource rich in several phytochemical constituents that have ethnomycological and pharmaceutical relevances (Osemwegie et al., 2014; Bach et al., 2019). In view of the aforementioned, Sum et al. (2019) utilized molecular techniques using ribosomal internal transcribed spacer (ITS) DNA in the characterization of *Echinochaete*, *Inonotus*, *Antrodia*, *Fomitiporia*, *Polyporus*, *Ganoderma*, *Hexagonia*, *Perenniporia*, *Favolaschia*, and *Skeletocutis*. The active secondary metabolites produced by their fungal mycelia were obtained using ZM½ (sugar-malt) liquid media and yeast malt agar media, and Q6½ (cottonseed). The active secondary metabolites were extracted using ethyl acetate while the organic solvent was then evaporated in order to obtain the crude extract. The crude extract was then studied for antimicrobial potential against *Candida albicans*, *Bacillus subtilis*, *Mucor plumbeus*, and *Escherichia coli*. Nine out of the 12 fungal strains tested showed antimicrobial activity. Furthermore, *Favolaschia calocera* and *Skeletocutis nivea* showed a more pronounced antimicrobial activity against *Cystopteri tenuis* and *B. subtilis*. The minimum inhibitory concentration (MIC) of the crude extract of *F. calocera* and *S. nivea* at <2.34 and 4.69 µg/mL concentrations, respectively, competed well with synthetic antibiotics like nystatin and ciprofloxacin with similar MIC concentration. Similarly, the crude extract of *Inonotus pachyphloeus* and *Hexagonia* sp. inhibited the growth of *E. coli* at 3.00 µg/mL while *S. nivea* and *F. calocera* suppressed the growth of *M. plumbeus* at 3.00 µg/mL, and 3.75 µg/mL, respectively.

Different accounts on the antimicrobial and ethnomycological potential of many African mushrooms exist in literature. Anyanwu et al. (2016) observed antimicrobial effect of *Pleurotus tuber-regium* sclerotium crude ethanol and aqueous extracts on *Pseudomonas aeruginosa*, *Candida albicans*, *Staphylococcus aureus*, *Aspergillus niger*, *Streptococcus* sp., and *E. coli*. While the target microorganisms showed varied sensitivity to the different crude extracts, the ethanolic extracts was more effective than the aqueous extracts and standard antibiotics used as controls in the study. The effectiveness of the ethanolic extract of *P. tuber-regium* might be due to the occurrence of alkaloids, polyphenols, glycosides, flavonoids, saponins, and tannins which were detected in the extracts. This knowledge places *P. tuber-regium* fruit body and its sclerotium on the list of alternative biogenetic resources that could be utilized as a permanent replacement to synthetic antibiotics.

The mycelia from many medicinal fungi have also been demonstrated to have antimicrobial property. This was corroborated by Tahany et al. (2016) who studied the antimicrobial effect of *Pleurotus ostreatus* mycelia extract against one pathogenic fungal species and four bacterial species using susceptibility assays. It was however observed that the use of macrodilution technique yielded lower MIC and minimum bactericidal concentration (MBC) when compared to the agar diffusion method. Furthermore, it was deduced from previous studies that the tenfold dilution produces a better inhibitory result than the bifold dilution technique. While there is presently no clear scientific explanation for the observed inconsistency in the antimicrobial performance of extracts of fungi relative to sensitivity techniques (serial dilution, well diffusion, Etest, spectrophotometry, disc diffusion, genotype, etc.), it is hypothetical to assume that this might be a reflection of the limitation and accuracy of each technique. It could also be associated with the innate resistance characteristics (inactivation of the porin channel, modification of antimicrobial receptors, enzyme neutralization of antimicrobial biomolecules) of test microbes (Khan et al., 2019) or the extract's physicochemical nature (Khan et al., 2019). A mixture of *P. ostreatus* mycelia extract and Norfloxacin also enhanced the antimicrobial effectiveness of the mushroom against *B. subtilis* (fractional inhibitory concentration index (FICI) = 0.011) using macrodilution technique, while the synergetic effect of the mixture of Terbinafine and *P. ostreatus* mycelia extract showed higher inhibitory effect against *C. albicans* (FICI = 0.10001). This study further strengthened the suggestion that the bioactive molecules from fungi in the form of extract do interact complementarily to activate a cascade of enzymatic and protein reactions that could mediate microbial sensitivity (Llauradó et al., 2015). Renuga and Krishnakumari (2015) reported numerous pharmacological values of mushrooms to include antimicrobial, anticancer, antidiuretic, antidiabetic, and anti-inflammation properties. Some scientists have also affirmed the potential of mushrooms as sustainable resource alternatives for biopharming antimicrobials, polysaccharides, and other beneficial bioactive compounds required for variable industrial applications (Akpi et al., 2017; Gebrevohannes et al., 2019; Oli et al., 2020; Balaji et al., 2020; Ogidi et al., 2020).

Udu-Ibiam et al. (2014) compared the antimicrobial effectiveness of two species (garlic and ginger) and seven edible mushrooms against *Pseudomonas aeruginosa*, *Candida albicans*, *Streptococcus pyogenes*, *Escherichia coli*, and *Staphylococcus aureus*. The antimicrobial constituents present in the two species (garlic and ginger) and seven edible mushrooms were extracted using diethyl ether, cold aqueous, ethanol, and hot aqueous solvents. The antimicrobial effectiveness was performed using agar well diffusion techniques and the result showed that the hot aqueous extract of *Psalliota campestris*, *Trichaptum* sp., *Tricholoma* sp., *Boletus* sp., and *Flammulina* sp., exhibited antimicrobial activity against *P. aeruginosa*, *C. albicans*, *S. aureus*, *E. coli*, and *S. pyogenes*. Furthermore, it was observed that the cold aqueous water obtained from *P. campestris* demonstrated the maximum inhibitory effect of 25.0 mm against *P. aeruginosa* while the zone of inhibition of 25.0 mm was detected for ginger. Also, the hot aqueous extract followed by the cold aqueous extract obtained from the experimental plant species and mushroom exhibited more antimicrobial effectiveness when compared to the extraction obtained from diethyl ether and ethanol extract. Their study further suggested that the antimicrobial properties responsible for all the observed effectiveness should be harnessed and patented for mass production as drugs to treat various ailments combating humanity.

Waithaka et al. (2017) performed a study on the antimicrobial effect of two mushrooms utilizing the metabolites of *Trametes gibbosa* and *Agaricus bisporus*, respectively. The antimicrobial activity was carried out against the following plant pathogens *Ralstonia solanacearum* and *Erwinia* spp. The result observed from their study shows that the active metabolites obtained from the fungi prevented the growth of all the tested pathogens with the exception of gram positive bacteria. Moreover, there was no significant difference in the rate of killing exhibited by the mushrooms extracts against the bacterial growth ($F = 1.92$, $P = 0.09$) even though fungal extracts exhibited a high level of inhibition against the bacteria ($F = 12$, $P = 0.00001$). Moreover, no significant difference was exhibited in the growth of the pathogenic fungus when the crude extract from the mushroom was tested ($F = 1.44$, $P = 0.24$). Conversely, a high level of significance was observed in the rate of inhibiting all the fungal pathogens by the fungal extracts ($F = 2.88$, $P = 0.025$) while there was a high level of significance in the rate of killing the mycopathogen by the fungal and mushroom extracts. This study shows that fungi and mushrooms have the capability to produce active metabolites that could be utilized in the treatment of diseases that have become a menace to the well-being of animal, plant, and human being.

Chelela et al. (2014) tested the antifungal and antibacterial effects of the crude extract derived from wild mushrooms such as *Lactarius densifolius*, *L. gymnocarpoides*, *Lactarius* sp., *Amanita phalloides*, *Russula kivuensis*, and *Amanita muscaria*, respectively. The result obtained indicated that the petroleum ether extraction of *A. muscaria* (MS1PE) exhibited slight inhibitory effect against *Klebsiella oxytoca* and *Shigella flexneri* with MIC of 1.56 mg/mL while the MIC of 3.13 mg/mL was obtained from ethanol (MS1E), petroleum ether (MS1PE), and chloroform (MS1C) extracts of *A. muscaria* were observed to be effective against *Streptococcus pyogenes* and *Vibrio cholera*. It was also noted that the *Mycobacteria* exhibited more resistance to the extract derived from the wild mushroom extract. Furthermore, the MIC of 0.78 mg/mL was observed from the crude extract containing MS1E against *Candida albicans*, while 12.5 mg/mL was observed against *Cryptococcus neoformans*. Also, the MIC of 1.56 mg/mL with slight activity was observed from MS1C and MS1PE against *C. neoformans* and *C. albicans*, while a weak inhibitory effect with a MIC of 6.25 mg/mL was observed against *C. neoformans* and *C. albicans*, respectively. Suffices to say that all the wild mushroom species evaluated demonstrated an inhibitory effect against all the tested fungi and bacteria pathogens suggesting that they possess some pharmacological potentials.

Udu-Ibiam et al. (2015) evaluated the antimicrobial effect of *Psychotria microphylla* which is an herbal plant and *Pleurotus* species a mushroom against five different clinical isolates. The antibacterial effect of the herb and mushroom using ethanol and aqueous solvents were evaluated. This was achieved by agar well diffusion technique modified from Kirby Bauer disk diffusion and tested against all the isolates.

Ayodele and Idoko (2011) appraised the antimicrobial effectiveness of four wild Nigerian edible mushrooms such as *Coprinellus micaceus*, *Lentinus squarrosulus*, *Volvorella volvacea*, and *Psathyrella atroumbonata*. It was observed that the filtrate of the wild mushrooms exhibited an inhibitory effect against all the tested pathogens but the *C. micaceus* and *P. atroumbonata* showed contrary result against *Penicillum notatum*. Moreover, *P. atroumbonata* and *L. squarrosulus* were established to demonstrate the best antibacterial activity against all the tested isolates while *L. squarrosulus* and *V. volvacea* exhibited the best antifungal activity. The study showed that the biologically active component present in these mushrooms could be utilized for the mass production of antimicrobial agents useful for the management of fungi and bacterial infections.

Ndyetabura et al. (2010) investigated the antimicrobial effect of the ethyl acetate crude extracts of *Coprinus cinereus* from Tanzania against *Aspergillus niger*, *Candida albicans*, and *Escherichia coli*, respectively. Their result indicated the presence of some bioactive constituents in the Tanzanian *C. cinereus* which might be a new source of antimicrobial agent that has the potential to be utilized for disease management.

Gbolagade and Fasidi (2005) tested the antimicrobial effectiveness of the methanol extracts of five Nigerian mushrooms which entail *Tricholoma lobayensis*, *Auricularia polytricha*, *Daedalea elegans*, *Corilopsis occidentalis*, and *Daldinia concentrica*. The antimicrobial effectiveness was demonstrated against *Bacillus cereus, Proteus vulgaris, Escherichia coli, Staphylococcus aureus*, and *Klebsiella pneumoniae,* utilizing filter paper disc and hole diffusion technique. The result obtained showed that the mushroom extracts exhibited an inhibitory effect against all the tested isolates with the exception of *Pseudomonas aeruginosa* that display some level of resistance to all the mushroom samples except *Tricholoma lobayensis.* Moreover, the antifungal influence carried out showed that the crude extract obtained from the mushroom exhibit a slight inhibitory effect against *Microsporum boulardii, Aspergillus niger, Candida albicans,* and *Aspergillus flavus*, respectively. Furthermore, it was observed that the MIC ranged between 10.50 and 17.50 mg/mL for fungi, while 1.25 and 9.00 mg/mL were observed for bacteria.

Mirfat et al. (2014) reported the antimicrobial actions of *Schizophyllum commune* which belong to the Basidiomycetes. It is well known for its medical and health promoting benefits. Hence, the authors evaluated the antimicrobial activity utilizing the well diffusion method. Common pathogenic bacteria such as *Pseudomonas aeruginosa, Streptococcus sanguis, Streptococcus mutans, Shigella* sp., *Streptococcus mitis, Shigella flexneri, Salmonella* sp., *Salmonella typhi, Plesiomonas shigelloides, Enterobacter faecalis*, and many others were used as targets. Also, the pathogenic fungi verified are *Candida albicans, Saccharomyces pombe*, and *Candida parapsilosis.* The antimicrobial actions of ethyl acetate, methanol, water and dichloromethane extracts of *S. commune* were quantitatively and qualitatively evaluated by the diameter and presence of inhibition zones. Antibiotic discs were compared as a positive control to the crude extract. The results indicated that the crude extract of *S. commune* contained an antimicrobial property against all the tested pathogens used for the study. Table 12.1 shows the list of some African mushrooms that possess antimicrobial activity.

## 12.3 NUTRITIONAL BENEFITS OF MUSHROOMS

The use of mushrooms as food is an important cultural heritage that dated back to many centuries back. Humanity's attraction to mushrooms equates to their knowledge heritage on the discernment of edible wild mushrooms, ethnomycological applications, nutritional and culinary values, and in recent decade, biotechnological versatility. This growing interest has led to a huge global market for medicinal and edible mushrooms as well as an explosion of the mushroom cultivation business and edible strains emergence. Furthermore, it has lowered mycophobic trends or mushroom abhorrence while promoting their positive use in interior decorations, dye production, animal feeds formulation, single cell protein production, and other areas of an expanding mushroom biotechnology (Chang, 2008). Mushrooms are consumed in different forms (pickled, raw, cooked, roasted/grilled/fried, and smoked) in most cultures primarily due to their organoleptic, medicinal, and rich nutrients properties. Although only fewer wild edible mushrooms (Boa, 2004) have been successfully cultivated and identified, the comprehensive understanding of their nutrients, metabolites, glycoproteins, and polysaccharides compositions is disproportionate to the total documented edible species. Mushrooms are rich source of proteins, particularly the nine essential amino acids, nutrients like phosphorus, iron, and vitamins (B1, B2, B3, C, D), and a variety of bioactive substances. This is why in recent times mushrooms have been upgraded by the Food and Agriculture Organization (FAO) to the list of

**TABLE 12.1**
**African Mushrooms Reported in Literature with Antimicrobial Potential**

| Mushroom | Health Benefit | References |
|---|---|---|
| *Pleurotus tuber-regium* (Fr.) Singer | Antimicrobial, antidiabetic | Osemwegie et al. (2002) |
| *Pleurotus tuber-regium* (Fried) Singer | Antibacterial activity | Ezeronye et al. (2005) |
| *Lycoperdon pusillum* Batsch and *Lycoperdon giganteum* (Pers.) Batsch | Antimicrobial activities | Jonathan and Fasidi (2003) |
| *Trametes* sp. (Aphyllophoromycetideae) | Antimicrobial activity | Ofodile et al. (2008) |
| Four wild mushrooms, *Termitomyces robustus* (Beeli) R. Heim, *Lenzites* sp., *Lentinus subnudus* Berk., and *Termitomyces clypeatus* R. Heim | Free radical scavenging and antimicrobial properties | Oyetayo (2009) |
| *Pleurotus pulmonarius* LAU 09 | Antimicrobial and anti-inflammatory potential of polysaccharide | Adebayo et al. (2012) |
| *Pleurotus tuber-regium* (Fr.) Singer | Antimicrobial effect | Anyanwu et al. (2016) |
| *Pleurotus tuber-regium* (Fr.) Singer (Agaricomycetideae) | Antagonistic/antifungal activities of medicinal mushrooms | Badalyan et al. (2008) |
| *Psalliota campestris* (L.) Quel., *Trichaptum* sp., *Tricholoma* sp., *Boletus* sp., and *Flammulina* sp. | Antimicrobial activity | Udu-Ibiam et al. (2014) |
| *Coprinellus micaceus* (Bull.) Vigalys, Hopple & Jacq., *Lentinus squarrosulus* Mont., *Volvorella volvacea* (Bull.) Singer, and *Psathyrella atroumbonata* Pegler | Antimicrobial activity | Ayodele and Idoko (2011) |
| *Tricholoma lobayensis* R.Heim, *Auricularia polytricha* (Mont.) Sacc., *Daedalea elegans* Spreng., *and Daldinia concentrica* (Bolton) Ces. & De Not. | Antimicrobial activity | Gbolagade and Fasidi (2005) |

therapeutic and functional foods affirming their health benefit-promoting potentials that exceed nutrient and hunger satiety (Raghavendra et al., 2018).

The phytochemicals present in mushrooms are non-nutritive substances responsible for the health promotion and protection of consumers against diseases. Anjali and Vinita (2017) corroborated this observation when they attributed the health merits gained by the general public to the rich mix of mushroom in several Indian local diets. Further studies have shown that mushrooms have the tendency to increase the immune system, promote organs' functionality, and stimulate antibodies and antioxidant to limit cancer in humans. This opened the door of research into the use of mushrooms in the bioaugmentation and biofortification of foods. One major example is the fortification of potato pudding with mushroom powder to optimize their sensory, protein, fat, fiber, and carbohydrate values (Valverde et al., 2015).

Also, the level of mineral (Zn, Cu, and Fe) contents in large samples of Iranian mushrooms studied across seasons was reported by Parisa and Haineh (2013) to be improved by fried techniques and reduced considerably with freezing techniques after frying. Majesty et al. (2019) related the widespread attraction to the consumption of *Pleurotus ostreatus*. They noted that oyster mushrooms including *Pleurotus ostreatus are highly priced diets* due to their richness in crude fats, protein, vitamins A, fatty acids, C, K, and B complex, potassium, calcium, sodium, copper, zinc, phosphorus, amino acid, and manganese mineral contents. It was also confirmed that they have alkaloids, saponins, tannins, oxalate, and phytate which corroborated the reduced lipid profile ration and intoxication observed in the experimental Wistar rats. The hematological, electrolyte ion concentration, and enzyme activity (aspartate and alanine aminotransferases) parameters obtained from the study further supported this observation.

James et al. (2016) wrote extensively on medicinal mushrooms used generally for health booster in Chinese traditional medicine for several decades. These mushrooms and some herbal medications have recently gained attractiveness, principally because of the growing number of researches that suggest their important medical values. Studies have shown that they are found to impact immunomodulatory roles via alterations in the production of some precise cytokines in blood, together with distinct activities and numbers of immune cell subsets. The authors reported that *Ganoderma lucidum* and *Coriolus versicolor* displayed significant immunomodulatory effects in both adaptive and innate immunity.

Oyetayo (2011) elaborated on the exploitation of mushrooms as a crucial source of food with nutrients and pharmacological compounds that are necessary for mitigating several diseases affecting human beings. The author emphasized on the ethnomedicinal benefits of various medicinal mushrooms such as *Pleurotus tuber-regium* which was utilized for the treatment of constipations, headache, fever, cold, and stomach pain; *G. applanatum* for the treatment of diabetes, inflammation, fungal, and bacterial infections, as well as a source of antioxidants; *Termitomyces microcarpus* for managing gonorrhea; *Calvatia cyathiformis* for treating barreness, as well as leucorrhea; and *G. resinaceum* for the treatment of liver diseases.

Khatua et al. (2017) explained that *Laetiporus sulphureus* which is an example of macrofungus has been identified as a typical example of a nutritional food that is also valuable in traditional medicine for the treatment of coughs, rheumatism, pyretic diseases, and gastric cancer. Hence, the species is mostly well-thought-out as a basin of drug for therapy, and thus it has become progressively more popular in the scientific world. The authors revealed that nutritional sciences have experienced a justifiable food supply boost to the rising population via the growing culinary acceptance of mushrooms because of their protein (isoleucine, histidine, lysine, leucine, threonine, methionine), carbohydrate (trehalose> mannitol> fructose), vitamins (D, B, E), minerals (phosphorus, calcium, sodium, magnesium, iron, potassium, manganese, zinc, copper). This is besides some of their biological actions like antihyperglycemic, anti-inflammatory, antimicrobial, antioxidant, immunomodulation, and antitumor influence. It was therefore suggested that a thorough research on this fungus will be necessary.

Vandana et al. (2019) reviewed many studies on plant-based diets that capable of lowering the risk of developing gestational diabetes mellitus. Medicinal mushrooms contain a lot of bioactive compounds and have been utilized conventionally as antidiabetic food across the globe. The bioactive compounds are proteins, polysaccharides, lectins, alkaloids, dietary fibers, sterols, lactones, terpenoids, and polyphenolic compounds, all of which are known to offer various health benefits.

Zhang et al. (2016) wrote extensively on the bioactivities and medicinal effects of many mushrooms which have been used as medicines and foods for decades. The authors reported that mushrooms contain polysaccharides, polyphenols, riboflavin, selenium, niacin, ergothioneine, minerals, and vitamins. Diverse studies carried out on mushrooms revealed high level of bioactivities, like anti-inflammatory, antioxidant, anticancer, antimicrobial, immunomodulatory, antidiabetic, and hepatoprotective properties, making it to attract attention in scientific communities, for its possible role as medicines or functional food in the management and stoppage of several chronic diseases, like cardiovascular illnesses, cancer, neurodegenerative ailments, and diabetes mellitus. The cultivation of *P. ostreatus* is already happening in many African nations. However, the observation of Poongkodi et al. (2015) on the marked differential in the concentration of trace minerals (nitrogen, calcium, iron, inorganic phosphorus, manganese, copper, nickel, magnesium, phosphorus, silicon, and titanium) and nutrients between *P. ostreatus* and *Agaricus bisporus* from India may shift the current paradigm in Africa. The structural precursors from the environment bioassimilated during the mushroom morphogenesis might hypothetically, undetectably, influence their nutrient and non-nutrient status, respectively, or cause variations in the physiochemical compounds diversity between same species from different geographical and climate divides or their distribution between the cap and stipe of a matured species. This realization has boost world production of cultivated edible mushrooms with *Lentinus edodes* supplying 22% of the annual world production. Experimentations using imported spawns of temperate mushrooms and their yield comparison with the indigenous varieties are now major attraction to mushroom biologists as the drive to alleviated hunger and malnutrition heightens in many African regions.

Arini and Tajul (2016) discussed extensively on the medical potential of the future in the management of diseases. Based on the great medical importance of mushrooms, many authors therefore studied the antioxidant properties and nutritional value of *Pleurotus sajor-caju* with the purpose of generating facts on the nutritional constituents of the fresh oyster mushroom. Proximate investigation of the nutritional constituents of gray oyster was carried out and the results showed high moisture content, followed by crude fiber content, protein content, and carbohydrates content, respectively. The lowest nutritional content was shown to be fat content and ash content, respectively. Also, the antioxidant activities were measured by total phenolic content using 2, 2-diphenyl-1-picrylhydrazyl radical scavenging assay and Folin-Ciocalteu reagent method. The results obtained shows that could serve as a functional food because of the presence of antioxidants and so many health benefiting compounds possible consumed in the food matrix.

## 12.4 UTILIZATION OF MUSHROOMS AS ANTICANCER AGENTS

Siwulski et al. (2015) reported that *Ganoderma lucidum* possess physiological and pharmacological effect such as immunomodulatory, antiviral, antidiabetic, antiulcer and antiaging, anti-infection, antiasthmatic, antiallergic, anti-inflammatory, antidiabetic and anticancer.

Wu et al. (2007) revealed that mushrooms possess antiviral, antitumor, and antibacterial properties which may result into their capability to modify cell's immunological purposes. Though, several studies have assessed the modulatory consequence of administering mushrooms' extracts, little or nothing has been reported regarding the immunological purposes of a dietary intake of white button mushrooms, which constitutes 90% of mushrooms eaten in the U.S. The authors performed an experiment by feeding C57BL/6 rats with a diet having 0, 2, or 10% (wt/wt) white button mushroom powder for 10 weeks experimental period and analyzed native and cell-mediated immunity parameters. The results revealed that the addition of mushroom improved natural killer (NK) cell activity, tumor necrosis factor-a (TNFa) generation, but in splenocytes it increased interleukin, IL-2 generation. They further revealed that the addition of mushroom doesn't alter the change of macrophage production of prostaglandin E2, IL-6, TNFa, $H_2O_2$, and nitric oxide, nor did it affect cytotoxic, the percentage of total T cells, suppressive T cells, helper T cells, macrophages, regulatory T cells, NK cells, and total B cells in spleens. They therefore concluded that improved consumption of white button mushrooms could encourage innate immunity against tumors and viruses through the improvement of NK activity which might affect mainly cytokine fabrication.

Fu-Qiang et al. (2013) wrote on cancer biology as an important cause of death globally. The authors revealed that in recent times, the search for safer and effective healing agents for the management and treatment of cancer through chemoprevention has increased tremendously. *Inonotus obliquus* is a white rot fungus which is priced as a medicinal agent. Thorough molecular research has revealed that *I. obliquus* possesses various types of secondary metabolites, such as lanostane-type triterpenoids, phenolic compounds, and melanins. Also *I. obliquus* contain several active constituents utilized for immunomodulatory role, antitumoral, antioxidant, and antiviral activities. Moreover, the anticancer activities of *I. obliquus* has been confirmed by the authors as a major reference point of recent, even when their modes of action are still unclear. Some of the bioactive compounds isolated from *I. obliquus* have been confirmed to arrest cancer growth at G0/G1 phase of cell cycle phase and subsequently induce cell death or differentiation. The polysaccharides derived from *I. obliquus* exhibited anticancer through the activation of the immune system and antioxidant ability which prevents the generation of free radical known to induce cancer cells.

## 12.5 UTILIZATION OF MUSHROOMS FOR THE MANAGEMENT OF SEVERAL DISEASES

Raphael and Chijioke (2015) investigated the indigenous understanding and the usage of edible medicinal mushrooms in different parts of Anambra state. They surveyed 11 local government areas of the state with a well-structured questionnaire.

The authors discovered that the people of Anambra state had local knowledge and understanding of the uses of mushrooms. Also results discovered that occupants of Anambra state normally eat edible mushrooms as staple food and medicine. They revealed that more than 85% of respondents cross-examined had eaten edible mushrooms as food, while only 2% of them used some of the mushrooms as medicine for treatment and management of different diseases. The investigation indicated that mushrooms served as an alternative source of income to rural people of Anambra state. The authors established that the information collected from the people living in these regions on the usage of edible and medicinal mushrooms would form a baseline studies on the application of ethnomedicinal practices among people of Nigeria and other nations of Africa.

Rathee et al. (2012) revealed that mushrooms are healthy food rich in protein, carbohydrate, and fat but could also serve as a good source of bioactive compounds of medicinal value which include antiviral, anticancer, immune-booster, hepatoprotective, and hypocholesterolemic agents.

Choi et al. (2012) showed that many edible mushrooms could reduce risk factors of cardiovascular disease like high low-density lipoprotein cholesterol, total cholesterol, low high-density lipoprotein cholesterol, atherosclerosis, high blood pressure, and oxidative stress and inflammatory impairment.

Ahmed et al. (2016) showed that oyster mushrooms are generally rich source of both micro and macro nutrients. The authors evaluated the yield of newly introduced oyster mushroom strains, viz. *Pleurotus flabellatus* (FLB), *Pleurotus sajor-caju* (PSC), *Pleurotus ostreatus* (PO2 and PO3), *Pleurotus florida* (FLO), *Pleurotus geesteranus* (PG1 and PG3), and *Pleurotus ostreatus* (HK-51), and also to justify their nutritional significant. Their results indicated that these strains possess high level of mineral elements and nutrients to battle socioeconomic encounters such as malnutrition and diseases associated with malnutrition, agricultural diversity, and poverty prevention.

Solomon (2014) did comprehensive assessment on the recent improvements, challenges, suggestions, and future perspectives of medicinal mushroom science. Fungi and medicinal mushrooms are discovered to retain about 130 medicinal roles, such as immunomodulating, antiparasitic, antitumor, anti-hypercholesterolemic, cardiovascular, antioxidant, antibacterial, antifungal, hepatoprotective, antidiabetic, antiviral, and detoxication effects. Many higher Basidiomycetes have bioactive molecules in their fruit bodies, cultured broth, and cultured mycelium. The author focused on the numerous polysaccharides and useful secondary metabolites derived from 700 species of higher homo- and heterobasidiomycetes. Various bioactive polysaccharide–protein complexes or polysaccharides from the medicinal mushrooms seem to have been affirmed to possess antitumor and cell-mediating and induce innate immune responses in animals as well as in human. Though the modes of action involved in the antitumor activity is yet unknown, modulation and stimulation of important host immune responses by these mushroom molecules seem significant in its activation. Low molecular weight secondary metabolites and polysaccharides obtained from these mushrooms also perform a noteworthy role in the control of other diseases and cancers. Similarly, Tang et al. (2016) associated the presence of several biological compounds to the various biological activities observed with mushrooms. Some of the compounds include Ganodermic acids—triterpenes, Ling Zhi-8 protein, Eritadenine, Ergosterol, Polysaccharides—spachyman and packymaran, Cyathane derivatives, Beta-glycoprotein-Proflamin, Beta (1–6)-D-glucan, Lovastatin, Ergopsterol, Beta and hetero-Beta-glucans, Cordycepin, Sterols, Galactomannans, Beta (1,3)-D-glucan, Beta (1–4)-D-glucan, and Proteoglucans. However, there are several species of edible mushrooms which are either cultivated or harvested from the wild. These mushrooms are richer in nutritional values as well as appreciated for their medicinal properties such as anti-hypertensive, neuritogenesis, and antioxidative which are exceedingly valuable to the well-being of humanity.

## 12.6 MUSHROOM FARMING CULTURE IN AFRICA

In Africa, majority of medicinal and edible mushrooms are obtained from the wild because of the slow emergence of mushroom farming culture. Many people from history practice mushroom hunting and this has been handed down from one generation to another. This practice is however common among the women and children (Okhuoya, 1997). In Nigeria, about 25 various edible mushroom species have been handed down generational lines, most especially through verbal communication. Most of these mushrooms that might be obtained from farmlands, plantations, and forests are normally cooked when they are still fresh after the application of some ingredient, oil, salt while some people might sun-dry the mushroom to preserve for consumption at a later date.

While mushroom cultivation is still less developed with many documented edible species yet to be cultivated, numerous people involved in mushroom farming relied on several agricultural wastes as substrates for their mass production (Adetunji and Adejumo, 2017, 2019). Consequently, the advancement of mushroom cultivation, associated technologies, and extension of utilizable substrate biomass beyond agro-wastes to domestic and industrial wastes may offer a partial alternative option to managing the wastes menace of the growing population. The world population has been reported to increase drastically and consequently forcing the need to increase food production to match the ever-increasing population. Therefore, adopting the concept of commercial mushroom production in Nigeria and Africa is critical to increasing access to the nutritional and medicinal benefits derivable from mushroom. There are other economic benefits associated with developed mushroom production culture that include international and local trade of both edible and medicinal mushrooms; household income generation; widespread alternative bioresources for pharmaceutical as well as industrial applications. There are several factors that could enhance the mass production of edible mushrooms in a developing nation like Nigeria that might include the national will to foster linkages between relevant stakeholders such as mushroom farmers, marketers, researchers, government, women, and youths. This is in addition to the sensitization of the member of the public through scientific presentation on numerous electronic and print media, workshops, and seminars as well as proposing a sustainable model that could enhance the availability of spawn, most especially, to the farmers/home-growers. The widespread dissemination of the mushroom spawn and cultivation technologies together with the cross-fertilization of mushroom growing technologies among developed and developing nations is fundamental for the domestication of the large number of wild economic mushrooms of ethnomycological relevance (Saguy, 2011).

Moreover, adequate relationship with the local branch of the mushroom growers association may be equally central to the long-term development of regional mushroom germplasm banks for future cultivar availability and accessibility. This would eventually lead to the emergence of sustainable research centers that could help in the storage of extinct and extant mycoresources, maintain mushroom genetic stability, and undertake breeding experiment as well as the introduction of quality control of the spawn and mushroom culture collections. Also, adequate investigation into the successful exploitation and mushroom prospecting of locally available wild mushroom is necessary for establishing strategies for the effective conservation of the myco-diversity of indigenous mushroom resource, taxonomic, and ethnomycological heritages. It could also play a significant role in the emergence of more novel application or derived product that could potentiate the economic values of mushrooms in Nigeria (Okhuoya and Akpaja, 2005; Akpaja et al., 2003, 2005; Osemwegie et al., 2006).

## 12.7 CONCLUSION AND FUTURE RECOMMENDATION TO KNOWLEDGE

This chapter has provided a comprehensive and detailed information on the biological potential of mushrooms laying emphasis on their bioactivities. Moreover, there is a need to intensify efforts on the application of metabolomics approaches for the structural elucidation of bioactive and beneficial components in African indigenous mushroom diversity. Furthermore, there is need to develop an intentional documentation that will entails all the unique biological attributes of different mushrooms or their ethnomycological applications after their in vitro and in vivo scientific potential have been validated. The popularity of mushrooms in various spheres of human endeavors should necessitate policies, driven by positive political initiatives that could promote their cultivation, biotechnological applications in food, industrial, pharmaceutical, and waste management efforts. Understanding their primary role in ecological cycle could also give impetus to research that focus on their exploitation beyond crop management and crop preservation to developing novel products, analogs, and pharmaceutics. Moreover, there is a need to implement policy that will discourage mushroom hunting and reduce the incidences of mycosis (or mycetisma) among the locals but rather promotes the use of mushrooms in the conversion of waste to wealth, while converted products are tested as organic manure for gardening, mycomeat for animal feeds, single cell protein basic, and packing for biofiltration. Many local farmers may also need to be trained about the current technologies that increase mushroom yield and maximize profits especially in the Sub-Sahara Africa. Finally, the knowledge of bioinformatics techniques for designing more potent drugs with active component is expanding to mushrooms with rudimentary efforts observed in Sub-Sahara Africa. More research needs to be carried out on the post-harvest management of mushrooms in order to enhance their shelf life after harvest most especially in the region where farmers cannot afford to control and have access to atmospheric storage facilities. This review therefore suggests that mushrooms are versatile biogenetic resource with a regenerative capacity for transformative change in the global effort to alleviation of food insecurity (SDG 2), boost sustainable development of good health and well-being (SDG 3), and reasonable consumption (SDG 13).

## REFERENCES

Adetunji, C.O. and Adejumo, I.O. (2017). Nutritional assessment of mycomeat produced from different agricultural substrates using wild and mutant strains from *Pleurotus sajor-caju* during solid state Fermentation. *Animal Feed Science and Technology* 224: 14–19. doi:10.1016/j.anifeedsci.2016.12.004.

Adetunji, C.O. and Adejumo, I.O. (2019). Potency of agricultural wastes in *Pleurotus sajor-caju* biotechnology for feeding broiler chicks. *International Journal of Recycling of Organic Waste in Agriculture* 1–9. doi:10.1007/s40093-018-0226-6.

Adetunji, C.O. and Anani, O.A. (2021a). Recent advances in the application of genetically engineered microorganisms for microbial rejuvenation of contaminated environment. In: Adetunji, C.O., Panpatte, D.G., Jhala, Y.K. (eds). *Microbial Rejuvenation of Polluted Environment. Microorganisms for Sustainability*, vol 27. Springer, Singapore. https://doi.org/10.1007/978-981-15-7459-7_14

Adetunji, C.O., Anani, O.A. and Panpatte D. (2021). Mechanism of actions involved in sustainable ecorestoration of petroleum hydrocarbons polluted soil by the beneficial microorganism. In: Panpatte, D.G., Jhala, Y.K. (eds). *Microbial Rejuvenation of Polluted Environment. Microorganisms for Sustainability*, vol 26. Springer, Singapore. https://doi.org/10.1007/978-981-15-7455-9_8

Adetunji, C.O. and Anani O.A. (2021b). Utilization of microbial biofilm for the biotransformation and bioremediation of heavily polluted environment. In: Panpatte, D.G., Jhala, Y.K. (eds). *Microbial Rejuvenation of Polluted Environment. Microorganisms for Sustainability*, vol 25. Springer, Singapore. https://doi.org/10.1007/978-981-15-7447-4_9

Anani, O.A., Mishra, R.R., Mishra, P., Enuneku, A.A., Anani, G.A. and Adetunji, C.O. (2020). Effects of toxicant from pesticides on food security: Current developments. In: Mishra, P., Mishra, R.R., Adetunji, C.O. (eds). *Innovations in Food Technology*. Springer, Singapore. https://doi.org/10.1007/978-981-15-6121-4_22

Anani, O.A. and Adetunji, C.O. (2021c). Bioremediation of polythene and plastics using beneficial microorganisms. In: Adetunji, C.O., Panpatte, D.G., Jhala, Y.K. (eds). *Microbial Rejuvenation of Polluted Environment. Microorganisms for Sustainability*, vol 27. Springer, Singapore. https://doi.org/10.1007/978-981-15-7459-7_13

Ahmed, M., Abdullah, N. and Nuruddin, M.M. (2016). Yield and nutritional composition of oyster mushrooms: An alternative nutritional source for rural people. *Sains Malaysiana* 45(11): 1609–1615.

Akpaja, E.O., Isikhuemhen, O.S. and Okhuoya, J.A. (2003). Ethnomycology and usage of edible and medicinal mushrooms among the Igbo people of Nigeria. *International Journal of Medicinal Mushrooms* 5: 313–319.

Akpaja, E.O., Okhuoya, J.A. and Eliwer Heferere, B.A. (2005). Ethnomycology and indigenous uses of mushrooms among the Bini-speaking people of Nigeria: A case study of Aihuobabekun community near Benin City, Nigeria. *International Journal of Medicinal Mushroom* 7(3): 373–374.

Anjali, V. and Vinita, S. (2017) Formulation and quality evaluation of mushroom (Oyster mushroom) powder fortified potato pudding. *Asian Journal of Dairy and Food Research* 36(1): 72–75.

Anyanwu, N.G., Mboto, C.I., Solomon, L. and Frank-Peterside, N. (2016). Phytochemical, proximate composition and antimicrobial potentials of *Pleurotus tuber-regium* sclerotium. *New York Science Journal* 9(1): 35–42. doi:10.7537/marsnys09011606.

Arini, N.M.R. and Tajul, A.Y. (2016). Nutritional and Antioxidant Values of Oyster Mushroom (*P. Sajor-caju*) Cultivated on Rubber Sawdust. *International Journal on Advanced Science Engineering Information Technology* 6(2): 161–164.

Ayodele, S.M. and Idoko, M.E. (2011) Antimicrobial activities of four wild edible mushrooms in nigeria. *International Journal of Science and Nature* 2(1): 55–58.

Adebayo, E.A., Oloke, J.K., Majolagbe, O.N., Ajani, R.A., and Bora, T.C. (2012). Antimicrobial and anti-inflammatory potential of polysaccharide from *Pleurotus pulmonarius* LAU 09. *African Journal of Microbiology Research* 6(13): 3315–3323.

Balaji, P., Madhanraj, R., Rameshkumar, K., Veeramanikandan, V., Eyini, M., Arun, A., Thulasinathan, B., Al Farraj, A.D., Elshikh, S.M., Alokda, M.A., Mahmoud, H.A., Tack, C.J. and Kim J.H. (2020). Evaluation of antibiotic activity of Pleurotus pulmonarius streptozotocin-nicotinamide induced diabetic wistar albino rats. *Saudi Journal of Biological Sciences* 27: 913–924.

Boa, E.R. (2004). *Wild Edible Fungi: A Global Overview of Their Use and Importance to People* (No. 17). FAO, Rome.

Bach, F., Zielinski, F.A.A., Helm, V.C., Maciel, M.G., Pedro, C.A., Staffussa, P.A., Avila, S. and Haminiuk, I.W.C. (2019). Bio-compounds of edible mushrooms: *In vitro* antioxidant and antimicrobial activities. *LWT-Food Science and Technology* 107: 214–220.

Badalyan, S., Isikhuemhen, O.S. and Gharibyan, G.M. (2008). Antagonistic / antifungal activities of medicinal mushrooms *Pleurotus tuberregium* (Fr.) Singer (Agaricomycetideae) against selected filamentous fungi. *International Journal of Medicinal Mushrooms* 10: 155–162.

Chang, S.T. (2008). Overview of mushroom cultivation and utilization as fuctional foods. In: Cheung, K.C. (ed.). *Mushrooms as Functional Foods*. John Wiley and Sons Ltd., Hoboken, NJ, pp. 1–29.

Chelela, B.L., Chacha, M. and Matemu, A. (2014). Antibacterial and antifungal activities of selected wild mushrooms from Southern Highlands of Tanzania. *American Journal of Research Communication* 2(9): 58–68.

Choi, E., Ham, O., Lee, S., Song, B., Cha, M., Lee, C.Y., Park, J., Lee, J., Song, H. and Hwang, K. (2012). Mushrooms and cardiovascular disease. *Current Topics in Nutraceutical Research* 10(1): 43–52.

Dwivedi, N., Dwivedi, S. and Adetunji, C.O. (2021). Efficacy of microorganisms in the removal of toxic materials from industrial effluents. In: Adetunji, C.O., Panpatte, D.G., Jhala, Y.K. (eds). *Microbial Rejuvenation of Polluted Environment. Microorganisms for Sustainability*, vol 27. Springer, Singapore. https://doi.org/10.1007/978-981-15-7459-7_15

Ezeronye, O.U., Daba, A.S., Okwujiako, A.I. and Onumajuru, I.C. (2005). Antibacterial of crude polysaccharide extracts from sclerotium and fruitbody (sporophore) of Pleurotus tuber-regium (Fried) Singer on some clinical isolates. *International Journal of Molecular Medicine and Advance Sciences* 1(3): 202–205.

Fu-Qiang, S., Ying, L., Xiang-Shi, K., Wei, C. and Ge, S.G. (2013). Progress on understanding the anticancer mechanisms of medicinal mushroom: *Inonotus Obliquus*. *Asian Pacific Journal of Cancer Prevention* 14(3): 1571–1578.

Gbolagade, J.S. and Fasidi, I.O. (2005). Antimicrobial Activities of Some Selected Nigerian Mushrooms. *African Journal of Biomedical Research* 8(8): 3–87.

Gezar, K., Duru, M.E., Kivrak, I., Turkoglu, A., Mercan, N., Turkoglu, H. and Gulcan, S. (2006). Free-radical scavenging capacity and antimicrobial activity of wild edible mushroom from Turkey. *Africa Journal of Biotechnology* 5: 1924–1928.

Isikhuemhen, O.S. and LeBauer, D.S. (2004). Growing *Pleurotus tuber-regium.* Oyster Mushroom Cultivation: *Mushworld.* 270–281.

Islam, S., Thangadurai, D, Adetunji, C.O., Nwankwo, W., Kadiri, O., Makinde, S., Michael, O.S., Anani, O.A. and Adetunji, J.B. (2021). Nanomaterials and nanocoatings for alternative antimicrobial therapy. In: Kharissova, O.V., Martínez, L.M.T., Kharisov, B.I. (eds). *Handbook of Nanomaterials and Nanocomposites for Energy and Environmental Applications.* Springer, Cham. https://doi.org/10.1007/978-3-030-11155-7_3-1

James, N.W., Cho, W.C. and Sze, D.M.Y. (2016). The use of medicinal mushroom or herb as effective immunomodulatory agent. *Herb Medicine* 2: 1.

Jong, S.C., Birmingham, J.M. and Pai, S.H. (1991). Immunomodulatory substances of fungal origin. *Journal of Immunology and Immunopharmacology* 3: 115–122.

Jonathan, S.G. and Fasidi, I.O. (2003). Antimicrobial activities of two Nigeria edible macrofungi: Lycoperdon pusilum (Bat. Ex) and *Lycoperdon giganteum* (Pers.). *African Journal of Biomedical Research* 6: 85–90.

Karaman, F., Gulluce, M., Outcu, H., Engulf, M. and Adyguzen, A. (2003). Antimicrobial activity of Braziliam propolis against *Paenibacillus. Journal of Ethnopharmacology* 77: 568–571.

Khan, A.Z., Siddiqui, E.M. and Park, S. (2019). Current and emerging methods of antibiotic susceptibility testing. *Diagnostics* 9: 49. doi:10.3390/diagnostics9020049.

Khatua, S., Ghosh, S. and Acharya, K. (2017). *Laetiporus sulphureus* (Bull.: Fr.) Murr. as Food as Medicine. *Pharmacognosy Journal* 9(6s): s1–s15.

Lekgari, L. (2010). Extraction of Phytochemicals. *Biotech online Articles: Biology.* www.bionlinearticles.

Llauradó, G., Morris, J.H., Ferrera, L., Camacho, M., Castán, L., Lebeque, Y., Beltrán, Y., Cos, P. and Bermúdez, C.R. (2015). *In vitro* antimicrobial activity and complement/macrophage stimulation effect of a hot-water extract from mycelium of the oyster mushroom *Pleurotus* sp. *Innovative Food Science and Emerging Technologies.* doi:10.1016/j.ifset.2015.05.002.

Majesty, D., Ijeoma, E., Winner K. and Prince, O. (2019). Nutritional, anti-nutritional and biochemical studies on the oyster mushroom, *Pleurotus ostreatus. EC Nutrition* 14 (1): 36–59.

María, E.V., Hernández-Pérez, T. and Octavio, P.-L. (2015). Edible mushrooms: Improving human health and promoting quality life. *International Journal of Microbiology* 2015: 1–14. Article ID 376387. doi:10.1155/2015/376387.

Mirfat, A.H.S., Noorlidah, A. and Vikineswary, S. (2014). Antimicrobial activities of split gill mushroom *Schizophyllum commune* Fr. *American Journal of Research Communication* 2(7): 113–124. www.usa-journals.com, ISSN: 2325-4076.

Mishra, P., Mishra, R. R., & Adetunji, C. O. (Eds.). (2020). *Innovations in Food Technology: Current Perspectives and Future Goals.* Springer Nature Singapore Pte Ltd. 2020. Springer, Singapore. Pp. 143-162. 1st Edition. DOI: 10.1007/978-981-15-6121-4.

Ndyetabura, T., Lyantagaye, S.L. and Mshandete, A.M. (2010). Antimicrobial activity of ethyl acetate extracts from edible Tanzanian *Coprinus cinereus* (Schaeff) S. Gray s.lat. cultivated on grasses supplemented with cow dung manure. *ARPN Journal of Agricultural and Biological Science* 5(5): 79–85.

Ofodile, L.N., Simmons, S.J., Grayer, R.J. and Uma, N.U. (2008). Antimicrobial activity of two species of the genus *Trametes* Fr. (Aphyllophoromycetideae) from Nigeria. *Journal of Medicinal Mushroom* 10(3): 265–268.

Ogidi, O.C., Ubaru, M.A., Ladi-Lawal, T., Thonda, A.O., Aladejana, M.O. and Malomo, O. (2020). Bioactivity assessment of exopolysaccharides produced by *Pleurotus pulmonarius* in submerged culture with different agro-waste residues. *Heliyon* 6:e05685.

Oli, N.A., Edeh, A.P., Al-Mosawi, M.R., Mbachu, A.N., Al-Dahmoshi, M.O.H., Al-Khafaji, N., Ekuma, O.U., Okezie, U.M. and Saki, M. (2020). Evaluation of the phytoconstituents of Auricularia auricula-judae mushroom and antimicrobial activity of its protein extract. *European Journal of Integrative Medicine* 38: 101176.

Okhuoya, J.A. (1997). Mushroom cultivation: The Nigerian experience. In: Dirar, A. (ed.). *Food Processing Technologies for Africa - Emerging Technologies Series.* UNIDO, pp. 153–168.

Okhuoya, J.A. and Akpaja, E.O. (2005). Mycomedicine and ethnomycology: The Nigerian experience. *International Journal of Medicinal Mushrooms* 7(3): 439–440.

Osemwegie, O.O., Eriyaremu, E.G. and Abdulmalik, J. (2006). A survey of macrofungi in Edo/Delta region of Nigeria, their morphology and uses. *Global Journal of Pure and Applied Science* 12(2): 149–157.

Osemwegie, O.O., Isikhuemhen, O.S., Onuoha, O.J. and Okhuoya, A.J. (2002). Cultivation of selected sporophores- only-producing strains of the edible and medicinal mushroom. *Pleurotus tuber-regium* (fr) Singer (Agaricomycetideae) on waste paper and plantain peelings. *International Journal of Medicinal Mushrooms* 4: 343–348.

Osemwegie, O.O., Okhuoya, A.J. and Dania, T.A. (2014). Ethnomycological conspectus of West African mushrooms: An awareness document.

Olaniyan, O.T. and Adetunji, C.O. (2021). Biochemical role of beneficial microorganisms: An overview on recent development in environmental and agro science. In: Adetunji, C.O., Panpatte, D.G., Jhala, Y.K. (eds). *Microbial Rejuvenation of Polluted Environment. Microorganisms for Sustainability*, vol 27. Springer, Singapore. https://doi.org/10.1007/978-981-15-7459-7_2

Oyetayo, O.V. (2011). Medicinal uses of mushrooms in Nigeria: Towards full and sustainable exploitation. *African Journal of Traditional Complementary and Alternative Medicine* 8(3): 267–274.

Oyetayo, V.O. 2009. Free radical scavenging and antimicrobial properties of extracts of wild mushrooms. *Brazilian Journal of Microbiology* 40: 380–386.

Poongkodi, G.K., HarithraPriya, G.P. and Harshitha, P.G.P. (2015) Nutrient contents of edible mushrooms, *Agaricusbisporus* and *Pleurotusostreatus. International Journal of Modern Chemistry and Applied Science* 2(2): 78–86.

Raphael, N.O. and Chijioke, M.N. (2015). Ethnostudy and usage of edible and medicinal mushrooms in some parts of Anambra State, Nigeria. *Natural Resources* 6: 79–89. doi:10.4236/nr.2015.61008.

Raghavendra, B.V., Venkitasamy, C., Pan, Z. and Nyayak, C. (2018). Functional foods from mushrooms. In: Gupta, K.V., Treichel, H., Shapaval, O.V., de Oliveira, A. and Tuohy, G.M. (eds.). *Microbial Functional Foods and Nutraceuticals.* John Wiley and Sons Ltd., Hoboken, NJ, pp. 39–65.

Rathee, S., Rathee, D., Rathee, D., Kumar, V. and Rathee, P. (2012) Mushrooms as therapeutic agents. *Revista Brasileira de Farmacognosia Brazilian Journal of Pharmacognosy* 22(2): 459–474.

Renuga, D.M. and Krishnakumari, S. (2015). Quantitative estimation of primary and secondary metabolites in hot aqueous extract of *Pleurotus sajor caju. Journal of Pharmacognosy and Phytochemistry* 4(3): 198–202.

Sadler, M. and Saltmarsh, M. (eds.) (1998). *Functional Foods: The Consumer, The Products and the Evidence.* Royal Society of Chemistry, Cambridge.

Saguy, S.I. (2011). Paradigm shifts in academia and food industry required to meet innovation challenges. *Trends in Food Science and Technology* 22: 467–475.

Siwulski, M., Sobieralski, K., Golak-Siwulska, I., Sokół, S. and Sękara A. (2015) *Ganoderma lucidum* (Curt.: Fr.) Karst. – health-promoting properties. *A Review* 61 (3): 105–118.

Solomon, P.W. (2014) Medicinal mushroom science: Current perspectives, advances, evidences, and challenges. *Biomedical Journal* 37(6): 345–356. doi:10.4103/2319-4170.138318.

Sum, W.C., Indieka, S.A., and Matasyoh, J.C. (2019). Antimicrobial activity of Basidiomycetes fungi isolated from a Kenyan tropical forest. *African Journal of Biotechnology* 18(5): 112–123. doi:10.5897/AJB2018.16660.

Tahany, M.A., Abd, E., Rasha, M.N., Sayed, M.S., Alaa, M.S., Aliaa, M.A., Amina, F.A., Fatma, R.W. and Linah, M.H. (2016). Combination antimicrobial efficacy between the mushroom (*Pleurotus ostreatus*) and some commercial antibiotics. *International Journal of Advanced Research in Engineering and Applied Sciences* 5(7): 56–72.

Tang, C., Hoo, P.C.X., Tan, L.T.H., Pusparajah, P., Khan, T.M., Lee, L.H., Goh, B.-H. and Chan, K.-G. (2016). Golden needle mushroom: A culinary medicine with evidenced-based biological activities and health promoting properties. *Frontiers in Pharmacology* 7: 474. doi:10.3389/fphar.2016.00474.

Thatoi, H. and Singdevsachan, S.K. (2014) Diversity, nutritional composition and medicinal potential of Indian mushrooms: A review. *African Journal of Biotechnology* 13(4): 523–545.

Thangadurai, D., Dabire, S.S., Sangeetha, J., Said Al-Tawaha, A.R.M., Adetunji, C.O., Islam, S., Shettar, A.K., David, M., Hospet, R. and Adetunji, J.B. (2021). Greener composites from plant fibers: Preparation, structure, and properties. In: Kharissova, O.V., Martínez, L.M.T., Kharisov, B.I. (eds). *Handbook of Nanomaterials and Nanocomposites for Energy and Environmental Applications.* Springer, Cham. https://doi.org/10.1007/978-3-030-11155-7_21-1

Udu-ibiam, O.E., Ogbu, O., Ibiam, U.A. and Nnachi, A.U. (2015). Synergistic Antibacterial Activity of *Pleurotus* Species (Mushroom) and *Psychotria microphylla* (Herb) against Some Clinical Isolates. *British Journal of Pharmaceutical Research* 7(1): 1–8.

Udu-Ibiam, O.E., Ogbu, O., Nworie, O., Ibiam, U.A., Agah, M.V., Nnachi, A.U., Ogbu, K.I. and Chukwu, O.S. (2014). Antimicrobial activities of some selected edible mushrooms and spices against clinical isolates from federal university teaching hospital Abakaliki (FETHA), Ebonyi State, Nigeria. *International Journal of Scientific and Technology Research* 3(5): 251–255.

Vandana, G., Mansi D.S. and Pankaj, G. (2019). Role of mushrooms in gestational diabetes mellitus. *AIMS Medical Science* 6(1): 49–66. doi:10.3934/medsci.

Valverde, M.E., Hernández-Pérez, T. and Paredes-López, O. (2015). Edible mushrooms: improving human health and promoting quality life. *International Journal of Microbiology* 2015:376387. doi: 10.1155/2015/376387. Epub 2015 Jan 20. PMID: 25685150; PMCID: PMC4320875.

Waithaka, P.N., Gathuru, E.M., Githaiga, B.M. and Onkoba, K.M. (2017). Antimicrobial activity of mushroom (*Agaricus Bisporus*) and fungal (*Trametes Gibbosa*) extracts from mushrooms and fungi of egerton main campus, Njoro Kenya. *Journal of Biomedical Sciences* 6(3): 1–6.

Wu, D., Pae, M., Ren, M., Guo, Z., Smith, D. and Meydani, S.N. (2007). Dietary supplementation with white button mushroom enhances natural killer cell activity in C57BL/6 Mice. *Journal of Nutrition* 137: 1472–1477.

Yasin, H., Zahoor, M., Yousaf, Z., Aftab, A., Saleh, N., Riaz, N. and Shamsheer, B. (2019). Ethnopharmacological exploration of medicinal mushroom from Pakistan. *Phytomedicine* 54: 43–55.

Zhang, J., Li, Y., Zhou, T., Xu, D., Zhang, P., Li, S. and Li, H. (2016). Bioactivities and health benefits of mushrooms mainly from China. *Molecules* 21: 938. doi:10.3390/molecules21070938.

Ziarati, P. and Rabizadeh, H. (2013). Safety and nutritional comparison of fresh, cooked and frozen mushroom (*Agaricus bisporus*). *International Journal of Farming and Allied Sciences* 2(24): 1141–1147.

# Index

**Note:** **Bold** page numbers refer to tables; *italic* page numbers refer to figures.

Printed and bound by CPI Group (UK) Ltd, Croydon, CR0 4YY
12/09/2024
01034038-0006